Undergraduate Texts in Mathematics

Editorial Board
S. Axler
K.A. Ribet

For other titles Published in this series, go to
http://www.springer.com/series/666

Glenn H. Hurlbert

Linear Optimization

The Simplex Workbook

Springer

Glenn H. Hurlbert
School of Mathematical and Statistical Sciences
Arizona State University
Tempe, AZ 85287-1804
USA
hurlbert@asu.edu

Editorial Board:
S. Axler
Mathematics Department
San Francisco State University
San Francisco, CA 94132
USA
axler@sfsu.edu

K. A. Ribet
Mathematics Department
University of California at Berkeley
Berkeley, CA 94720
USA
ribet@math.berkeley.edu

ISBN 978-0-387-79147-0   e-ISBN 978-0-387-79148-7
DOI 10.1007/978-0-387-79148-7
Springer New York Dordrecht Heidelberg London

Library of Congress Control Number: 2009936080

Mathematics Subject Classification (2000): Primary: 90-01, Secondary: 05-01, 15-01, 52-01, 91-01

© Springer Science+Business Media, LLC 2010
All rights reserved. This work may not be translated or copied in whole or in part without the written permission of the publisher (Springer Science+Business Media, LLC, 233 Spring Street, New York, NY 10013, USA), except for brief excerpts in connection with reviews or scholarly analysis. Use in connection with any form of information storage and retrieval, electronic adaptation, computer software, or by similar or dissimilar methodology now known or hereafter developed is forbidden.
The use in this publication of trade names, trademarks, service marks, and similar terms, even if they are not identified as such, is not to be taken as an expression of opinion as to whether or not they are subject to proprietary rights.

Printed on acid-free paper

Springer is part of Springer Science+Business Media (www.springer.com)

Dedication

To Karen, for her constant love and support.
To Calvin and Kate, for patiently waiting for Daddy to play.
To Uncle Frank, who would have loved to read it.
To Mom, my teaching model.

Preface

The Subject

A little explanation is in order for our choice of the title *Linear Optimization*[1] (and corresponding terminology) for what has traditionally been called *Linear Programming*. The word *programming* in this context can be confusing and/or misleading to students. Linear programming problems are referred to as optimization problems but the general term linear programming remains. This can cause people unfamiliar with the subject to think that it is about programming in the sense of writing computer code. It isn't. This workbook is about the beautiful mathematics underlying the ideas of optimizing linear functions subject to linear constraints and the algorithms to solve such problems. In particular, much of what we discuss is the mathematics of Simplex Algorithm for solving such problems, developed by George Dantzig in the late 1940s.

The word *program* in linear programming is a historical artifact. When Dantzig first developed the Simplex Algorithm to solve what are now called linear programming problems, his initial model was a class of resource allocation problems to be solved for the U.S. Air Force. The decisions about the allocations were called 'Programs' by the Air Force, and hence the term. Dantzig's article[2] is a fascinating description of the origins of this subject written by the person who originated many of the ideas. Included is a description of how Tjalling Koopmans (who won a Nobel Prize in economics for his work in decision science) suggested shortening Dantzig's description 'programming in a linear structure' to 'linear programming' during a walk on the beach with Dantzig. Also included is a note that, at the time, 'code' was the word used for computer instructions and not 'program'.

To be clear to potential and current students that this is a mathematics course requiring a background in writing proofs and not a computer coding class, we prefer the terminology *linear optimization*. We do look at computer algorithms but focus on the underlying mathematics. A small amount of computer coding (for example, simple MAPLE) programs will be very useful, but writing code is not the central purpose of the course. The shorthand LP has been used to refer to both the general subject of

---

[1] I'm not that original — there are at least a dozen books that use this terminology.
[2] G. Dantzig, Linear Programming, *Operations Research* **50** (2002), 42–47.

linear programming as well as specific instances of linear programs. To distinguish, in our notation, LO (linear optimization) refers to the general class of optimizing linear functions subject to linear constraints while LOP (linear optimization problem) refers to specific instances of such problems. Furthermore, using the optimization term brings the subject in line with other, closely related fields that are increasingly called Optimization (Nonlinear, Quadratic, Convex, Integer, Combinatorial).

The Simplex Algorithm is the focus of study in this book. In particular, we do not discuss Karmarkar's Algorithm or Khachian's Ellipsoid Algorithm or more general interior-point approaches to solving LOPs. The main reason for this is that, as I hope you'll experience, the Simplex Algorithm leads to richer connections with linear algebra, geometry, combinatorics, game theory, probability, and graph theory. Furthermore, in the post-optimality analysis that occurs in economic modeling and in Integer Optimization, the Simplex Algorithm plays a central role.

Terminology

Besides the usage of LOP and ILOP, we introduce other quirks into the language, mostly for handiness and consistency and occasionally for fun. For example, we discuss four kinds of linear combinations, based on whether or not the extra affine and conic conditions hold, so it makes sense to use the similar notations lspan, aspan, nspan, and vspan for linear, affine, conic, and convex (both affine and conic) combinations, respectively. In particular, lspan makes more sense in this scheme than does span, the more common term found in linear algebra texts. Geometric hulls get the same treatment, with lhull, ahull, nhull, and vhull, respectively. For fun we use the term FLOP (Fractional LOP) when we need to distinguish a LOP from being an ILOP. Indeed, the first step in solving an ILOP is to relax the integer constraints to allow for rationals and find the resulting floptimal solution, which is used as a first approximation to the iloptimal solution. The term BLOP refers to an ILOP whose variables are binary (either 0 or 1).[3] Also, when we discuss game theory, we talk about the GLOP (Game LOP) derived from a game. We don't go too much farther down the self-parody road — hopefully there is no SLOP in the book.

I think we're also the first LO book to use the term *parameter* in place of *nonbasic variable*. That must be worth some kind of award, right? Actually, I stole it from virtually every linear algebra text ever written.

Chapter Flow

Outlandish as it may seem, someone studying from this text will need to start with Chapter 1, followed by Chapter 2. After that, there are many directions of travel.

If and when you wish to study geometry, you'll want to learn Chapter 3 before Chapter 8, but you can pretty much learn them whenever you

---

[3]Thus it is quite natural for many academics to be interested in BO.

want. No other chapter uses Chapter 3 explicitly, although it does offer very beneficial intuition that permeates almost every other chapter. This is why it is placed so early. Chapter 8, however, isn't necessary for much (unless one continues on to study graduate level optimization), but does use material from Chapter 7 (and Section 8.4 needs Chapter 6) — and is kind of fun.

Chapter 4 is really the heart of the course. Everything feeds off of duality. This is why we derive the dual immediately in Section 1.1. Spending extra time here pays dividends later, as everything thereafter depends critically upon it.

The material of Chapter 5 is useful for learning the tip of the how-this-is-done-in-the-real-world iceberg. The ability to state and work with everything in matrix form is a very useful skill in general, and in particular comes in handy in Chapter 12. Thus it does not need to be studied before any other chapter (if at all, as it is not essential material otherwise — in fact, Chapter 5 is only crucial to proving Theorem 12.1.4).

Chapter 6 puts Chapter 4 in general context, and is required for all subsequent chapters.

Chapter 7 contains material that is necessary for Chapter 8. Chapter 9 is not needed by anyone who doesn't want to have a good time. Chapter 10 is required by Chapter 11, and Section 12.4 is key for Chapter 13. Chapters 7–13 offer the greatest flexibility for studying your favorite topics within a semester's time. Of course, you could slow down, spend extra time on the exercises, and complete the whole book in two semesters. Have at it!

A visual description of the above dependencies is given below.

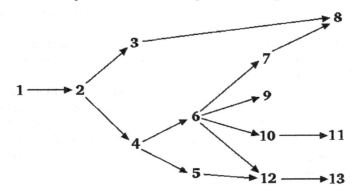

Book Format and Usage

It is not difficult to notice that this text is different from most, and not just because I can't resist even the lamest of jokes,[4] so it may be worth some discussion on why this is so, what benefits this may have for you, and how best to take advantage of the new format.

First and foremost, a great deal of information is missing, information that typically is included in mathematics texts. Having spent most of your

---

[4] Most of these require my age to understand anyway.

life reading from such texts, you may be used to being fed facts and algorithms, and well used to memorizing them in order to reproduce them on exams. But I believe that you are capable of much more: of deriving results, in fact, discovering them through experimentation, of making conjectures, of proving theorems, of solving problems and checking them yourself, and of asking creative questions. That's what this format is all about, giving your professor the opportunity to lead you through the kinds of experiences that will develop your skills in each of these areas, helping you become a highly critical thinking machine,[5] able to wrestle with complex problems in all areas of society, rather than just someone who went to college and remembers a few math facts.

You may find that your classroom environment may also differ from the norm. Many professors who use this will ask their students to participate more in discussion, rather than simply listen to lectures. Some may even ask their students to make daily presentations of the theorems they've proven, the exercises and workouts they've solved, and the algorithms they've written, in order to create an environment in which the students become responsible for their own learning, questioning each other for understanding, while the professor acts as a facilitator. Such an environment may be upsettingly abnormal to you initially, but I promise you will warm to it (indeed, embrace it) in time.

This form of discourse centers more on the learning than the teaching, and those who engage in it deeply are affected (infected?) for life. My hope is that you are enticed enough, not only by the material, but by the interesting problems, challenging questions, and your professor's invitation to question, challenge, wonder, experiment, guess, and argue, to throw your energies in this new direction so as to be stirred to the point that it transforms the way you think about everything, from mathematics and science, to politics and religion, to sports and fine arts. Be inquisitive, be skeptical, be critical, be creative, and keep thinking.

Keep in mind, this isn't some crazy, new, experimental pedagogy. This approach is as old as it gets, predating all forms of formal classroom teaching. It has come to be known by many names through the years: the Socratic Method, Discovery-Based Teaching, the Moore Method, and Inquiry-Based Learning, among others. There are maybe two central tenets that identify the philosophy.

- A thing isn't true because someone says so or because it's written in a book, but because it is reasoned to be true.

- One doesn't master something by hearing or seeing it, but by doing it.

So what else can you do outside of the classroom in order to master this material? Some students find that keeping a journal, separate from their class notes, that holds all their proofs and solutions, can be quite useful.

---

[5] You can even wear a cape: leap tall buildings, etc.

Preface

This is especially true if one uses a word processor (LaTeX is certainly best, whether using AMSTeX on Linux, MiKTeX through WinEdt or Scientific Word on Windows, or even with TeXShop on MacOS, although Microsoft Word, with its Equation Editor should suffice), since the work can continue to be edited and organized to follow the text. Also, do the workouts as you read along, such as the very first one:

Workout 0 *What pattern is there regarding Section 3 of each chapter?*

The point of doing them as you go is partly to help you learn the material by making sure you read the stuff (I can remember undergraduate books that I wouldn't read at all, instead just solving the exercises required for homework by mimicking the examples — what kind of comprehension does that foster?), and partly to help you learn the process of experimenting, conjecturing, clarifying, strategizing, proof writing, and generalizing. That will serve you long after you forget the Complementary Slackness Theorem.

On the other hand, you can simply distract yourself with the trivia contest that floats throughout the book, signified by the o in the margins. Of course, you have to swear not to Google anything if you want to win.

Index

Just a small note about the Index. I believe that index entries should signify what's memorable, not important — importance should be indicated by the page references instead. So in this era of inclusivity, I've thrown in the kitchen sink. The main reason for this is that I don't know how your memory works. It doesn't help you if you can't find *tractor pull* because it's only listed under *theological tractor pull* and you couldn't remember the obvious connection to theology. So the index is fattened by including various permutations of words. Also, even though *tractor pull* is not central to the theory of linear optimization, in 25 years when you're trying to show your kids a nice example of how the theory works and all you can remember is that awesome example with the tractor pull, you'll thank me. Plus, this adds an extra page to the book, which increases its cost by .037¢. With about 2.7 million readers annually, my 1% commission generates an extra deluxe cheeseburger per year, which in turn helps me satisfy the requirements of Problem 1.1.1 for that day. While the entries are many, I did make an effort to restrict the page references of the most common terms to their most important instances.

With regard to the fonts you'll encounter, theorems (in their full names) are in *italics*, and page numbers that refer to definitions are in **bold**. (You'll notice that theorems and definitions find their way into the margins for easy location.) *Italicized* page numbers denote appearances in theorems, while sans serif numbers signal inclusion in workouts and exercises, and roman fonts cite regular occurrences of the term. With regard to the order of terms, mathematical symbols come first (numbers, then capital letters, then lower case letters), followed by As,[6] and then standard words.

---

[6] Short for Acronyms.

Thanks

I wish to express my gratitude to many people who have helped bring this project to fruition. First and foremost is Garth Isaak, who really helped me get the ball rolling in many ways and contributed a number of great ideas at the start. Rob Hochberg and Nate Dean were excellent sounding boards for crazy ideas, some of which found their way onto these pages and some of which didn't. The National Science Foundation was instrumental for funding the time for me to get the bulk of it up and going. Several semesters worth of students tested early versions of the material, and Josh Maximoff and Ben Hester in particular combed through many exercises carefully. Other students, Sriram Penumatcha, Jake Hawkes, and Josh Wolfe, developed WebSim with me, and Jennifer Broatch organized the assessment of the book's methods for the NSF. Much appreciation goes to Harry Lucas and the Educational Advancement Foundation for supporting conferences and workshops dedicated to the pedagogical approach of R.L. Moore, in particular, the Legacy of Moore Conferences at the University of Texas, and the Inquiry-Based Learning Workshops run by Stan Yoshinobu, Ed Parker, Carl Leinart, and Jenny Smith. I will even thank Mike Starbird for many a conversation more fruitful to me than he might imagine. Francis Su's MAA PREP Workshop on Combinatorial Geometry and Mike Jones's MAA Short Course on Game Theory were quite beneficial as well. Several colleagues piloted versions of the course at various stages of development and offered significant feedback, including Dan Biebighauser, Nancy Childress, Steven Dunbar, Mark Ellingham, Gary Gordon, Donovan Hare, and Attila Sali. Many thanks are due to those who helped me with foreign language translations: Hélène Barcelo, Airat Bekmetjev, Anthony Chambers, Gil Kalai, Vikram Kamat, Irina Long, Rose Sau Lugano, Mkamburi Lyabaya, Faris Odish, Andrea Richa, Derar Serhan, Nandor Sieben, and Jennifer Tom, and especially to Klaus Lagally, Dominik Wujastyk, and Rajiv Monsurate for their assistance with LaTeX language packages (and to Renate Mittelmann for loading things for me). Of course, I really also need to thank Don, not only for creating TeX in the first place, but also for the excellent sense of taste and humor displayed in his own texts, not that I stole any ideas or anything. The Coffee Buzz, Lilo's Coffee Haus, The Harem, Bunna Coffee & Tea Market, Steve's Espresso, Extreme Bean Coffee Co., and Gold Bar Espresso deserve much credit for comfortable and creative writing environments (sorry if your copy has any coffee stains on it). Finally, the tremendous group at Springer deserves a raise for putting up with me: Vaishali Damle, Frank Ganz, Marcia Bunda, Frank McGuckin, and players to be named later.[7]

Feedback

Thanks to the many students and professors and editors (in particular, Christopher Curioli) who combed through this book before its printing, the

---
[7] I might trade for some draft picks.

text would have been perfect if not for me. Any errors in typography or content that remain are clearly the fault of some guy I met on the bus one day who distracted me, scattering errors of varying types throughout the book. If you tell me about them I'll be sure to let him know if I ever happen to see him again. I will also try to correct them before the 7th edition. From my web site (search for `Hurlbert math homepage`), click on the `LinOpt` logo and scroll down to the `Submit Feedback` link. You can search to see if your idea has already been posted, and otherwise post it by Chapter, Section, and Type of Error for others to see, and it will be on queue for the next revision.

Disclaimer

This material is based upon work supported by the National Science Foundation under Grant No. 0443087. Any opinions, findings, and conclusions or recommendations expressed in this material are those of the author and do not necessarily reflect the views of the National Science Foundation.

Contents

Preface   **vii**

1 Introduction   **1**
    1.1 The Diet Problem 1
    1.2 The Matching Problem 4
    1.3 Un Problema de la Práctica 7
    1.4 Standard Form and the Dual 8
    1.5 Exercises 12

2 The Simplex Algorithm   **29**
    2.1 Geometric Lens 29
    2.2 Algebraic Lens 32
    2.3 مثال آخر 37
    2.4 Infeasible Basis 38
    2.5 Shortcut Method 41
    2.6 Infeasibility 45
    2.7 Unboundedness 47
    2.8 Cycling 48
    2.9 The Fundamental Theorem 49
    2.10 Exercises 50

3 Geometry   **59**
    3.1 Extreme Points 59
    3.2 Convexity 61
    3.3 小试牛刀 64
    3.4 Carathéodory's Theorem 65
    3.5 Exercises 67

4 The Duality Theorem   **73**
    4.1 Primal-Dual Relationship 73
    4.2 Complementary Slackness Conditions 77
    4.3 Jizoezi, Jizoezi, Jizoezi 79
    4.4 Finding Optimal Certificates 80
    4.5 Exercises 82

5 Matrix Environment — 89
- 5.1 Format and Dictionaries 89
- 5.2 Simplex Phases and Advantages 91
- 5.3 קוַת תרגול 97
- 5.4 Basic Coefficients 98
- 5.5 Exercises 100

6 General Form — 107
- 6.1 Nonstandard Duals 107
- 6.2 General Simplex and Phase 0 110
- 6.3 Plus de Pratique 113
- 6.4 General Duality and Slackness 113
- 6.5 Exercises 115

7 Unsolvable Systems — 119
- 7.1 Infeasible Certificates 119
- 7.2 Inconsistency 122
- 7.3 इसका अध्ययन करो 124
- 7.4 Unsolvable Subsystems 124
- 7.5 Exercises 126

8 Geometry Revisited — 129
- 8.1 Helly's Theorem 129
- 8.2 Permutation Matrices 131
- 8.3 Pratique de Novo 134
- 8.4 Cones 134
- 8.5 Exercises 137

9 Game Theory — 145
- 9.1 Matrix Games 145
- 9.2 Minimax Theorem 147
- 9.3 Bitte Praxis 150
- 9.4 Saddles 151
- 9.5 Exercises 155

10 Network Environment — 163
- 10.1 Shipping 163
- 10.2 Trees 167
- 10.3 Nilai! 172
- 10.4 Integrality 173
- 10.5 Exercises 174

11 Combinatorics — 183
- 11.1 Matchings 183
- 11.2 Covers 185
- 11.3 もっと練習しましょう 188
- 11.4 Systems of Distinct Representatives 189

	11.5 Exercises	190

12 Economics — 195
	12.1 Shadow Prices	195
	12.2 Reduced Costs	200
	12.3 Gyakoroljon egy Kicsit	202
	12.4 Dual Simplex	203
	12.5 Exercises	206

13 Integer Optimization — 209
	13.1 Cutting Planes	209
	13.2 Branch-and-Bound	215
	13.3 Последняя Практика	221
	13.4 Integer Certificates	222
	13.5 Exercises	223

A Linear Algebra Review — 231

B Equivalence of Auxiliary and Shortcut Methods — 235

C Complexity — 241
	C.1 P versus NP	241
	C.2 Examples	243
	C.3 LO Complexity	244

D Software — 247
	D.1 WebSim	247
	D.2 Algorithms	248
	D.3 MAPLE	250

Index — 257

Chapter 1

Introduction

1.1 The Diet Problem

Problem 1.1.1 *Imagine that your entire class is allowed to eat from the following menu. A hamburger, a chicken sandwich, a fish sandwich, and a deluxe cheeseburger (water will be your only beverage). The game is that your meal must satisfy certain percentages of the USRDA of Vitamin A, Vitamin C, Calcium, and Iron, and the winner will be the person who consumes the fewest calories. The table below contains all the necessary information regarding percentages of USRDA and number of calories for each item and for the requirements. Keep in mind that, while you must buy whole number amounts of each item, you are allowed to eat fractional amounts of each item. To add excitement, the winner will receive what Carol Merrill is hiding behind door number 3.*

	%A	%C	%Calc	%Iron	Calories
Hamburger	4	4	10	15	250
Chicken	8	15	15	8	400
Fish	2	0	15	10	370
Cheeseburger	15	6	30	20	490
Requirements	10	10	15	15	

For example, if one ate the chicken and the cheeseburger, that would amount to the percentages of 23, 21, 45, and 28, along with 890 calories. Another could eat exactly half of that, but then would fail the Iron requirement, needing 2 more percentage points. Eating an extra one tenth of the cheeseburger would make up that difference, increasing her total calorie intake to 494, a great improvement over 890. Still better, a third person could decide on eating two fifths of each of the hamburger, chicken, and cheeseburger. This satisfies the four requirements with only 456 calories.

G. H. Hurlbert, *Linear Optimization*, Undergraduate Texts in Mathematics,
DOI: 10.1007/978-0-387-79148-7_1, © Springer Science+Business Media LLC 2010

Workout 1.1.2 *Find a menu that satisfies the requirements while consuming fewer than 456 calories.*

How low can one go? Is it possible to consume fewer than 400 calories under these conditions? Less than 350? Before we try to answer this, let us first try to set the problem in more precise mathematical terms.

If we set y_1 through y_4 to be the amounts consumed of hamburger through cheeseburger, respectively, then we come upon the following observations. The total percentage of Vitamin A eaten is $4y_1 + 8y_2 + 2y_3 + 15y_4$. By the above discussion, this quantity should be at least 10. We can carry out the same analysis on the other nutrients as well, and in fact, the quantity we would like to minimize is $250y_1 + 400y_2 + 370y_3 + 490y_4$, the number of calories consumed. So we can succinctly display our challenge mathematically as follows.

Problem 1.1.3

$$\begin{aligned} \text{Minimize} \quad w &= 250y_1 + 400y_2 + 370y_3 + 490y_4 \\ \text{subject to} \quad 4y_1 &+ 8y_2 + 2y_3 + 15y_4 \geq 10 \\ 4y_1 &+ 15y_2 + 6y_4 \geq 10 \\ 10y_1 &+ 15y_2 + 15y_3 + 30y_4 \geq 15 \\ 15y_1 &+ 8y_2 + 10y_3 + 20y_4 \geq 15 \\ \text{and} \quad y_1, \; y_2, \; & y_3, \; y_4 \geq 0 \end{aligned}$$

non-negativity/ problem constraint

LOP

objective function

Notice that we haven't forgotten how difficult (actually unsightly) it is to consume a negative amount — these **nonnegativity constraints** are quite common in many similar problems and shouldn't be neglected. The other constraints, called **problem constraints**, are all linear inequalities (linear referring to the absence of terms like $y_1 y_3$ and y_4^2 in the sum), the only types of inequalities we will consider here. The final building block of this (and every) linear optimization problem (**LOP**) is what we call the **objective function**, and we often reserve a separate variable for it, say w.

As we begin to formalize the LOP in this way, it becomes easier to think about answering the problem of how low we can go with calories. For example, if we multiply the third constraint by 15 we see that $150y_1 + 225y_2 + 225y_3 + 450y_4 \geq 225$. This leads to the following lower bound.

$$\begin{aligned} w &= 250y_1 + 400y_2 + 370y_3 + 490y_4 \\ &\geq 150y_1 + 225y_2 + 225y_3 + 450y_4 \\ &\geq 225. \end{aligned}$$

Carefully, let's think of why we can reason that $w \geq 150y_1 + 225y_2 + 225y_3 + 450y_4$. It is not simply because $250 \geq 150$, and so on, but because we know that $y_1 \geq 0$, and so on. Because of *both* of these two facts, we have $250y_1 \geq 150y_1$, and so on.

1.1. The Diet Problem

We can improve our lower bound to 350 by multiplying the first, second, and fourth constraints each by 10, and then adding them together. Check closely to see that these next inequalities hold true.

$$\begin{aligned} w &= 250y_1 + 400y_2 + 370y_3 + 490y_4 \\ &\geq 230y_1 + 310y_2 + 120y_3 + 410y_4 \\ &= 10(4y_1 + 8y_2 + 2y_3 + 15y_4) \\ &\quad + 10(4y_1 + 15y_2 + 6y_4) \\ &\quad + 10(15y_1 + 8y_2 + 10y_3 + 20y_4) \\ &\geq 10(10) + 10(10) + 10(15). \\ &= 350. \end{aligned}$$

With just a minor adjustment (use 11 in place of the second and third 10s), 375 becomes a slightly better lower bound. Can you push the lower bound above 400?

Workout 1.1.4 *Find a lower bound that is greater than 375 calories.*

At this point, it seems we are splashing in the same kind of water we started in. It is nice to know the answer lies somewhere between 375 and 456, but we are resorting to hit-or-miss guessing. It works up to a point, say, right about here. Certainly, if the LOP involved many more variables or constraints, we'd have had to abort much sooner. The aim of this scroll is to consider precisely this kind of analysis and see where it leads us. Soon, it will lead us to an algorithmic solution by the famous Simplex Algorithm, to the theory of Duality, and to a myriad of fascinating and powerful applications. While we were considering lower bounds, we were in the process of building the very similar LOP, known as its dual, below.

Problem 1.1.5

$$\begin{aligned} \text{Maximize} \quad z = 10x_1 + 10x_2 + 15x_3 + 15x_4 \end{aligned}$$

subject to

$$\begin{aligned} 4x_1 + 4x_2 + 10x_3 + 15x_4 &\leq 250 \\ 8x_1 + 15x_2 + 15x_3 + 8x_4 &\leq 400 \\ 2x_1 \quad\quad\quad + 15x_3 + 10x_4 &\leq 370 \\ 15x_1 + 6x_2 + 30x_3 + 20x_4 &\leq 490 \end{aligned}$$

and $x_1, x_2, x_3, x_4 \geq 0$

Notice the similarities, as well as the subtle differences, between Problems 1.1.3 and 1.1.5. We will discuss their dual relationship in detail in Chapter 4. For now be amazed that if z^* is the maximum z under these conditions, and w^* is the minimum w subject to its constraints, then $z^* = w^*$! This impressive result is called the Strong Duality Theorem 4.1.9 and it plays a most central role in this course. It has surprising and powerful implications in fields as dissimilar as Game Theory, Linear Algebra, Combinatorics, Geometry, and Economics.

Workout 1.1.6 (Weak Duality Theorem) *Suppose that the set of y_i satisfies the constraints of Problem 1.1.3 and produces the objective value w. Suppose also that the set of x_j satisfies the constraints of its dual Problem 1.1.5 and produces the objective value z. Use <u>all</u> the constraints together to prove that $z \le w$.*

Now to satisfy your curiosity, let's present the solution. The winner will consume roughly 424.156 calories (exactly $w^* = 766450/1807$) by eating $y_1^* = 535/1807$ of the hamburger, $y_2^* = 810/1807$ of the chicken sandwich, $y_3^* = 0$ of the fish sandwich, and $y_4^* = 630/1807$ of the cheeseburger deluxe (see Exercise 2.10.6). (We use the star superscript to connote optimal values.) One of the charming qualities of Linear Optimization is that we can be easily convinced of the minimality of this particular solution without showing any of the details which led to its discovery. We simply multiply the first constraint by $x_1^* = 27570/1807$, the second by $x_2^* = 24880/1807$, and the fourth by $x_4^* = 16130/1807$, and then add them up (we could say that we also multiply the third constraint by $x_3^* = 0$). How these figures were obtained is for future lectures.

certificate   What is significant is that, while the y_is offer a *proposed optimal* solution, the x_js provide a **certificate** of their optimality. The existence of certificates is a hallmark of Linear Optimization. While finding optimal solutions may be time consuming, checking their optimality is trivial. The same cannot be said of Calculus, for example — how do you know your answer is correct without redoing the problem?

One final note: observe that the diet solution included only rational values for the y_i^*s, x_j^*s and $w^* = z^*$. This is no coincidence. While polynomials of degree at least two in one variable with integer coefficients can have irrational roots (e.g., $x^2 - 2$), linear functions can only have rational roots. Likewise, Cramer's Rule (recall your Linear Algebra here!) gives a formula for the solution of a multivariable linear function in terms of ratios of determinants. Thus if the coefficients are integers only then so are the determinants, and hence the solutions can only be rational. How this relates to solutions of LOPs we will see in Chapter 5.

1.2 The Matching Problem

Problem 1.2.1 *The Mathematics Department has 10 courses it would like to offer during a particular time slot. Luckily, there are 10 professors available to teach at that time. They, and their qualifications are: Aguilera, Math 401, 402, 407, and 409; Backman, Math 400, 403, 404, 405, and 408; Carter, Math 404 and 406; Dykstra, Math 404 and 406; Elster, Math 400, 402, 403, 405, and 408; Fernandez, Math 402, 404, and 406; Gooden, Math 401, 406, 407, and 409; Hernandez, Math 404 and 406; Innis, Math 402 and 406; and Johnson, Math 402 and 404. How many of these courses can the department actually offer at that time?*

It is a bit difficult to digest all of that information in one gulp. One

1.2. The Matching Problem

way of presenting the information visually is as a bipartite graph. A **graph** is a pair (V, E), where V is a set of **vertices** and E is a set of **edges**, each edge being an unordered pair of two vertices. The graph is **bipartite** if V can be split into two parts with every edge containing a vertex from each part. Those of you familiar with that term may also be familiar with this Matching Problem. For others, you simply list the professors in a vertical column on the left side of a sheet of paper (go ahead and do it!), list the courses in a vertical column on the right side, and draw a line connecting each professor to each of the classes he or she can teach. It's still a mess, isn't it? What if, by sheer fancy, you listed the professors in the order B, E, H, J, D, F, I, C, A, and G, and the courses in the order 5, 8, 0, 3, 4, 2, 6, 7, 9, and 1? (Try it!) Now something interesting might pop out at you. Notice that the four courses 5, 8, 0, and 3 have among them only the 2 professors B and E who are qualified to teach them, so some two of those courses cannot be offered. Also, A and G are the only professors able to teach 7, 9, and 1, so another course will be lost. So far we know at least three classes cannot be offered, and without too much trouble you can probably find a way to match 7 different professors to 7 different courses, thus answering the question. Such a pairing of professors and courses corresponds to what we call a **matching** in the corresponding bipartite graph: a set of edges that share no endpoints.

graph
vertex
edge
bipartite graph
matching

Workout 1.2.2 *Find a matching of size 7 in the above bipartite graph.*

Likewise, one can see that, among the 6 professors H, J, D, F, I, and C, the only courses they are qualified to teach are 4, 2, and 6, so some three of these professors will not be able to teach. We call the set {B, E, A, G, 4, 2, 6} a **cover** because every pair (X,j), where professor X is qualified to teach course $400 + j$, involves at least one of the members of that set. (Technically, a set of vertices is a cover if every edge has at least one of its endpoints in the set.) It is no coincidence that the size of this set is 7. Rather it is a consequence of the König–Egerváry Theorem 11.2.8, which itself is a consequence of the Duality Theorem, surprise, surprise. We will see this theorem later in the course as well.

cover

Workout 1.2.3 *Use the definitions of matching and cover to prove that the size of any matching in a bipartite graph is at most the size of any cover.*

Of course, with a larger problem, it may not be so easy to spot the kind of thing we spotted here. Let's see if there is another way to represent the information we are given. Instead of listing the courses vertically, list them horizontally so that we can define a 10×10 matrix C, with $C(i,j) = 1$ if the professor whose initial is the ith letter of the alphabet is qualified to teach course $400 + j$, and $C(i,j) = 0$ otherwise.

C	0	1	2	3	4	5	6	7	8	9
1	0	1	1	0	0	0	0	1	0	1
2	1	0	0	1	1	1	0	0	1	0
3	0	0	0	0	1	0	1	0	0	0
4	0	0	0	0	1	0	1	0	0	0
5	1	0	1	1	0	1	0	0	1	0
6	0	0	1	0	1	0	1	0	0	0
7	0	1	0	0	0	0	1	1	0	1
8	0	0	0	0	1	0	1	0	0	0
9	0	0	1	0	0	0	1	0	0	0
10	0	0	1	0	1	0	0	0	0	0

We know the total number of courses a particular professor teaches at this time is at most 1. Let's define a set of variables $x_{i,j}$ ($1 \leq i \leq 10$, $0 \leq j \leq 9$) which take on the value $x_{i,j} = 1$ if professor i ends up teaching course $400 + j$, and $x_{i,j} = 0$ otherwise. Then we can translate the above observation into the linear inequality $x_{1,1} + x_{1,2} + x_{1,7} + x_{1,9} \leq 1$. But this interpretation doesn't rule out $x_{1,4} = 1$, for example.

Let's write this inequality more generally as $c_{1,0}x_{1,0} + c_{1,1}x_{1,1} + \ldots + c_{1,9}x_{1,9} \leq 1$, with the interpretation that professor i teaches course $400 + j$ if and only if both $x_{i,j} = c_{i,j} = 1$. In this way, each row of C determines a constraint, and by similar considerations, each column does the same. Likewise, if we add up every possible product $c_{i,j}x_{i,j}$, we get the total number of courses that will be taught, and this is what we want to maximize. Thus we have modeled the problem by the following LOP.

Problem 1.2.4

$$\text{Max.} \quad z = \sum_{i=1}^{10} \sum_{j=0}^{9} c_{i,j} x_{i,j}$$

$$\text{s.t.} \quad \sum_{j=0}^{9} c_{i,j} x_{i,j} \leq 1 \quad \text{for } 1 \leq i \leq 10$$

$$\sum_{i=1}^{10} c_{i,j} x_{i,j} \leq 1 \quad \text{for } 0 \leq j \leq 9$$

$$\& \quad x_{i,j} \geq 0 \quad \text{for } 1 \leq i \leq 10, 0 \leq j \leq 9.$$

Notice that the values $x_{i,j} = 1/6$ for each variable yield a feasible solution to Problem 1.2.4 producing an objective value of $31/6$. How can anyone teach a sixth of a class, and how can 5.166... classes be offered? It seems there should be one last constraint included in the statement of the problem, that being each variable $x_{i,j}$ should be an integer (that $x_{i,j} \leq 1$ is already implied).

Integer Optimization is a subject which deals with the same programs as Linear Optimization, but with the added constraint that every variable must be an integer. We will not deal with integer linear optimization problems (**ILOPs**) here, but you can imagine in calculus trying to maximize a continuous function of one variable, but restricting your attention to integer

ILOP

values in the domain. The true maximum of 1000 might occur at $x = 1/2$, but the function near $1/2$ might be so spiked that it quickly drops below zero, so fast that $f(x) < 0$ for every $.1 < |x - 1/2| < 50$. Finally, the function rises to an eventually constant level of 6. In this case, maybe the integer maximum over the interval $[0, 100]$ is 6, far lower than 1000, and maybe it occurs at $x = 100$, nowhere near $1/2$. Although we are looking only at linear objective functions, we are also involving many (sometimes thousands of) variables, so it should not be surprising that the integer optimum can be so unrelated to the true optimum. For example, consider the following simple LOP below.

Problem 1.2.5

$$\begin{array}{rrcrcr}
\text{Max.} & z & = & x_1 & + & 1000x_2 \\
\text{s.t.} & & & x_1 & + & 625x_2 \leq 500 \\
\& & & & x_1 & , & x_2 \geq 0
\end{array}$$

Workout 1.2.6 *Prove that Problem 1.2.5 attains its linear maximum of 800 at $(0, .8)$, but its integer maximum of only 500 is far off at $(500, 0)$. [HINT: Would graphing help?]*

Have we set this up enough so that what we are about to say is truly amazing? It turns out to be a consequence of the Integrality Theorem (Theorem 10.4.1) that our Matching Problem (Problem 1.2.4) has a linear maximum solution involving only integer-valued $x_{i,j}$ — that is, in this case, because this problem has a particularly nice structure, the Integrality Theorem says that the integer constraints are quite unnecessary! How different in nature from the Diet Problem this is. We will see this theorem toward the end of the course, and again, it will be a simple observation of the workings of the Simplex Algorithm in a particular setting.

1.3 Un Problema de la Práctica

Consider the following LOP.

Problem 1.3.1

$$\begin{array}{rrcrcr}
\text{Max.} & z & = & x_1 & + & x_2 \\
\text{s.t.} & & & 3x_1 & + & 5x_2 \leq 90 \\
& & & 9x_1 & + & 5x_2 \leq 180 \\
& & & & & x_2 \leq 15 \\
\& & & & x_1 & , & x_2 \geq 0
\end{array}$$

Workout 1.3.2

a. Graph the constraints of Problem 1.3.1.

b. Recall from Calculus where in the domain maxima are allowed to occur. Which of these can be ruled out in this case?

c. Use parts a and b and a little algebra to find the values of x_1^*, x_2^* and z^*.

1.4 Standard Form and the Dual

standard form

Now let's discuss some terminology. Problem 1.3.1 is one which we will refer to as being in **standard form**. There are three reasons for this. First, it is a maximization rather than minimization problem. Second, all the linear inequalities have the linear combination of variables less than or equal to a constant, as opposed to greater than or equal to, or equal to. Third, every variable is nonnegative. As the course progresses, we will consider variations on each of these themes; minimization problems, equality constraints, and variables which can take on negative values. We will always group the components of the problem as above; the objective function, followed by the problem constraints, followed by the nonnegativity constraints.

Minimization LOPs are not significantly different from maximization LOPs for the simple reason that a change of variable converts one to the other. For example, let $v = -z$. Then max $z = -$ min v. Likewise, any inequality $\sum a_j x_j \geq b$ can rewritten as $\sum -a_j x_j \leq -b$, so reversing an inequality is not difficult. In fact, restricting our attention solely to equalities (still with all variables nonnegative) may seem like a qualitative change in our format, but this too is not so different. With the introduction of another inequality, one can convert an equality to standard form as follows. Given $\sum a_j x_j = b$, we can replace it by the pair $\sum a_j x_j \leq b$ and $\sum a_j x_j \geq b$, the latter of the pair then reversed as discussed. Any inequality $\sum a_j x_j \leq b$ can be converted to the equality $\sum a_j x_j + s = b$ slack/ with the introduction of what we call a **slack variable** s, which itself problem must be nonnegative. (Variables which are original to the LOP are called variable **problem variables**.)

So far, it seems that any problem we start out with can be put into standard form without much trouble. Is this true? Any explicit upper bound on a nonnegative variable, such as in Problem 1.3.1, can be considered merely as one of the problem constraints. An explicit lower bound of $x_j \geq l$ can be converted to $x_j' \geq 0$ via the substitution $x_j' = x_j - l$ (in which case the substitution must be carried out in the objective function and problem constraints as well). If a variable x_j has no explicit lower bound, but does have an explicit upper bound $x_j \leq l$, then we use the substitution $x_j' = l - x_j \geq 0$. In any of these cases the variable x_j is called restricted/ **restricted**, while if x_j has neither an explicit upper or lower bound then it free is called free. In this last case, we can use the substitution $x_j = x_j^+ - x_j^-$, variable

1.4. Standard Form and the Dual

with $x_j^+ \geq 0$ and $x_j^- \geq 0$. Thus it is true that any linear problem can be altered so as to be in standard form.

Workout 1.4.1 *Convert Problem 1.3.1 into the form of equalities with free variables. [MORAL: The form of equalities with free variables is not as general as the standard form of inequalities with restricted variables.]*

Problem 1.4.2

$$\begin{aligned} \text{Min.} \quad z &= x_2 \\ \text{s.t.} \quad -2x_1 + 3x_2 &\geq 12 \\ 6x_1 + 5x_2 &= 30 \\ \& \quad x_2 &\geq 0 \end{aligned}$$

Workout 1.4.3 *Write Problem 1.4.2 in standard form.*

Workout 1.4.4 *Solve Problem 1.4.2. [HINT: It may be easiest to do in its original form.]*

We will often use vectors and matrices to convey information, and unless otherwise stated, all vectors will be of the column variety. Thus every problem in standard form (written in summation form),

$$\begin{aligned} \text{Maximize} \quad z &= \sum_{j=1}^{n} c_j x_j \\ \text{subject to} \quad \sum_{j=1}^{n} a_{i,j} x_j &\leq b_i \quad \text{for } 1 \leq i \leq m \\ \text{and} \quad x_j &\geq 0 \quad \text{for } 1 \leq j \leq n \end{aligned} \quad (1.1)$$

can be written in matrix form as

$$\begin{aligned} \text{Max.} \quad z &= \mathbf{c}^\mathsf{T} \mathbf{x} \\ \text{s.t.} \quad \mathbf{A}\mathbf{x} &\leq \mathbf{b} \\ \& \quad \mathbf{x} &\geq \mathbf{0}, \end{aligned} \quad (1.2)$$

where $\mathbf{c}^\mathsf{T} = (c_1, \ldots, c_n)$, $\mathbf{x} = (x_1, \ldots, x_n)^\mathsf{T}$, $\mathbf{b} = (b_1, \ldots, b_m)^\mathsf{T}$, $\mathbf{0}$ is the $(n \times 1)$ column vector of all zeros, and $\mathbf{A} = [a_{i,j}]$ is the $(m \times n)$ matrix of coefficients in the problem constraints.

Workout 1.4.5 *Write the matrix $\mathbf{A}$ and vectors $\mathbf{c}$, $\mathbf{x}$ and $\mathbf{b}$ for the standard form version of Problem 1.3.1.*

(in)feasible point/ solution

infeasible/ unbounded problem

feasible objective value

Before we try to solve Problem 1.3.1, let's continue our habit of looking for estimates. We call a point (vector, solution) **x feasible** if it satisfies all the problem and nonnegativity constraints, and **infeasible** otherwise. (An **infeasible problem** is one that has no feasible points — its constraints are unsolvable. An **unbounded problem** is a feasible problem with no optimum.) Thus, we can say that we are trying to maximize the function $z = z(\mathbf{x})$ over all feasible points $\mathbf{x}$. (We will also call a particular objective value **feasible** if it is equal to $z(\mathbf{x})$ for some feasible $\mathbf{x}$.) If we denote this maximum by z^*, then $z^* \geq z(\mathbf{x})$ for all feasible $\mathbf{x}$. In Problem 1.3.1, since the point $\mathbf{x} = (10, 12)^\mathsf{T}$ is feasible and $z(10, 12) = 22$, we get $z^* \geq 22$. In contrast, it doesn't help to compute $z(10, 13) = 23$ because the point $(10, 13)^\mathsf{T}$ is infeasible. It cannot, then, help us to determine if $z^* \geq 23$ or if $z^* \leq 23$.

In order to get an upper bound we might notice that $z = x_1 + x_2 \leq 12x_1 + 11x_2$, because from $x_1, x_2 \geq 0$ follow both $12x_1 \geq x_1$ and $11x_2 \geq x_2$. We didn't just pull 12 and 11 out of a hat: $12x_1 + 11x_2 = (3x_1 + 5x_2) + (9x_1 + 5x_2) + (x_2)$. From here, we can use the problem constraints to say $(3x_1 + 5x_2) + (9x_1 + 5x_2) + (x_2) \leq 90 + 180 + 15 = 285$. All of this implies $z = z(x) \leq 285$ for all feasible x, which means $z^* \leq 285$. A better upper bound would be found by lowering the estimate $12x_1 + 11x_2$ on z, for example,

$$\begin{aligned} z = x_1 + x_2 &= \frac{1}{12}[12x_1 + 12x_2] \\ &\leq \frac{1}{12}[(3x_1 + 5x_2) + (9x_1 + 5x_2) + 2(x_2)] \\ &\leq \frac{1}{12}[300] = 25, \end{aligned}$$

so that $z^* \leq 25$.

dual multipliers

Here, we just used the multipliers $y_1 = \frac{1}{12}$, $y_2 = \frac{1}{12}$, and $y_3 = \frac{1}{6}$, respectively, on the problem constraints 1, 2, and 3. Because each of these multipliers was nonnegative, none of the inequalities got turned around. We could hope to use the following analysis as a general upper bound.

$$\begin{aligned} z = x_1 + x_2 &\leq (3y_1 + 9y_2)x_1 + (5y_1 + 5y_2 + y_3)x_2 \\ &= y_1(3x_1 + 5x_2) + y_2(9x_1 + 5x_2) + y_3(x_2) \\ &\leq y_1(90) + y_2(180) + y_3(15) = w. \end{aligned}$$

For this we would need the coefficients in the first inequality to either increase or stay the same, that is $1 \leq 3y_1 + 9y_2$ and $1 \leq 5y_1 + 5y_2 + y_3$. Under these conditions, we would be clever to find which nonnegative multipliers would produce the smallest upper bound w. In other words, we have constructed the following LOP.

1.4. Standard Form and the Dual

Problem 1.4.6

$$\text{Min. } w = 90y_1 + 180y_2 + 15y_3$$

$$\text{s.t. } \begin{array}{rcl} 3y_1 + 9y_2 & \geq & 1 \\ 5y_1 + 5y_2 + y_3 & \geq & 1 \end{array}$$

$$\&\quad y_1,\ y_2,\ y_3 \geq 0$$

Problem 1.4.6 is called the **dual** to Problem 1.3.1, which we will often refer to as the **primal** problem. The variables belonging to each will also be named primal and dual, respectively.[1] In general, we can write the primal and dual problems, written in summation form,

primal/ dual problem/ variable

Primal		Dual	
Max. $z = \sum_{j=1}^{n} c_j x_j$		Min. $w = \sum_{i=1}^{m} b_i y_i$	z, c_j, x_j
s.t. $\sum_{j=1}^{n} a_{i,j} x_j \leq b_i$ $(1 \leq i \leq m)$		s.t. $\sum_{i=1}^{m} a_{i,j} y_i \geq c_j$ $(1 \leq j \leq n)$	w, b_i, y_i
& $x_j \geq 0$ $(1 \leq j \leq n)$		& $y_i \geq 0$ $(1 \leq i \leq m)$	$a_{i,j}$

in matrix form as

Max. $z = \mathbf{c}^\mathsf{T} \mathbf{x}$	Min. $w = \mathbf{b}^\mathsf{T} \mathbf{y}$	**c, x**
s.t. $\mathbf{A}\mathbf{x} \leq \mathbf{b}$	s.t. $\mathbf{A}^\mathsf{T} \mathbf{y} \geq \mathbf{c}$	**b, y**
& $\mathbf{x} \geq \mathbf{0}$	& $\mathbf{y} \geq \mathbf{0}$.	**A**

Then, for every primal-feasible x and dual-feasible y, we have

Inequality 1.4.7

Weak Duality Theorem

$$z = \sum_{j=1}^{n} c_j x_j \leq \sum_{j=1}^{n}\left(\sum_{i=1}^{m} a_{i,j} y_i\right) x_j = \sum_{i=1}^{m}\left(\sum_{j=1}^{n} a_{i,j} x_j\right) y_i \leq \sum_{i=1}^{m} b_i y_i = w .$$

This means that $z^* \leq w^*$, because the inequality holds for every pair of feasible values z and w, z^* and w^* being one particular case. This primal-dual relationship can be expressed more succinctly as

$$z = \mathbf{c}^\mathsf{T}\mathbf{x} \leq \mathbf{y}^\mathsf{T} \mathbf{A} \mathbf{x} \leq \mathbf{y}^\mathsf{T} \mathbf{b} = w .$$

It is important that we remember that these inequalities can only be used when <u>both</u> **x** and **y** are feasible (in particular, nonnegative — see Exercise 1.5.9)!

---

[1] While the meaning of the primal variables is given in the problem, the meaning of the dual variables is not as transparent, but is discussed in Chapter 12.

It is also instructive to note that, because of weak duality, if we ever come across **x** and **y** that satisfy $\mathbf{c}^\mathsf{T}\mathbf{x} = \mathbf{y}^\mathsf{T}\mathbf{b}$, then we know we have found both z^* and w^*. While such **x** and **y** can be found by solving the respective primal and dual LOPs, we will see in Chapter 4 that, once $\mathbf{x}^*$ has been found, we can produce the certificate of optimality $\mathbf{y}^*$ without solving the dual (see Exercise 1.5.10).

1.5 Exercises

Practice

1.5.1 *Write each of the following LOPs in standard form.*

a.

$$\text{Max. } z = 2x_1 - 3x_2$$
$$\text{s.t. } \quad x_1 + 2x_3 \leq 5$$
$$-4x_1 + x_2 + 3x_3 \leq 8$$
$$-x_1 + 4x_2 - 9x_3 \geq 3$$
$$3x_2 + 3x_3 \leq 5$$
$$\& \quad x_1, \; x_2, \; x_3 \geq 0$$

b.

$$\text{Min. } w = -5y_1 + y_2$$
$$\text{s.t. } \quad 3y_1 + 4y_2 \leq -3$$
$$y_1 - 2y_2 \leq -2$$
$$6y_1 \leq 1$$
$$\& \quad y_1 \geq 0$$

c.

$$\text{Min. } w = 4y_1 - 6y_2 - 2y_3$$
$$\text{s.t. } -2y_1 + 3y_3 = 5$$
$$y_2 - 2y_3 \leq 7$$
$$\& \quad y_1, \; y_2, \; y_3 \geq 0$$

1.5. Exercises

1.5.2 Consider the following LOP P.

$$\text{Max. } z = 4x_1 + 8x_2$$

s.t.
$$\begin{aligned}
x_1 - x_2 &\leq -1 \\
2x_1 - x_2 &\leq 5 \\
-3x_1 - x_2 &\leq -3 \\
2x_1 + x_2 &\leq 7 \\
10x_1 + x_2 &\leq 20
\end{aligned}$$

& $x_1, x_2 \geq 0$

Decide whether each of the following solutions are feasible or infeasible.

a. $\mathbf{x} = (2,3)^\mathsf{T}$;
b. $\mathbf{x} = (1,4)^\mathsf{T}$;
c. $\mathbf{x} = (0,7)^\mathsf{T}$.
d. Graph the set of points that satisfy the constraints of P.

1.5.3 Consider the following LOP P.

$$\text{Max. } z = 3x_1 + 3x_2 + 3x_3$$

s.t.
$$\begin{aligned}
-5x_1 + x_3 &\leq -1 \\
4x_1 + 5x_2 + 6x_3 &\leq 14 \\
-3x_3 &\leq 1 \\
x_1 - 7x_2 - x_3 &\leq -5
\end{aligned}$$

& $x_1, x_2, x_3 \geq 0$

Decide whether each of the following solutions are feasible or infeasible.

a. $\mathbf{x} = (73, 34, 0)^\mathsf{T}/33$;
b. $\mathbf{x} = (55, 62, 166)^\mathsf{T}/109$;
c. $\mathbf{x} = (1, 1, 1)^\mathsf{T}$.
d. Draw the set of points that satisfy the constraints of P. [HINT: First reduce the number of constraints by discarding redundant ones (constraints that are implied by a collection of other constraints).] [MORAL: A little 3-dimensional drawing and visualization never hurt anyone!]

1.5.4 Consider the following LOP P.

$$\text{Max. } z = 2x_1 - 3x_2 + 4x_4 - x_5$$

s.t.
$$\begin{aligned}
x_1 + x_2 + 3x_5 &\leq 2 \\
2x_1 + x_4 - x_5 &\leq 6 \\
x_3 + 2x_4 + 3x_5 &\leq 4
\end{aligned}$$

& $x_1, x_2, x_3, x_4, x_5 \geq 0$

Decide whether each of the following solutions are feasible or infeasible.

  a. $\mathbf{x} = (11, 0, 0, 0, 4)^\mathsf{T}/3$;

  b. $\mathbf{x} = (2, 0, 0, 2, 0)^\mathsf{T}$;

  c. $\mathbf{x} = (0, 2, 0, 2, 0)^\mathsf{T}$.

1.5.5 *Consider the following LOP P.*

$$\begin{aligned}
\text{Max.} \quad z &= 3x_1 + 4x_2 \\
\text{s.t.} \quad 2x_1 &- 3x_2 \leq 3 \\
4x_1 &+ x_2 \leq 6 \\
x_1 &+ x_2 \leq 5 \\
\& \quad x_1 &, x_2 \geq 0
\end{aligned}$$

  a. *Find a primal feasible solution $\mathbf{x}$ and its corresponding objective value $z = z(\mathbf{x})$.*

  b. *Write the LOP D that is dual to P.*

  c. *Find a dual feasible solution $\mathbf{y}$ and its corresponding objective value $w = w(\mathbf{y})$.*

  d. *What upper and lower bounds do parts a and c produce for z^*?*

1.5.6 *Repeat Exercise 1.5.5 N times, each with a different modestly sized LOP of your own design. [MORAL: Many exercises in the book can be repeated by the reader simply by making up a new LOP.]*

1.5.7 *Consider the following LOP P*

$$\begin{aligned}
\text{Max.} \quad z &= 5x_1 + 5x_2 + 5x_3 \\
\text{s.t.} \quad x_1 &+ 2x_2 + 3x_3 \leq 4 \\
4x_1 &+ 3x_2 + 2x_3 \leq 1 \\
\& \quad x_1 &, x_2, x_3 \geq 0
\end{aligned}$$

  a. *Write the matrices $\mathbf{A}, \mathbf{b}, \mathbf{c}$ which correspond to P.*

  b. *Write the LOP D that is dual to P.*

  c. *Prove, for this particular P and D, that $z \leq w$, as in Inequality 1.4.7. Explain each step.*

1.5.8 *Repeat Exercise 1.5.7 N times, each with a different modestly sized LOP of your own design. Try LOPs for which the number of variables is different from the number of problem constraints.*

1.5. Exercises

1.5.9 *Consider the following LOP.*

$$\text{Max. } z = x_1 + 2x_2 + 3x_3 + 4x_4$$

$$\text{s.t.} \quad \begin{aligned} x_1 + 3x_2 + x_3 + 4x_4 &\leq 6 \\ x_1 + 7x_2 + 3x_3 + 9x_4 &\leq 12 \\ x_1 + 6x_2 + 5x_3 + 9x_4 &\leq 8 \end{aligned}$$

$$\& \quad x_1, \ x_2, \ x_3, \ x_4 \geq 0$$

a. *Find the dual multipliers that yield the following constraint.*

$$x_1 + 2x_2 + 3x_3 + 4x_4 \leq 2$$

b. *Does it follow from part a that $z^* \leq 2$?*

[MORAL: Multipliers of standard form LOP constraints *must* be nonnegative!]

1.5.10 *Consider the following LOP P.*

$$\text{Max. } z = -12x_1 - 11x_2 - 13x_3$$

$$\text{s.t.} \quad \begin{aligned} -x_1 + x_2 &\leq -2 \\ -x_2 - x_3 &\leq -3 \\ x_1 + x_3 &\leq 5 \end{aligned}$$

$$\& \quad x_1, \ x_2, \ x_3 \geq 0$$

a. *Write the LOP D that is dual to P.*
b. *Consider the point* $\mathbf{x} = (2, 0, 3)^\mathsf{T}$.
  (i) Show that $\mathbf{x}$ is P-feasible.
  (ii) Find $z(\mathbf{x})$.
c. *Consider the point* $\mathbf{y} = (12, 13, 0)^\mathsf{T}$.
  (i) Show that $\mathbf{y}$ is D-feasible.
  (ii) Find $w(\mathbf{y})$.
d. *Use parts b and c to find z^*.*
e. *Consider the point* $\mathbf{y} = (13, 14, 1)^\mathsf{T}$.
  (i) Show that $\mathbf{y}$ is D-feasible.
  (ii) Find $w(\mathbf{y})$. [MORAL: a given LOP may have several optimal solutions.]

1.5.11 *Consider the following LOP P.*

$$\text{Max. } z = 5x_1 + 3x_2$$

$$\text{s.t.} \quad \begin{aligned} x_1 + 2x_2 &\leq 14 \\ 3x_1 - 2x_2 &\leq 18 \\ x_1 - 2x_2 &\geq -10 \end{aligned}$$

$$\& \quad x_1, \ x_2 \geq 0$$

a. Graph the system of constraints of P.

b. Draw the following lines on your graph in part a. (These are usually referred to as **level curves** in Calculus, or **contour lines** in cartography.)

level curves/ contour lines

   (i) $z = 12$ (i.e., $5x_1 + 3x_2 = 12$);
   (ii) $z = 24$;
   (iii) $z = 36$.

c. Plot the following points on your graph in part a:
   (i) $\mathbf{x} = (2,3)^\mathsf{T}$;
   (ii) $\mathbf{x} = (4,6)^\mathsf{T}$;
   (iii) $\mathbf{x} = (8,3)^\mathsf{T}$.

d. Find $\mathbf{x}^*$ and z^*.

e. Write the LOP D that is dual to P.

f. Find $\mathbf{y}^*$. [HINT: Find dual-feasible $\mathbf{y}$ such that $w(\mathbf{y}) = z^*$.]

1.5.12 Repeat Exercise 1.5.11 N times, each with a different LOP of your own design, having 2 variables and 3 constraints. [NOTE: You may want to reverse-engineer this — that is, derive the constraints from a graph you draw.] You will need to choose your own objective lines to draw (two are enough), and find which points are important to plot.

1.5.13 Write the dual of the following LOP.

$$\begin{array}{rl} \text{Max.} \quad z = & \sum_{j=1}^{n} c_j x_j \\ \text{s.t.} & \sum_{j=1}^{n} a_j x_j \leq b \\ & x_j \leq 1 \quad (1 \leq j \leq n) \\ \& & x_j \geq 0 \quad (1 \leq j \leq n) \end{array}$$

Challenges

1.5.14 Consider the following LOP P.

$$\begin{array}{rl} \text{Max.} \quad z = & -2x_2 \\ \text{s.t.} & x_1 + 4x_2 - x_3 \leq 1 \\ & -2x_1 - 3x_2 + x_3 \leq -2 \\ & 4x_1 + x_2 - x_3 \leq 1 \\ \& & x_1,\ x_2,\ x_3 \geq 0 \end{array}$$

a. Write the dual D of P.

b. Show that $\mathbf{y} = (2, 3, 1)^\mathsf{T}$ is D-feasible.

1.5. Exercises

    c. Use part b to show that P is infeasible.

    d. Show that $\mathbf{y}(t) = (2t, 3t, t)^\mathsf{T}$ is D-feasible for all $t \geq 0$.

    e. Use part d to prove that D is unbounded.

[MORAL: Simple certificates like $\mathbf{y}$ are very powerful tools in LO.]

1.5.15 *Repeat part a N times, each with a different LOP of your own design.*

    a. Let P be a LOP in standard max form and let D be its dual LOP. Write D in standard max form as D' and let Q' be its dual. Finally, write Q' in standard max form as Q. Compare P and Q.

    b. Prove a statement about the relationship between P and Q in general.

1.5.16 *Consider the problem: Max. $\mathbf{c}^\mathsf{T}\mathbf{x}$ s.t. $\mathbf{A}\mathbf{x} = \mathbf{b}$ (note that $\mathbf{x}$ is free).*

    a. Let $V = \{\mathbf{v}^i\}_{i=1}^k$ be a basis for the nullspace of $\mathbf{A}$. Write all solutions to $\mathbf{A}\mathbf{x} = \mathbf{b}$ in terms of V. [HINT: Recall Gaussian elimination.]

    b. Use part a to prove that if $\mathbf{c}^\mathsf{T}\mathbf{v}^i = 0$ for all $1 \leq i \leq k$ then every feasible point is optimal.

    c. Prove that if $\mathbf{A}\mathbf{x} = \mathbf{b}$ is feasible then there is some $\mathbf{v}$ and feasible $\mathbf{x}^0$ such that $\mathbf{x}^0 + t\mathbf{v}$ is feasible for all $t \in \mathbb{R}$.

    d. Use part c to prove that if $\mathbf{c}^\mathsf{T}\mathbf{v}^i \neq 0$ for some $1 \leq i \leq k$ then the problem is unbounded.

    e. Use parts b and d to prove that every such problem is either infeasible, optimal or unbounded.

1.5.17 *Write pseudocode for an algorithm that takes as input a LOP with n variables and m constraints and outputs its dual LOP.*

Modeling

1.5.18 *A factory manufactures two products, each requiring the use of three machines. The first machine can be used at most 70 hours; the second machine at most 40 hours; and the third machine at most 90 hours. The first product requires 2 hours on machine 1, 1 hour on machine 2, and 1 hour on machine 3; the second product requires 1 hour on machines 1 and 2 and 3 hours on machine 3. The profit is \$40 per unit for the first product and \$60 per unit for the second product. Write a LOP that will compute how many units of each product should be manufactured in order to maximize profit.*

1.5.19 *A birchwood table company has an individual who does all its finishing work and it wishes to use him in this capacity at least 36 hours each week. By union contract, the assembly area can be used at most 48 hours each week. The company has three models of birch tables, T_1, T_2 and T_3. T_1 requires 1 hour for assembly, 2 hours for finishing, and 9 board feet of birch. T_2 requires 1 hour for assembly, 1 hour for finishing and 9 board feet of birch. T_3 requires 2 hours for assembly, 1 hour for finishing and 3 board feet of birch. Write a LOP that will compute how many of each model should be made in order to minimize the board feet of birchwood used.*

1.5.20 *The State of Florida must make two types of ballots, A and B, which they will use for an election, using three types of materials: construction paper, tissue paper, and ink. Ballot A uses 210 cm^2 of construction paper, 35 cm^2 of tissue paper, and 3 tsp of ink, and generates 7 chads. Ballot B uses 190 cm^2 of construction paper, 55 cm^2 of tissue paper, and 2 tsp of ink, and generates 3 chads. The State must generate at least 70,000 chads, but only 2.8 million cm^2 of construction paper, .63 million cm^2 of tissue paper, and 35 thousand tsp of ink are available. Moreover, construction paper costs 23 cents per 100 cm^2, tissue paper costs 2 cents per 100 cm^2, and ink costs 15 cents per tsp. Assuming that the State of Florida wishes to minimize its cost, write down the associated LOP.*

1.5.21 *Farmer Brown has 50 acres on which she can grow arugula or broccoli. Each acre of arugula requires $10 in capital costs and uses 5 hours of labor. Each acre of broccoli requires $7 in capital costs and uses 3 hours of labor. Labor is $6 per hour. The sale of arugula yields $200 per acre and the sale of broccoli yields $80 per acre.*

> a. *Suppose that $1950 is available for capital expenses and labor. Formulate a LOP whose solution determines how much of each should be planted in order to maximize profit.*
>
> b. *Reformulate if, in addition, only $450 of the funds in (a) are available for capital costs.*
>
> c. *Which of the LOPs in parts a or b will have a larger maximum? Why?*
>
> d. *Write down the duals for each of a and b.*
>
> e. *Which of the duals will have a larger minimum? Why?*

1.5.22 *Eumerica makes bottled air at three plants in Vienna, Athens, and Moscow, and ships crates of their products to distributors in Venice, Frankfurt, and Paris. Each day the Athens plant produces 25 thousand crates, while Vienna can produce up to 18 thousand, and Moscow can produce up to 15 thousand. In addition, Venice must receive 14 thousand and Paris must receive 22 thousand crates, while Frankfurt can receive up to 19 thousand. The company pays Arope Trucking to transport their products at the following per-crate Eurodollar costs.*

120	Vienna to Frankfurt		240	Venice to Paris
100	Frankfurt to Athens		250	Paris to Venice
120	Athens to Frankfurt		290	Frankfurt to Venice
150	Frankfurt to Paris		270	Venice to Moscow
130	Paris to Athens		280	Moscow to Frankfurt
160	Athens to Vienna			

Eumerica would like to tell Arope which shipments to make between cities so as to minimize cost. Write the ILOP that solves this problem.

1.5.23 CarbonDating.com keeps a database of their clients and their love interests (for unrealism, we assume symmetry: if A loves B then B loves A). Annette loves David, John and Warren, Kathy loves Bill, John and Regis, Monica loves Bill, David and Warren, Teresa loves Bill, John and Regis, and Victoria loves David, Regis and Warren. A marriage of a woman and a man is good if the couple love each other. The company would like to find as many pairwise disjoint, good marriages as possible. Write the corresponding ILOP.

1.5.24 The Police Chief of Gridburg decides to place 15 policemen at the 15 street corners of his town, as shown in the map below. A policeman has the ability to see the activity of people on the streets leading from his street corner, but only as far as one block. The Mayor fires the Chief for spending over budget, and hires Joseph Blough to make sure that every street can be seen by some policeman, using the fewest possible policemen. Write Joe's ILOP that solves this problem.

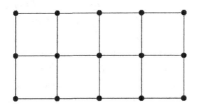

1.5.25 The Commerce Secretary of Gridburg decides to place 3 hot dog vendors on the 3 East-West streets of Gridburg (see the map above). The Hotdogger's Union requires that no two vendors can be on street blocks that share an intersection. The Mayor fires the Secretary for not generating enough commerce in town, and hires Anna Benannaugh to place the maximum number of vendors, subject to union rules. Write Anna's ILOP.

1.5.26 Consider the following six committees of students. Committee 1: Clifford, Kara; Committee 2: Ben, Donyell, Jake, Rebecca; Committee 3: Clifford, Kara, Sue; Committee 4: Ben, Jake, Kara, Nykesha, Rebecca; Committee 5: Clifford, Sue; Committee 6: Kara, Sue. Each committee must choose a representative to send to the school Senate, but the Senate requires that no person represent more than one committee. Write the ILOP that computes whether or not this is possible.

1.5.27 Öreg MacDonald owns 1,000 acres of land and is contemplating conserving, farming, and/or developing it. His annual considerations are as follows. It will only cost him $1 per acre in registration fees to own conservation land, and he will reap $30 per acre in tax savings. Farming will cost him $50 per acre for seeds, from which he can earn $190 per acre by selling vegetables. He can earn $290 per acre by renting developed land, which costs $85 per acre in permits. Öreg has only $40,000 to use, and is also bound by having only 75 descendants, each of whom can work at most 2,000 hours. How should he apportion his acreage in order to maximize profits, if conservation, farming, and development uses 12, 240, and 180 hours per acre, respectively? Write the appropriate LOP that solves this.

1.5.28 Biff has infestations of crickets, ants, and moths in his house. He estimates that there are 50 ounces of crickets, 20 ounces of ants, and 15 ounces of moths, and realizes that they must be removed before his girlfriend Muffy arrives in an hour. He could buy cockroaches, a pound of which would cost 48¢ and eat 5 ounces of ants and 3 ounces of moths in an hour. He could also purchase a pound of black widow spiders that would eat 6 ounces of crickets, 3 ounces of ants, and 2 ounces of moths per hour for 73¢. A pound of scorpions would eat 8 ounces of crickets and 4 ounces of ants per hour and would cost Biff 56¢. His final choice is to spend 93¢ for a pound of rough green snakes that would eat 11 ounces of crickets and 1 ounce of moths per hour. Formulate Biff's LOP.

1.5.29 Aussie Foods Co. makes three different emu pet foods in 10-kg bags. The Premium bag is a mixture of 5 kgs of kiwi fruit, 2 kgs of wattle leaves, 2 kgs of boab seeds, and 1 kg of ground diamond weevil, and AFC makes a profit of 91¢ per bag sold. The Regular bag mixes 4, 4, 0, and 2 kgs of kiwi, wattle, boab, and diamond, respectively, making 84¢ for AFC. The corresponding numbers for the Bargain bag are 1, 2, 3, 4, and 73¢, respectively. The weekly supply available to AFC is 1,000 kgs of kiwi fruit, 1,200 kgs of wattle leaves, 1,500 kgs of boab seeds, and 1,400 kgs of ground diamond weevil. Write the ILOP that Aussie Foods should solve in order to maximize their weekly profit. (Would the LOP have relevance to them?)

1.5.30 Consider the matrix $\mathbf{A}$, below.

$$\begin{pmatrix} 3 & 0 & -2 \\ -1 & 4 & 0 \\ 2 & -3 & 1 \\ 0 & -2 & 1 \end{pmatrix}$$

During a probability calculation, Carlos realizes he needs to approximate the vector $\mathbf{Ax}$ by something simpler, for every probability vector $\mathbf{x}$. In particular, he needs to satisfy $\mathbf{Ax} \geq x_0 \mathbf{J}_4$, where $\mathbf{J}_k$ is the vector of k ones. Write a LOP that finds the maximum of all such x_0 that that Carlos can use.

1.5. Exercises

1.5.31 *Kingsbury's butcher is asked to grind up several cuts of meat to form a blend of equal parts of proteins and fats. The butcher, being conscientious, wishes to do this at least cost per pound of meat purchased. The following table gives fat and protein contents and costs in dollars.*

	Rib	Thigh	Breast	Rump	Calf	Forearm	Neck
%Protein	19	20	16	17	19	16	17
%Fat	16	18	25	23	11	28	20
cost/lb	.69	.98	1.39	1.29	1.19	1.50	1.65

a. *Write a LOP that will compute the amounts of meat and how much the butcher should charge.*

b. *Usually the butcher has extra fat available free per pound. How does this alter the LOP? (Does this LOP have an obvious solution?)*

1.5.32 *The Hendrix factory buys bags of sand and produces sand castles. Each sand castle requires one bag of sand and the factory has a production capacity of 3,000 sand castles per quarter year. However, sand is available in different amounts and sand castles are required for sale or distribution in different amounts each quarter. Furthermore, storing sand castles is expensive and carrying them over from one quarter to the next is to be minimized. At the beginning of the year, 3,000 sand bags are available and at least this many must be left over at the end of the year. The availability of sand bags and requirements for sand castles per quarter is as follows:*

quarter	sand bags available for purchase	sand castles required for sale
1	5,000	1,000
2	3,000	4,000
3	1,000	3,000
4	2,000	1,500

There is storage room available for 10,000 sand bags or 2,000 sand castles or any combination in this ratio. (That is, in quarter q, if B_q and C_q respectively represent the number of sand bags and sand castles on hand at the end of the quarter, then $B_q + 5C_q \leq 10,000$. Here we ignore bottlenecks during a quarter.)

Write down an ILOP that will compute purchases of sand bags and the number of sand castles made for each quarter, minimizing carryover of sand castles, subject to the availability and requirement constraints.

1.5.33 *Hal's Refinery can buy two types of gasoline. Boosch Oil has available, at $60 per barrel, 130,000 barrels of 92 octane gasoline with vapor pressure 4.6 psi and sulfur content 0.58%. Chayni Oil has available, at $70 per barrel, 140,000 barrels of 85 octane gasoline with vapor pressure 6.5 psi and sulfur content 0.40%. Hal needs to blend these two to produce at least*

200,000 barrels of a mixture with octane between 87 and 89, with vapor pressure at most 6.0 psi and sulfur content at most 0.50%. Formulate a LOP to determine the proportions of each type he should use to minimize his cost.

1.5.34 The system of equations

$$\begin{aligned} x_1 + 4x_2 - x_3 &= 2 \\ -2x_1 - 3x_2 + x_3 &= 1 \\ -3x_1 - 2x_2 + x_3 &= 0 \\ 4x_1 + x_2 - x_3 &= -1 \end{aligned}$$

has no solution. A 'best' approximate solution minimizes the error according to some measure. For a given (x'_1, x'_2, x'_3) the error e_1 in the first equation is $e_1 = 2 - x_1 - 4x_2 + x_3$ and similarly for the errors e_2, e_3, e_4 for the other rows.

a. Write a LOP that will compute the best L_1 approximation, which minimizes the sum of the absolute values of the errors.

b. Write a LOP that will compute the best L_∞ approximation, which minimizes the maximum absolute value of an error.

[HINT: Creating a new variable that is an upper bound on both e_i and on $-e_i$ makes it an upper bound on $|e_i|$.]

○ **1.5.35** The streets of Old Yorktown are set up in an orthogonal grid, with parallel North-South streets every tenth of a mile and parallel East-West streets every tenth of a mile. Three subway systems eminate from the center, taking passengers about the town in the following manner. The Red trains travel in the direction of the vector $(11, 4)^\mathsf{T}$ and its negative, making station stops at integer multiples of it (in tenths of a mile units). The Yellow and Blue trains travel along integer multiples of $(9, 13)^\mathsf{T}$ and $(8, -5)^\mathsf{T}$, respectively. The subway lines are coordinated so that every color can stop at each station. For example, one could take any combination of 3 Red, -2 Yellow (meaning the opposite direction), and 5 Blue trains in order to reach the station 5.5 miles East and 3.9 miles South of center (coordinates $(55, -39)^\mathsf{T}$). Kate lives at $(-35, 6)^\mathsf{T}$, her brother Calvin at $(20, 23)^\mathsf{T}$, and they are planning to meet for lunch at Le Café Barphe in Central Station for lunch.

a. Their reservation allows them only enough time to travel for at most 6 stops. Set up the ILOPs that must be solved in order to minimize the walking distances along the streets from their houses to nearby subway stations.

b. Suppose it is Sunday, so they have no time constraints and the Yellow subway isn't running. Set up the ILOP to solve this case.

1.5.36 *Melissa has 9 Christmas packages to deliver to her neighbors. Three are wrapped in red paper, three in green, and three in blue. She enlists her three children to help carry the packages and, of course, they insist on each having a package of each color. Furthermore, whoever has the heaviest packages will complain, so Melissa wants to split up the packages so that the heaviest and lightest set of three differ in weight by as little as possible. Write the ILOP she must solve in order to achieve this, supposing that the weights of the red packages are 50oz, 37oz, and 33oz, the weights of the green packages are 62oz, 55oz, and 24oz, and the weights of the blue packages are 48oz, 44oz, and 29oz. [HINT: Consider a BLOP.]*

1.5.37 *Two warehouses have canned tomatoes on hand and three stores require more in stock, as described in the table below.*

Warehouse	Cases on hand	Store	Cases required
I	100	A	75
II	200	B	125
		C	100

The cost (in cents) of shipping between warehouses and stores per case is given in the following table.

	A	B	C
I	10	14	30
II	12	20	17

a. *Set up an ILOP to minimize the total shipping cost.*

b. *Reformulate part a, assuming the cases required at Store B are only 60 and introducing a disposal activity at the warehouses at a loss of 5¢ per case disposed.*

c. *Reformulate part a, assuming that the cases available at Warehouse I are only 90 and introducing a purchase activity from outside sources at a cost of 45¢ per case.*

d. *Write down the duals to the LOPs obtained by ignoring the integrality constraints (the LOP relaxations) for each of parts a, b and c.*

e. *Generalize part a to r warehouses and s stores with cost c_{kl} to ship from Warehouse k to Store l. Denote the cases on hand at Warehouse k by b_k^W and the requirements at Store l by b_l^S.*

f. *What happens in part e if the total cases on hand in the warehouses is not equal to the total requirements at the stores?*

1.5.38 *Curly, Larry, and Moe have to pay Huey, Dewey, and Louie monies* of varying amounts, which are under dispute. Below is a chart of what HDL claims CLM owe them.

	H	D	L
C	200	150	420
L	240	200	450
M	100	80	370

In order to settle the dispute, a judge decided that CLM will each pay one of HDL in such a way that each of HDL will receive something. Write an ILOP that solves how the payments should be made so that the amount of money paid in total is maximized. [HINT: Consider a BLOP.]

○ **1.5.39** When the gates open at Sidney Planet amusement park, the patrons all rush to the Spaced Out Center. One path goes through Mikey Moose Square. The path from the entrance to the square can handle up to 500 patrons and the path from the square to the center can handle at most 400 patrons. Patrons can also go from the entrance via Gumbo's Restaurant and then Large World Park with the path on the first leg handling up to 300 patrons, the second leg only 100 patrons and the third leg at most 400 patrons. There are also paths from Gumbo's to the square and from the square to the park handling no more than 100 and 300 patrons, respectively. Formulate an ILOP whose solution will give the maximum number of patrons who can get to the Center without exceeding path capacity. See if you can determine the answer by drawing a small diagram.

○ **1.5.40** In the game of Odds and Evens, Pete and Repete simultaneously put out either one or two fingers. If the total number of fingers shown is even then Repete pays Pete one dollar for each finger shown. If the total is odd then Pete pays Repete three dollars. The game will be played repeatedly. Pete will randomly show one finger with probability x_1 and two fingers with probability x_2. Formulate a LOP to determine the probabilities that will maximize the minimum expected gain for Pete. That is, determine x_1 and x_2 so that the minimum of the two expected gains, one for each of Repete's pure strategies, is as large as possible.

○ **1.5.41** Chamique, Diana, Lisa, Sheryl and Yolanda are to be seated at one of three tables (labeled 1, 2, and 3) for dinner. There are certain pairs who are not willing to sit together: Chamique and Diana, Diana and Lisa, Lisa and Sheryl, Sheryl and Yolanda, and Yolanda and Chamique. Formulate an integer system of inequalities with variables A, B, C, D, E taking on possible values 1,2,3 whose solutions correspond to feasible seatings. [HINT: You will need to introduce extra variables.]

○ **1.5.42** Pat has the following cash flow over the first four months of the year (negative indicates loss which must be covered).

Month	J	F	M	A
Cash Flow	−200	300	−50	100

1.5. Exercises

He or she can borrow up to $400 on an equity line of credit at a rate of 20% per month (due at the end of the month) and invest excess funds earning 10% per month. Formulate a LOP whose solution determines how much he or she should borrow and how much he or she should carry over in excess each month in order to maximize his or her worth at the end of the four months?

1.5.43 *Anthony wants to make a cassette tape of Sallie's favorite songs and compiles the following list.*

Time	Artist	Song
4:35	Sonny & Cher	I Got You, Babe
2:30	The Monkees	Your Auntie Grizelda
1:50	Tiny Tim	Tiptoe Through the Tulips
2:09	The Partridge Family	Come On Get Happy
2:15	Kermit the Frog	Being Green
2:35	The Archies	Bicycles, Roller Skates and You
2:07	The Sugar Bears	Happiness Train
2:26	Jimmy Osmond	Long Haired Lover From Liverpool
3:20	Tony Orlando and Dawn	Tie a Yellow Ribbon Round the Old Oak Tree
2:32	Paul Anka	Having My Baby
4:13	George Segal	If You Like-a-Me
1:41	Hee Haw Gospel Quartet	Turn Your Radio On
3:30	Van McCoy	Do the Hustle
2:56	Bay City Rollers	Saturday Night
3:05	Rick Dees	Disco Duck
5:45	Kajagoogoo	Too Shy
3:38	Madonna	Like a Virgin
4:16	New Kids on the Block	Hangin' Tough
4:31	Vanilla Ice	Ice Ice Baby
3:23	Billy Ray Cyrus	Achy Breaky Heart
2:50	Right Said Fred	I'm Too Sexy
1:00	Barney	I Love You
2:57	William Shatner	Lucy in the Sky with Diamonds
4:28	Hanson	Mmm...Bop
4:41	Celine Dion	My Heart Will Go On
3:41	Kenny Chesney	She Thinks My Tractor's Sexy
3:33	Britney Spears	Oops, I Did it Again
3:09	Bob the Builder	Can We Fix It
0:35	Ashlee Simpson	Pieces of Me (live SNL version)

a. Write the ILOP Anthony needs to solve in order to fit the most music onto a cassette with two 30-minute sides. [HINT: Consider a BLOP.]

b. Sammy reminds Anthony that he needs to add the 3:03 VeggieTales "Hairbrush Song" and that he should put it all on one 80-minute CD. Revise the above ILOP accordingly.

○ **1.5.44** *Anders Johnson is running for President of his homeowners association. There are three issues on the minds of other homeowners: yard maintenance, house colors, and swimming pool usage. Conservative interpretations of association rules outline such things as maximum grass height, allowable house colors, and owner-only use, while liberal interpretations allow for overgrown hedges, creative color combinations, and use by extended family and friends. When polled on a specific issue, individual association members rated their personal interpretation on a scale from -1 (most liberal) to 1 (most conservative), giving an overall profile $\mathbf{p} \in [-1, 1]^3$. Some of the other members have banded together along common attitudes to form special interest groups. For example, the 99-member Dolphin Boosters would be happy with a candidate in the $[-1, -.5] \times [-.7, -.2] \times [.3, .8]$ range because they like to have swim practice more than work on their houses, and enjoy having most of the pool to themselves. Likewise, the 44 Stepford Wives want the neighborhood properly color coordinated, and so would approve of a candidate in the $[-.7, .4] \times [.2, .7] \times [-.5, .5]$ range, while the 52 Gambinos would like a candidate in the $[-.6, 0] \times [0, 1] \times [-.8, -.3]$ range so they can invite the whole Family. Furthermore, the 89-member Inclusivity Club, the 59-member Gardeners Guild, and the 42 Aging Lappers would prefer a candidate in the $[0, .7] \times [-.3, .3] \times [-1, -.6]$, $[.3, 1] \times [-.5, .5] \times [-.7, .7]$, and $[-.6, .4] \times [-1, .3] \times [.5, 1]$ ranges, respectively. Assuming there to be no homeowner belonging to two of these groups, write the ILOP Anders should solve in order to figure out where he should position himself to earn the approval of the most members.*

Projects

1.5.45 *Write your own modeling exercise from an experience or situation in your own life (at a grocery store, at a baseball game, watching garbage trucks, getting on an airplane, surfing the web, putting away your clothes, registering for classes, driving, typing, etc.).*

1.5.46 *Present the role of the U.S. Air Force in the development of Linear Optimization.*

1.5.47 *Write a short biography on one of the following mathematicians, including their relationship with Linear Optimization: Kenneth Arrow, Evelyn Martin Lansdowne Beale, Garrett Birkhoff, Robert Bixby, Robert Bland, Constantin Carathéodory, Abraham Charnes, William Cooper, George Dantzig, René Descartes, Robert Dilworth, Jenö Egerváry, József Farkas, Lester Ford, Jean Fourier, Ragnar Frish, Ray Fulkerson, David Gale, Ralph Gomory, Philip Hall, Eduard Helly, Frank Hitchcock, Alan Hoffman, Leonid Kantorovich, Narendra Karmarkar, William Karush, Leonid Khachian, Victor Klee, Dénes König, Joseph Kruskal, Tjalling Koopmans, Harold Kuhn, Carlton Lemke, Wassily Leontief, Hermann Minkowski, George Minty, Oscar Morgenstern, Theodore Motzkin, John Nash, Arkady Nemirovsky,*

1.5. Exercises

John von Neumann, William Orchard-Hayes, Tyrell Rockafellar, Paul Samuelson, Lloyd Shapley, Naum Shor, Stephen Smale, George Stigler, Eva Tardos, Albert Tucker, Charles de la Vallé Poussin, Roger Wets, Philip Wolfe, Marshal Wood, Susan Wright.

1.5.48 *Present the problem of* P *versus* NP.

Chapter 2
The Simplex Algorithm

2.1 Geometric Lens

Refer again to Problem 1.3.1. In Figure 2.1 below we draw the region bounded by its 3 problem constraints and 2 nonnegativity constraints. This is called the **feasible region** (or **feasible set**) since it contains precisely all of the feasible points.[1] For any feasible region S, a feasible point $\mathbf{x}$ is in its **interior** if it is the center of some ball (of the same dimension as S) contained entirely in S (i.e., there is some small enough $\epsilon > 0$ so that every point within distance ϵ from $\mathbf{x}$ is in S). Otherwise, $\mathbf{x}$ is said to be on the **boundary** of S. A boundary point $\mathbf{x}$ is an **extreme point** of S if no line segment, with $\mathbf{x}$ as its center, has both its endpoints in S. Finally, $\mathbf{x}$ is **exterior** to S if it is infeasible. In Figure 2.1, $(10, 11)$ is in the interior, $(10, 12)$ and $(5, 15)$ are on the boundary, and $(10, 13)$ is an exterior point. Of these four, only $(5, 15)$ is an extreme point. We label a boundary line L_i if it arises from constraint i (note that the nonnegativity constraints are included, in order).

feasible region/set

interior/ boundary/ extreme/ exterior point

A feasible region S will be called **bounded** if there is a large enough integer K so that S is contained in the region defined by $|x_j| \leq K$ for all j. Otherwise, S is **unbounded**. In Figure 2.1, the value $K = 30$ suffices to show that the feasible region is bounded (20 is the smallest such K that works). We can see that, in two dimensions, every feasible region will be a polygon (including its interior), unless of course it is unbounded. That is because every constraint cuts $\mathbb{R}^2$ in half, so to speak. In fact, we call any region defined as the solution set of a single linear inequality a **half-space**. Similarly, in $\mathbb{R}^n$, that is, when a LOP involves n problem variables, each feasible region is what is known as a **polyhedron** ("many sides"), in particular a **polytope** if it is bounded. A polyhedron is more precisely defined as the intersection of finitely many half-spaces. We say that F is a

(un-)bounded region

half-space

polyhedron/ polytope

---

[1] While the terms *set* and *region* can be used interchangeably, we try to reserve the term set (resp. region) for the algebraic (resp. geometric) collection of vectors (resp. points) that satisfy the constraints — thus the region is the geometric realization of the set.

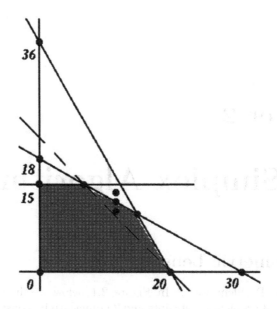

Figure 2.1: Feasible region for LOP 1.3.1

(k-)face/facet — **face** of a polyhedron P if there is some half-space $H = \{\mathbf{x} \mid \mathbf{a}^T\mathbf{x} \leq \mathbf{b}\} \supseteq P$ such that $F = P \cap \partial H$, where $\partial H = \{\mathbf{x} \mid \mathbf{a}^T\mathbf{x} = \mathbf{b}\}$ denotes the boundary of H. In this case H is called the **supporting hyperplane** of F. A **k-face** of P is a face of dimension k, and a **facet** is a face of dimension one less than that of P.

supporting hyperplane

Workout 2.1.1 *Consider the LOP P in Problem 1.4.2.*

    a. Draw the feasible region for P.

    b. Find $\mathbf{x}^*$ and z^*.

[MORAL: Although the boundedness of a linear problem's feasible region implies the existence of an optimal solution, such boundedness is not a requirement for optimality.]

We also have made note of the line $z = 20$, or $x_1 + x_2 = 20$, in Figure 2.1. The line $z = 15$ would be parallel, but closer to the origin, while the parallel line $z = 25$ doesn't intersect S, which implies that $z^* < 25$. Thus, we would like to think of the objective function as a whole family of parallel lines (or hyperplanes if we are in higher dimensions), and we could solve the problem if we knew which of them corresponded to the highest z-value while still intersecting S.

Workout 2.1.2 *Devise a LOP whose feasible region contains no extreme points and is*

    a. unbounded.

2.1. Geometric Lens

b. optimal.

[MORAL: Not every polyhedron has extreme points. In particular, it is possible for a LOP to be optimal not at an extreme point.]

Workout 2.1.3 *Let S be the feasible region of a LOP. Prove that if an interior point of S is optimal then every point of S is optimal.*

Note that the following more general statement is true (see Exercise 2.10.32): if $\mathbf{x}^0$ is an optimal point that is interior to the face F of S then every point of F is optimal.

Workout 2.1.4 *Let S be the feasible region of a LOP. Prove that if S is bounded then one of its extreme points is optimal. [HINT: Use the above generalized statement.]*

We also note that the hypothesis that S is bounded can be weakened by requiring only that S contains an extreme point (see Exercise 8.5.31). [MORAL: Except in degenerate cases one need only consider extreme points when looking for optima!]

Another way of thinking about this is through a calculus lens (the argument is not too dissimilar from above). For simplicity assume that S is bounded. Pick any line through S and parametrize it by the variable t. Then the objective function becomes a linear function of t as well, and because S is bounded we maximize the function $z = z(t)$ over the line segment L defined by the interval $t \in [\alpha, \beta]$. Bounded functions have maxima only at critical points and endpoints, and because continuous linear functions rule out nondifferentiable points, the maximum of z on L occurs at an endpoint (even in the case of stationary points). Such an endpoint is a boundary point of S. Thus we may assume that an optimal point is never an interior point of any line segment in S; i.e., an extreme point. We leave it to the reader to fill in more careful and general details.

From Figure 2.1, then, it is clear that $z^* = 24$ at $\mathbf{x}^* = (15, 9)^\mathsf{T}$. We will use the notation $\mathbf{x}^*$ in the general case to indicate the point (or one of the points) where z^* occurs. What we have discussed implicitly is a method (the **Graphic Method**) which reduces the search for the "best" point from an (uncountably) infinite set to a search from a finite set of extreme points.

Graphic Method

Workout 2.1.5 *Why is the set of extreme points of a polytope finite?*

If we have in hand the list of all extreme points of S, then we simply can check through them all to see which yields a maximum. Unfortunately, that list might be rather large (typically exponentially large, in terms of n), and so we may not have time to check them all. (In practical terms, even with 50 variables, never mind the tens of thousands or more encountered in common applications, we <u>don't</u> have time!) In addition, how is one to compute all the extreme points of S? (See Section 3.1.) Thus the geometric discussion doesn't as yet give us a real method for solving a linear problem, although one hopes you'll agree that it gives us plenty of insight into the nature of its solutions. The interplay between the algebraic Simplex Algorithm (hang on, it's coming) and its geometric underpinnings would make Descartes both excited and proud.

Problem 2.1.6

$$\text{Max.} \quad z = 226x_1 + 219x_2$$

$$\begin{align}
\text{s.t.} \quad 197x_1 + 185x_2 &\leq 9{,}650 \\
202x_1 + 178x_2 &\leq 9{,}595 \\
186x_1 + 190x_2 &\leq 9{,}502 \\
191x_1 + 196x_2 &\leq 9{,}781 \\
177x_1 + 205x_2 &\leq 9{,}661
\end{align}$$

$$\& \quad x_1, \; x_2 \geq 0$$

Workout 2.1.7 *Consider Problem 2.1.6. Draw its corresponding feasible region and use your drawing to find its maximum. [HINT: Would a* MAPLE *plot help?]*

Problem 2.1.8

$$\text{Max.} \quad z = 2x_1 + 3x_2 + x_3 + 2x_4$$

$$\begin{align}
\text{s.t.} \quad x_1 + 6x_2 + 5x_3 + 3x_4 &\leq 85 \\
4x_1 + 2x_2 + 6x_3 + x_4 &\leq 72 \\
7x_1 + 4x_2 + x_3 + 4x_4 &\leq 91 \\
3x_1 + x_2 + 5x_3 + 6x_4 &\leq 83
\end{align}$$

$$\& \quad x_1, \; x_2, \; x_3, \; x_4 \geq 0$$

Workout 2.1.9 *Consider Problem 2.1.8. Draw its corresponding feasible region and use your drawing to find its maximum.*

[MORAL: We need algebraic methods to solve linear problems.]

2.2 Algebraic Lens

We begin by rewriting the problem constraints of Problem 1.3.1 as equalities by introducing a new slack variable for each constraint. We call them slack variables because they pick up the slack, so to speak; that is, they take on whatever values are necessary to create equalities (notice that each slack variable must be nonnegative). In fact, it is a bit handier if we solve for each slack variable. Then Problem 1.3.1 can be written as follows.

Dictionary 2.2.1

$$\begin{aligned}
\text{Max.} \quad z &= 0 + x_1 + x_2 \\
\text{s.t.} \quad x_3 &= 90 - 3x_1 - 5x_2 \\
x_4 &= 180 - 9x_1 - 5x_2 \\
x_5 &= 15 - x_2 \\
\& \quad x_j &\geq 0 \qquad\qquad 1 \leq j \leq 5
\end{aligned}$$

We call this formulation a **dictionary**, and it has the property that the set of all variables is split in two, those on the left side (ignoring the nonnegativity constraints) appearing exactly once. Other than the objective variable z, the variables on the left-hand side of a given dictionary will be called **basic** (the set of which is called the **basis**), and those on the right will be called **nonbasic** (also referred to as **parameters**). One should notice that the initial basis of a LOP in standard form is always the set of slack variables. As we will see, further bases will be some mixture of problem and slack variables.

dictionary

(non)basic variable

(initial) basis

Another way of recording the same information is in a **tableau**, which is obtained from a dictionary by first rewriting each equation with all its variables on the left-hand side, constants on the right, and then writing the augmented matrix of coefficients of that system. By convention, the row for the objective function is written last. Thus, Problem 1.3.1 has the following tableau, corresponding to the above dictionary.

parameter

tableau

Tableau 2.2.2

$$\left[\begin{array}{cc|ccc|c} 3 & 5 & 1 & 0 & 0 & 90 \\ 9 & 5 & 0 & 1 & 0 & 180 \\ 0 & 1 & 0 & 0 & 1 & 15 \\ \hline -1 & -1 & 0 & 0 & 0 & 0 \end{array}\right]$$

Notice that we have ordered the columns in increasing fashion according to subscripts, with z coming last. We have included the dividing lines only as a visual aid to separate problem variables from slack, constraints from objective function, and left sides of the equations from right. The far right column will be referred to as the **b-column**, and the bottom row is called the **objective row**. One can typically spot the basic variables easily in a tableau by finding the simple columns (there are degenerate examples): the columns of the basis form a permutation of the columns of an identity matrix (or a positive multiple of one, as we will see later).

b-column

objective row

Workout 2.2.3 *Consider the linear problem P from Exercise 1.5.10.*

  a. *Write the initial dictionary for P.*
  b. *Write the initial tableau for P.*

(basic) solution

By a **solution** we will mean any set of values of the x_j that satisfy the problem constraints when written as equalities. Thus we can refer to either feasible or infeasible solutions, feasible being the case in which the nonnegativity constraints also hold. A solution is **basic** if it corresponds to the values one gets from a dictionary by setting all the parameters to zero. For example, Dictionary 2.2.1 yields the basic (feasible) solution $\mathbf{x} = (0, 0 \mid 90, 180, 15)^\mathsf{T}$, with value $z = 0$ (again, the divider distinguishes decision variables from slack).

Workout 2.2.4 *Consider the linear problem P from Exercise 1.5.10.*

    a. Find an infeasible basic solution to P.

    b. Find a feasible nonbasic solution to P.

β, π

(in)feasible basis/ dictionary/ tableau/ problem

Phase I/II

optimal basis/ dictionary/ tableau problem

When we need to discuss the distinction between basic and nonbasic variables, we will use the shorthand notations β for the set of subscripts of basic variables and π for the set of subscripts of parameters. For example, Dictionary 2.2.1 has $\beta = \{3, 4, 5\}$ and $\pi = \{1, 2\}$. We call a dictionary or tableau **feasible** if its corresponding basic solution is feasible (and **infeasible** otherwise). We say a linear problem is **feasible** if it has a feasible tableau (and **infeasible** otherwise). Whenever a tableau is infeasible we will say that we are in **Phase I** of the Simplex Algorithm; **Phase II** if the tableau is feasible. A basis or tableau or dictionary is **optimal** if the corresponding basic solution is optimal, and a problem is **optimal** if it has an optimal solution. The Simplex Algorithm, which recognizes such things, will halt at this stage (except for the chance of degeneracy — see Exercise 1.5.4 and Section 2.8).

One might notice from Dictionary 2.2.1 that an increase in x_1 from its basic value of zero would bring about a corresponding increase in the value of z. But since changes in x_3 and x_4 also would occur, we must be careful not to increase x_1 too much. If x_2 is held at zero, then because x_3 and x_4 must remain nonnegative we obtain the following restrictions on x_1.

$$90 - 3x_1 \geq 0 \quad \text{and} \quad 180 - 9x_1 \geq 0 .$$

The second restriction is the strongest, requiring $x_1 \leq 20$.

Thus, we might increase x_1 all the way up to 20, thereby decreasing x_4 all the way to 0. That has the ring of making x_1 basic and x_4 nonbasic, so we may as well solve the second equation for x_1 and substitute the result into the remaining equations. This produces the new basis $\beta^{(1)} = \{1, 3, 5\}$, parameter set $\pi^{(1)} = \{2, 4\}$, and dictionary below. (The superscripted (1), in general (k), serves to indicate the values after the first, in general k^{th}, modification; thus superscript (0) will denote original information.)

2.2. Algebraic Lens

Dictionary 2.2.5

$$\begin{aligned}
\text{Max.} \quad z &= 20 + .444x_2 - .111x_4 \\
\text{s.t.} \quad x_3 &= 30 - 3.333x_2 + .333x_4 \\
x_1 &= 20 - .555x_2 - .111x_4 \\
x_5 &= 15 - x_2 \\
\& \quad x_j &\geq 0 \qquad\qquad 1 \leq j \leq 5
\end{aligned}$$

Notice that we have rounded the fractions with denominator 9, so this dictionary is really only an approximation. We might rather maintain exactness by clearing the denominator and writing the following.

Dictionary 2.2.6

$$\begin{aligned}
\text{Max.} \quad 9z &= 180 + 4x_2 - x_4 \\
\text{s.t.} \quad 9x_2 &= 270 - 30x_2 + 3x_4 \\
9x_1 &= 180 - 5x_2 - x_4 \\
9x_5 &= 135 - 9x_2 \\
\& \quad x_j &\geq 0 \qquad\qquad 1 \leq j \leq 5
\end{aligned}$$

Here we have the basic solution $\mathbf{x}^{(1)} = (20, 0 \mid 30, 0, 15)^\mathsf{T}$ with $z^{(1)} = 20$. All of this, of course, corresponds to performing a **pivot operation**[2] (suitably modified to clear fractions) on the entry of 9 in row 2, column 1, of Tableau 2.2.2, resulting in Tableau 2.2.7. The notation we use to denote this is $1 \mapsto 4$, since x_1 replaces x_4 in the basis. The 9 is referred to as the **basic coefficient** since it is the coefficient of all the basic variables and z in the dictionary.

Tableau 2.2.7

$$\begin{bmatrix} 0 & 30 & 9 & -3 & 0 & 0 & 270 \\ 9 & 5 & 0 & 1 & 0 & 0 & 180 \\ 0 & 9 & 0 & 0 & 9 & 0 & 135 \\ 0 & -4 & 0 & 1 & 0 & 9 & 180 \end{bmatrix}$$

Workout 2.2.8 *Write the row operations that transformed Tableau 2.2.2 to Tableau 2.2.7.*

By similar analysis on Tableau 2.2.7, an increase in z is incurred by an increase in x_2, but only so far as $x_2 \leq \min\{270/30 = 9, 180/5 = 36, 135/9 = 15\}$, in order to maintain the feasibility of the next basic solution. Thus we arrive at Dictionary 2.2.9 and Tableau 2.2.10, with $\beta^{(2)} = \{1, 2, 5\}$, $\pi^{(2)} = \{3, 4\}$, $\mathbf{x}^{(2)} = (15, 9 \mid 0, 0, 6)^\mathsf{T}$, and $z^{(2)} = 24$.

---

[2]See Appendix A.

Dictionary 2.2.9

$$\text{Max.} \quad 30z = 720 - 4x_3 - 2x_4$$

$$\text{s.t.} \quad \begin{aligned} 30x_3 &= 270 - 9x_3 + 3x_4 \\ 30x_1 &= 450 + 5x_3 - 5x_4 \\ 30x_5 &= 135 + 9x_3 - 3x_4 \end{aligned}$$

$$\& \quad x_j \geq 0 \qquad 1 \leq j \leq 5$$

Tableau 2.2.10

$$\left[\begin{array}{cc|cccc|c} 0 & 30 & 9 & -3 & 0 & 0 & 270 \\ 30 & 0 & -5 & 5 & 0 & 0 & 450 \\ 0 & 0 & -9 & 3 & 30 & 0 & 180 \\ \hline 0 & 0 & 4 & 2 & 0 & 30 & 720 \end{array} \right]$$

Workout 2.2.11 *Write the row operations that transformed Tableau 2.2.7 to Tableau 2.2.10.*

Now, you may have noticed the wonderful way the objective function is written in Dictionary 2.2.9. It says that $z = (720 - 4x_3 - 2x_4)/30$, which means that if x_3 or x_4 takes on any value other than zero, then $z < 720/30 = 24$. Therefore z cannot be increased! Hence $z^* = 24$ and $\mathbf{x}^* = (15, 9 \mid 0, 0, 6)^\mathsf{T}$.

These are the workings of the Simplex Algorithm (Tableau Environment) in the simplest case (Phase II): given a feasible tableau, find a parameter whose increase from zero will increase z. That is, find a negative entry in the objective row (not including the rightmost, which can be negative at times), and pivot in the same column as that entry. This will be known as the **pivot column**, and if it is column j then x_j will be known as the **entering variable**, since it is entering the basis. In row i, if $a_{i,j}$ is the coefficient of x_j and b_i is the rightmost entry, then we know that $b_i/a_{i,j}$ is an upper bound on the value of the entering variable x_j whenever $a_{i,j} > 0$. (Notice that $b_i \geq 0$ because the tableau is feasible.) When $a_{i,j} < 0$, the quantity $b_i/a_{i,j}$ is a lower, rather than upper bound (when $a_{i,j} = 0$ the variable x_j is absent from the constraint corresponding to the current row i, so no bound of either type arises). Thus, we pivot in that row (**pivot row**) which forces the tightest restriction. That is, in pivot column j we pivot on $a_{i,j}$ when $a_{i,j} > 0$ and the "**b-ratio**" $b_i/a_{i,j}$ is minimum among positive $a_{i,j}$ (in the case of a tie we will choose that variable having the least subscript) — which we call the **smallest nonnegative ratio** and denote by $\mathrm{snr}(b_1/a_{1,j}, \ldots, b_m/a_{m,j})$. The basic variable corresponding to the pivot row will be called the **leaving variable**, since it leaves the basis during the pivot. An important observation to make is that the resulting tableau is also feasible because of this judicious choice. Therefore, we can continue to use the Phase II Algorithm.

It may be that we have several choices for a pivot column. One way to decide would be always to choose the leftmost (where the subscript j is least). Another would be always to choose the most negative of coefficients because that would force z to increase at the highest rate. A third might be actually to compute which of the allowable pivots results in the highest z-value, and pivot accordingly. We call these the **Least Subscript** (LS), **Most Negative** (MN), and **Greatest Increase** (GI) **Implementations**, respectively, and we will not discuss their comparative merits here. It turns out that none of them requires fewer pivots than any other on all problems, or even on average. We will use the Least Subscript Implementation in this text. It has one advantage of being the simplest, and another of guaranteeing termination of the algorithm (see Section 2.8).

Least Subscript/ Most Negative/ Greatest Increase Implementation

Workout 2.2.12 *For each of the implementations above, give an example of a LOP for which the chosen implementation requires fewer pivots than the other implementations. [HINT: One can consider two variable programs like that in Figure 2.1 geometrically. What determines the first feasible basis in each case.]*

It can be said that the general strategy of Phase II is to find, with respect to the basis, an entering variable that improves the objective value before finding an leaving variable that most restricts the entering one. That is, add a good variable to the basis before removing a bad one!

Workout 2.2.13 *Write the correct Simplex pivot operation $i \mapsto k$ for the following tableau.*

$$\begin{bmatrix} 0 & 5 & 0 & 2 & 0 & 0 & 8 & 7 & 0 & 6 \\ 0 & 0 & 0 & 21 & 0 & 7 & -3 & 0 & 0 & 35 \\ 7 & -2 & 0 & 0 & 0 & 0 & 6 & 0 & 0 & 13 \\ 0 & 13 & 0 & -5 & 7 & 0 & -2 & 0 & 0 & 0 \\ 0 & 7 & 7 & 15 & 0 & 0 & 1 & 0 & 0 & 25 \\ \hline 0 & 12 & 0 & -6 & 0 & 0 & -10 & 0 & 7 & -84 \end{bmatrix}$$

2.3 مثال آخر

Consider the following LOP.

Problem 2.3.1

$$\text{Max. } z = 4x_1 + 5x_2$$

$$\text{s.t. } \begin{aligned} 14x_1 + 11x_2 &\leq 154 \\ 7x_1 + 16x_2 &\leq 112 \end{aligned}$$

$$\& \quad x_1, \ x_2 \geq 0$$

Workout 2.3.2

  a. *Draw the feasible region for Problem 2.3.1.*

b. Write dictionary $D^{(0)}$, tableau $T^{(0)}$, basis $\beta^{(0)}$, parameters $\pi^{(0)}$, basic solution $\mathbf{x}^{(0)}$, and objective value $z^{(0)}$.

c. Write the first pivot operation $(i \mapsto j)$ and the corresponding row operations (avoiding fractions and keeping the basic coefficient constant).

d. Perform the first pivot and write $D^{(1)}$, $T^{(1)}$, $\beta^{(1)}$, $\pi^{(1)}$, $\mathbf{x}^{(1)}$, and $z^{(1)}$.

e. Write the second pivot operation and the corresponding row operations.

f. Perform the second pivot and write $D^{(2)}$, $T^{(2)}$, $\beta^{(2)}$, $\pi^{(2)}$, $\mathbf{x}^{(2)}$, and $z^{(2)}$.

g. Use part f to write $\mathbf{x}^*$ and z^*.

h. Plot the points $\mathbf{x}^{(k)}$ on your drawing from part a and draw an arrow from each $\mathbf{x}^{(k)}$ to $\mathbf{x}^{(k+1)}$.

Workout 2.3.3 What pattern emerges from the row operations found in Workouts 2.2.8, 2.2.11 and 2.3.2ce? Specifically, write the row operation R_k, used to modify row k while pivoting on entry $a_{i,j}$ $(k \neq i)$, where d is the basic coefficient.

Workout 2.3.4

a. Write an outline (or pseudocode) for the Phase II algorithm for feasible LOPs.

b. What can go wrong along the way?

2.4 Infeasible Basis

Let us return to the dual Problem 1.4.6, and rewrite it in standard form for maximization. Let $u = -w$ so that $\min w = -\max u$.

Problem 2.4.1

$$\begin{aligned}
\text{Max.} \quad u &= -90y_1 - 180y_2 - 15y_3 \\
\text{s.t.} \quad & -3y_1 - 9y_2 \leq -1 \\
& -5y_1 - 5y_2 - y_3 \leq -1 \\
\& \quad & y_1, \quad y_2, \quad y_3 \geq 0
\end{aligned}$$

This problem has the following initial tableau.

Tableau 2.4.2

$$\left[\begin{array}{ccc|ccc|c} -3 & -9 & 0 & 1 & 0 & 0 & -1 \\ -5 & -5 & -1 & 0 & 1 & 0 & -1 \\ \hline 90 & 180 & 15 & 0 & 0 & 1 & 0 \end{array}\right]$$

2.4. Infeasible Basis

Of critical importance is the fact that the initial basis $\beta^{(0)} = \{4, 5\}$ is infeasible. Indeed, $\mathbf{x}_4 = \mathbf{x}_5 = -1$. At first glance, we cannot discern whether or not Problem 2.4.1 is feasible. In order to solve this problem we resort to the **Auxiliary Method**. We already have discovered how Phase II of the Simplex Algorithm works, and we will use the Auxiliary Method to develop Phase I of the algorithm. Consider the following problem.

Auxiliary Method

Problem 2.4.3 (Auxiliary to Problem 2.4.1)

$$\text{Max.} \quad v = -y_0$$

$$\text{s.t.} \quad \begin{array}{rcrcrcrcr} -y_0 & - & 3y_1 & - & 9y_2 & & & \leq & -1 \\ -y_0 & - & 5y_1 & - & 5y_2 & - & y_3 & \leq & -1 \end{array}$$

$$\& \quad y_0, \ y_1, \ y_2, \ y_3 \geq 0$$

In general, the problem

$$\text{Max.} \quad z = \sum_{j=1}^{n} c_j x_j$$

$$\text{s.t.} \quad \sum_{j=1}^{n} a_{i,j} x_j \leq b_i \quad (1 \leq i \leq m) \tag{A}$$

$$\& \quad x_j \geq 0 \quad (1 \leq j \leq n)$$

has the following corresponding **Auxiliary Problem**.

Auxiliary Problem

$$\text{Max.} \quad v = -x_0$$

$$\text{s.t.} \quad -x_0 + \sum_{j=1}^{n} a_{i,j} x_j \leq b_i \quad (1 \leq i \leq m) \tag{B}$$

$$\& \quad x_j \geq 0 \quad (0 \leq j \leq n)$$

It is not difficult to see that Problem 2.4.1 is feasible if and only if Problem 2.4.3 is optimal at $v^* = 0$. In fact, such an auxiliary relationship holds in general.

Theorem 2.4.4 *Let P be a maximizing LOP in standard form whose initial basis is infeasible, and let Q be its corresponding Auxiliary Problem. Then P is feasible if and only if Q is optimal at 0.* ◇

Workout 2.4.5 *Verify Theorem 2.4.4 for the instance of Problem 2.4.1.*

Workout 2.4.6 *Prove Theorem 2.4.4 in general.*

The initial tableau for Problem 2.4.3 is as follows.

Tableau 2.4.7

$$\left[\begin{array}{c|ccc|ccc|c} -1 & -3 & -9 & 0 & 1 & 0 & 0 & -1 \\ -1 & -5 & -5 & -1 & 0 & 1 & 0 & -1 \\ \hline 1 & 0 & 0 & 0 & 0 & 0 & 1 & 0 \end{array}\right]$$

We can discover a feasible tableau with the pivot $0 \mapsto 4$, which results in the next tableau.

Tableau 2.4.8

$$\left[\begin{array}{c|ccc|ccc|c} 1 & 3 & 9 & 0 & -1 & 0 & 0 & 1 \\ 0 & -2 & 4 & -1 & -1 & 1 & 0 & 0 \\ \hline 0 & -3 & -9 & 0 & 1 & 0 & 1 & -1 \end{array}\right]$$

Workout 2.4.9 *Suppose that x_j is a (basic) slack variable in a given Auxiliary Problem. Show that the initial pivot $0 \mapsto j$ yields a feasible tableau if and only if x_j is a variable whose basic value in $\mathbf{x}^{(0)}$ is most negative.*

Since Tableau 2.4.8 is feasible we may resort to the Phase II Algorithm we know and love. The pivot $1 \mapsto 0$ yields the optimal tableau, below.

Tableau 2.4.10

$$\left[\begin{array}{c|ccc|ccc|c} 1 & 3 & 9 & 0 & -1 & 0 & 0 & 1 \\ 2 & 0 & 30 & -3 & -5 & 3 & 0 & 2 \\ \hline 3 & 0 & 0 & 0 & 0 & 0 & 3 & 0 \end{array}\right]$$

Now we know from Theorem 2.4.4 that Problem 2.4.1 is feasible since $v^* = 0$. To recover the corresponding tableau for Problem 2.4.1, we merely restate its objective variable in terms of the parameters y_2, y_3, and y_4 of Tableau 2.4.10. That is, we substitute $3y_1 = 1 - 9y_2 + y_4$ into $u = -90y_1 - 180y_2 - 15y_3$ to obtain $3u = -90 + 270y_2 - 45y_3 - 90y_4$. This produces Tableau 2.4.11, below.

Tableau 2.4.11

$$\left[\begin{array}{ccc|ccc|c} 3 & 9 & 0 & -1 & 0 & 0 & 1 \\ 0 & 30 & -3 & -5 & 3 & 0 & 2 \\ \hline 0 & -270 & 45 & 90 & 0 & 3 & -90 \end{array}\right]$$

At this point, we may as well finish the problem so we can compare the result with that of Problem 1.3.1, whose dual is Problem 1.4.6. The next and final pivot $2 \mapsto 5$ produces the following optimal tableau.

Tableau 2.4.12

$$\left[\begin{array}{ccc|ccc|c} 30 & 0 & 9 & 5 & -9 & 0 & 4 \\ 0 & 30 & -3 & -5 & 3 & 0 & 2 \\ \hline 0 & 0 & 180 & 450 & 270 & 30 & -720 \end{array}\right]$$

2.5. Shortcut Method

Thus, $u^* = -24$, and so $w^* = 24$, which agrees with $z^* = 24$. Here we have $\mathbf{y}^* = (4, 2, 0 \mid 0, 0)^T/30$. Compare these values with the optimal objective row in Tableau 2.2.10. Likewise, compare the values for $\mathbf{x}^* = (450, 270 \mid 0, 0, 180)^T/30$ with the optimal objective row in Tableau 2.4.12. The conclusions you draw will be used in the proof of the Duality Theorem (Theorem 4.1.9) in Chapter 4.

Workout 2.4.13 *Consider the LOP P from Exercise 1.5.10.*

a. *Use the Auxiliary Method to solve P. [HINT: First pivot in x_0 to replace the first slack variable.]*

b. *Write the LOP D that is dual to P and convert it to standard maximization form.*

c. *Use Phase II of the Simplex Algorithm to solve D.*

d. *Compare your results in parts a and c.*

2.5 Shortcut Method

Observe that we could have proceeded directly from Tableau 2.4.2 to Tableau 2.4.11 with the single pivot operation $1 \mapsto 4$. This operation is sort of a concatenation of the two operations $0 \mapsto 4$ and $1 \mapsto 0$ that were used to solve the Auxiliary Problem 2.4.3. We would like to generalize this observation in the hope that we could save the time and trouble of resorting to an auxiliary problem.

Let us consider a new problem, such as Problem 2.5.1, below.

Problem 2.5.1

$$\text{Max. } z = 28x_1 + 21x_2 + 26x_3$$

$$\text{s.t.} \quad \begin{aligned} -7x_1 + 2x_2 + 3x_3 &\leq -210 \\ 5x_1 - 8x_2 + x_3 &\leq -305 \\ 2x_1 + 4x_2 - 9x_3 &\leq -250 \end{aligned}$$

$$\& \quad x_1, \quad x_2, \quad x_3 \geq 0$$

If we are observant, we can notice two things about this problem. First, if $\mathbf{x} = (t, t, t)^T$, then $\mathbf{x}$ is feasible whenever $t \geq 152.5$. Second, such an $\mathbf{x}$ yields $z = 75t \to \infty$ as $t \to \infty$, and so the problem is unbounded. But let's disregard this quick analysis for the moment (we will discuss this trick in detail later — see Section 2.7), since it is difficult to perform on more general linear problems, and instead concentrate on how the Auxiliary Method handles this particular problem. Following that, we will discuss how we can replace the Auxiliary Method by a much simpler process. The Auxiliary to Problem 2.5.1 is the following.

Problem 2.5.2

Max. $v = -x_0$

s.t.
$$-x_0 - 7x_1 + 2x_2 + 3x_3 \leq -210$$
$$-x_0 + 5x_1 - 8x_2 + x_3 \leq -305$$
$$-x_0 + 2x_1 + 4x_2 - 9x_3 \leq -250$$

    & $x_0, \; x_1, \; x_2, \; x_3 \geq 0$

The corresponding sequence of auxiliary tableaux and pivots are as follows.

Tableaux 2.5.3 (Auxiliary)

	x_0	x_1	x_2	x_3	x_4	x_5	x_6	v	
Tableau 0:	-1	-7	2	3	1	0	0	0	-210
	-1	5	-8	1	0	1	0	0	-305
	-1	2	4	-9	0	0	1	0	-250
	1	0	0	0	0	0	0	1	0

Pivot 1: $0 \mapsto 5$

Tableau 1:	0	-12	10	2	1	-1	0	0	95
	1	-5	8	-1	0	-1	0	0	305
	0	-3	12	-10	0	-1	1	0	55
	0	5	-8	1	0	1	0	1	-305

Pivot 2: $2 \mapsto 6$

Tableau 2:	0	-114	0	124	12	-2	-10	0	590
	12	-36	0	68	0	-4	-8	0	3220
	0	-3	12	-10	0	-1	1	0	55
	0	36	0	-36	0	4	8	12	-3220

Pivot 3: $3 \mapsto 4$

Tableau 3:	0	-114	0	124	12	-2	-10	0	590
	124	274	0	0	-68	-30	-26	0	29930
	0	-126	124	0	10	-12	2	0	1060
	0	-274	0	0	68	30	26	124	-29930

Pivot 4: $1 \mapsto 0$

Tableau 4:	114	0	0	274	-36	-32	-46	0	28820
	124	274	0	0	-68	-30	-26	0	29930
	126	0	274	0	-47	-57	-22	0	32755
	274	0	0	0	0	0	0	274	0

2.5. Shortcut Method

Keep in mind that Pivot 1 was performed in order that Tableau 1 might be feasible. By pivoting x_0 into the basis, Tableau 1 will be feasible, according to Workout 2.4.9, if and only if the leaving variable is chosen to be that basic variable which is most negative. From that point on, we resort to the pivoting rules of Phase II, as you can see.

Now we wish to investigate the effect of these same pivots on the original Problem 2.5.1. But what does "same" mean? The pivots in Tableaux 2.5.3 are

$$0 \mapsto 5, \quad 2 \mapsto 6, \quad 3 \mapsto 4, \quad 1 \mapsto 0.$$

By ignoring variable x_0 coming into the basis at the start and leaving at the end, we convert those pivots into

$$2 \mapsto 5, \quad 3 \mapsto 6, \quad 1 \mapsto 4,$$

which we perform on Problem 2.5.1 in Tableaux 2.5.4, below.

Tableaux 2.5.4 (Shortcut)

	x_1	x_2	x_3	x_4	x_5	x_6	z	
Tableau 0:	-7	2	3	1	0	0	0	-210
	5	-8	1	0	1	0	0	-305
	2	4	-9	0	0	1	0	-250
	-28	-21	-26	0	0	0	1	0

Pivot 1: $2 \mapsto 5$

	x_1	x_2	x_3	x_4	x_5	x_6	z	
Tableau 1:	-46	0	26	8	2	0	0	-2290
	-5	8	-1	0	-1	0	0	305
	36	0	-68	0	4	8	0	-3220
	-329	0	-229	0	-21	0	8	6405

Pivot 2: $3 \mapsto 6$

	x_1	x_2	x_3	x_4	x_5	x_6	z	
Tableau 2:	-274	0	0	68	30	26	0	-29930
	-47	68	0	0	-9	-1	0	2995
	-36	0	68	0	-4	-8	0	3220
	-3827	0	0	0	-293	229	68	146615

Pivot 3: $1 \mapsto 4$

	x_1	x_2	x_3	x_4	x_5	x_6	z	
Tableau 3:	274	0	0	-68	-30	-26	0	29930
	0	274	0	-47	-57	-22	0	3275
	0	0	274	-36	-32	-46	0	28820
	0	0	0	-3827	-2869	2386	274	2275215

Let's spend a moment translating the rules we use in the Auxiliary Method into rules for the **Shortcut Method**. Recall that the intention of both methods is simply to produce a feasible tableau for the original problem. In both cases we start by choosing as pivot row that row whose **b**-column is most negative, in this case row 2.

In Auxiliary Pivot 2, x_2 is chosen as the entering variable because of the -8 in the objective row of Auxiliary Tableau 1, the first negative number we see when reading left-to-right (Least Subscript rule). This corresponds to the -8 in row 2 of both the Auxiliary and Shortcut Tableau 0, and it translates into a Shortcut rule that chooses the variable with the least subscript whose coefficient in the pivot row is negative.

Now take a gander at Shortcut Pivots 2 and 3 and ask yourself, could this Shortcut rule have determined these as well?

Problem 2.5.5

$$\begin{array}{rrrrrrrr}
\text{Max.} & z = & -4x_1 & +2x_2 & +16x_3 & +9x_4 & +9x_5 & +13x_6 \\
\\
\text{s.t.} & & 2x_1 & -x_2 & & +x_4 & & +x_6 & \leq & 2 \\
& & x_1 & +3x_2 & +4x_3 & +2x_4 & & +2x_6 & \leq & 5 \\
& & 4x_1 & & +x_3 & -3x_4 & -3x_5 & -2x_6 & \leq & -8 \\
\\
\& & & x_1, & x_2, & x_3, & x_4, & x_5, & x_6 & \geq & 0
\end{array}$$

Workout 2.5.6

  a. Write the LOP Q that is the Auxiliary to the LOP P in Problem 2.5.5.

  b. Use Phase II (including the special first pivot) to solve Q.

  c. Write the pivots used in part b and modify them as above to avoid the auxiliary variable x_0.

  d. Use the Shortcut Method on P.

  e. Write the pivots used in part d and compare them to the modified pivots in part c.

We will prove the following theorem in Appendix B.

Theorem 2.5.7 *The Auxiliary and Shortcut Methods are equivalent in the sense that, ignoring the auxiliary variable x_0, they make the same sequence of decisions for entering and leaving variables.*   ◇

If you are interested in trying to prove Theorem 2.5.7 yourself, here's a hint. Consider Auxiliary Pivot 2 of Tableaux 2.5.3. The variable x_6 is chosen to leave the basis because it is the variable that places the greatest restriction on the entering x_2. That is, its **b**-ratio of $\frac{55}{12}$ is the smallest nonnegative. Now consider Shortcut Pivot 2 of Tableaux 2.5.4. Here, x_6 is chosen to leave the basis because its basic value of -3220 is most negative. Your job, should you accept it, is to determine why these two conditions are equivalent (in the general case, of course).

The general strategy for finding a pivot in Phase I is the reverse of that which finds the pivot in Phase II in the following sense. We now choose a leaving variable that most violates its nonnegativity constraint before choosing an entering variable that will have a positive basic value. In other words, remove a bad variable from the basis before adding a good one!

Workout 2.5.8 *Write the correct Simplex pivot operation $i \mapsto k$ for the following tableau.*

$$\begin{bmatrix} -5 & 0 & 15 & 3 & 0 & 0 & 0 & 0 & 0 & -7 \\ 18 & 0 & 0 & -10 & 0 & 15 & 0 & 3 & 0 & 0 \\ -7 & 0 & 0 & 8 & 0 & 0 & 15 & 14 & 0 & -9 \\ 2 & 15 & 0 & -4 & 0 & 0 & 0 & 6 & 0 & 22 \\ 20 & 0 & 0 & 6 & 15 & 0 & 0 & -12 & 0 & -9 \\ \hline -31 & 0 & 0 & -47 & 0 & 0 & 0 & 63 & 15 & 206 \end{bmatrix}$$

Workout 2.5.9

    a. *Write an outline (or pseudocode) for the Phase I (Shortcut) Algorithm for LOPs with infeasible initial basis.*

    b. *What can go wrong along the way?*

2.6 Infeasibility

In putting together Phase I and Phase II Algorithms for a particular standard max form LOP, one first must determine which applies. If a given basis is feasible then we use Phase II, while otherwise we use Phase I. In the infeasible case we know that some variable is negative, and hence that some variable is most negative (in the case of a tie, we shall choose that most negative variable having the least subscript). This determines the pivot row in the corresponding tableau. If some entry in the pivot row is negative, then the first such determines the pivot. Now the question arises as to what it means if no such entry exists. Consider the following LOP, for example.

Problem 2.6.1

$$\begin{array}{rrrcrcr} \text{Max.} & z & = & 3x_1 & + & 8x_2 & \\ \text{s.t.} & & & 5x_1 & - & 2x_2 & \leq & 10 \\ & & & -2x_1 & + & 3x_2 & \leq & 6 \\ & & & -4x_1 & - & 5x_2 & \leq & -40 \\ \& & & & x_1 & , & x_2 & \geq & 0 \end{array}$$

One can see that after 2 pivots we obtain the following tableau.

Tableau 2.6.2

$$\begin{bmatrix} 0 & 33 & -4 & 0 & -5 & 0 & 160 \\ 0 & 0 & 22 & 33 & 11 & 0 & -22 \\ 33 & 0 & 5 & 0 & -2 & 0 & 130 \\ \hline 0 & 0 & -17 & 0 & -46 & 33 & 1670 \end{bmatrix}$$

At this point we find no negative entry in (the left side of) row 2 of Tableau 2.6.2. Recall that this row is merely the row of coefficients that represents the equality

$$22x_3 + 33x_4 + 11x_5 = -22 \, .$$

Because every variable is nonnegative, the left side of the equation is nonnegative, which contradicts the equality. Thus we can only surmise that Problem 2.6.1 is infeasible; that is, there are no values for its variables that satisfy both its problem and nonnegativity constraints.

Now someone who doesn't trust the arithmetic that produced Tableau 2.6.2 might then distrust this conclusion. Therefore, it would be beneficial to produce a certificate of this result. Since every row of any tableau is some linear combination of the rows $\{r_i^{(0)}\}$ of the original tableau (see Exercise 2.10.38), we should be able to produce the exact linear combination necessary.

Workout 2.6.3

   a. Find the linear combination of the rows of Tableau 0 of Problem 2.6.1 that produces row 2 of Tableau 2.6.2.

   b. Use part a to derive a contradiction directly from the problem and nonnegativity constraints of Problem 2.6.1.

   c. Where in Tableau 2.6.2 do you see the coefficients of the linear combination you found in part a?

Workout 2.6.4

   a. Generate a LOP P in standard form with 3 variables and 4 inequalities by writing random integers as coefficients and right-hand sides. (With fairly high probability the resulting LOP is infeasible.)

   b. Verify that the Simplex Algorithm outputs "Infeasible" on P. (If not, return to part a and repeat.)

   c. Find the linear combination of the rows of the initial tableau of P that produces the offending row of the final tableau from part b.

   d. Use part c to derive a contradiction directly from the problem and nonnegativity constraints of P.

   e. Where in the final tableau from part b do you see the coefficients of the linear combination you found in part c?

We will see in Chapter 7 a way to compute infeasibility certificates without using the infeasible tableaux that halt Phase I.

2.7 Unboundedness

Similar concerns arise in Phase II. If we see no negatives in the (left side of the) objective row of a feasible tableau, then we know that the tableau is optimal. We will see in Chapter 4 how to find the certificate of optimality without solving the dual LOP. Otherwise there is some first negative entry in the objective row, and that determines the entering variable. If the pivot column has a positive entry then we can find the leaving variable by the b-ratio test. But now we ask what meaning there is in the case that no such entry exists. Consider the following LOP, for example.

Problem 2.7.1

$$
\begin{array}{rrrrr}
\text{Max.} & z = & 3x_1 & + & 8x_2 \\
\text{s.t.} & & -5x_1 & + & 2x_2 & \leq & 10 \\
& & 2x_1 & - & 3x_2 & \leq & 6 \\
& & -4x_1 & - & 5x_2 & \leq & -10 \\
\& & & x_1 & , & x_2 & \geq & 0
\end{array}
$$

One can see that after 3 pivots we obtain the following tableau.

Tableau 2.7.2

$$
\left[\begin{array}{cc|ccccc|c}
-33 & 0 & 5 & 0 & 2 & 0 & & 30 \\
-11 & 0 & 3 & 2 & 0 & 0 & & 42 \\
-5 & 2 & 1 & 0 & 0 & 0 & & 10 \\
\hline
-46 & 0 & 8 & 0 & 0 & 2 & & 80
\end{array}\right]
$$

At this point we find no positive entries in column 1 of Tableau 2.7.2. This suggests that no variables place any restriction on the value of x_1, and since increasing x_1 increases z, we should find no upper bound on the LOP. Indeed, consider the following set of values

$$\mathbf{x}(t) = (2t, 10 + 5t \mid 0, 42 + 11t, 30 + 33t)^\mathsf{T} / 2 \ .$$

Note that $\mathbf{x}(t)$ satisfies all problem and nonnegativity constraints of Problem 2.7.1 whenever $t \geq 0$; i.e., $\mathbf{x}(t)$ is feasible for all $t \geq 0$. Moreover $\mathbf{x}(t)$ produces the (feasible) objective value $z(t) = (80 + 46t)/2$, which tends to infinity as $t \to \infty$. Thus $\mathbf{x}(t)$ is a certificate that shows that Problem 2.7.1 has no finite optimum. Such problems we call **unbounded**. Their corresponding tableaux and dictionaries are also called **unbounded**.

unbounded problem/ tableau/ dictionary

Problem 2.7.3

$$
\begin{array}{rrrrr}
\text{Max.} & z = & 31x_1 & + & 23x_2 \\
\text{s.t.} & & -7x_1 & + & 3x_2 & \leq & 21 \\
& & 4x_1 & - & 5x_2 & \leq & 20 \\
& & -9x_1 & - & 8x_2 & \leq & -72 \\
\& & & x_1 & , & x_2 & \geq & 0
\end{array}
$$

Workout 2.7.4 *Find a certificate for the unboundedness of Problem 2.7.3.*

2.8 Cycling

As is the case with all algorithms, we must be sure that the Simplex Algorithm halts in all cases. Does that mean we have to test every possible LOP? Not really. We merely have to imagine that we are given an arbitrary LOP, say P, and show that Simplex halts on it. So let P be a fixed LOP which, by the discussion of Section 1.4, we may assume is in standard max form, with n variables and m problem constraints.

First notice that the Simplex Algorithm as presented is deterministic. That is, for a particular tableau the pivot chosen by Simplex is determined by its rules, so if that same tableau turns up again later on then the same pivot will be chosen again. In particular, if that happens then Simplex will continue to make the same choices as before and the algorithm will cycle through a sequence of tableaux indefinitely. In fact, we argue that if Simplex doesn't halt on P then it will cycle in this manner. This is because the number of possible tableaux of P is finite (see Workout 2.8.1, below). Thus, if Simplex doesn't cycle then it will halt after a finite number of steps. Halting can be guaranteed, therefore, if cycling can be eliminated somehow.

Workout 2.8.1 *Why is the number of possible tableaux of P finite?* [HINT: *Why is the number of possible bases of P finite?*]

For the moment let us consider two different rules for choosing the leaving variable in Phase II. The first is the Least Subscript (LS) rule we have been using, and the second is the Lowest-Greatest (LG) rule. Both rules use the least subscript for entering variable choice: x_1 in Tableau 2.8.2 below. However, while LS uses least subscript for the breaking of ties in the choice of a leaving variable (x_4), LG uses the greatest subscript instead (x_7).

Tableau 2.8.2

$$\left[\begin{array}{cccc|c|ccc|c} -8 & 16 & 4 & 0 & 2 & -18 & 0 & 0 & 0 \\ 1 & -3 & -1 & 2 & 0 & 2 & 0 & 0 & 0 \\ 2 & 0 & 0 & 0 & 0 & 0 & 2 & 0 & 0 \\ \hline -44 & 186 & 42 & 0 & 0 & -48 & 0 & 2 & 0 \end{array}\right]$$

Now consider the following LOP. A certificate that Problem 2.8.3 is unbounded is given by $\mathbf{x}(t) = t(0, 5, 0, 11)^\mathsf{T}/11$ and $z(t) = 2793t$, as one can verify.

Problem 2.8.3

$$\text{Max. } z = 24x_1 + 288x_2 - 270x_3 + 123x_4$$

$$\text{s.t.} \quad x_1 + 11x_2 + 18x_3 - 5x_4 \leq 0$$
$$-x_1 - 3x_2 + 2x_3 - x_4 \leq 0$$

$$\& \quad x_1, \ x_2, \ x_3, \ x_4 \geq 0$$

Workout 2.8.4

a. Show that LS Simplex halts on Problem 2.8.3 in 2 steps.

b. Show that LG Simplex cycles on Problem 2.8.3 in 6 steps.

Note that any time cycling occurs the objective function value necessarily stays fixed. Pivots that do not change the objective value are called **degenerate**. Tableau, bases and extreme points associated with such a pivot are similarly named, as are the LOP and polyhedron involved. Degenerate pivots do not signal cycling, however; they sometimes occur in problems that don't cycle (see Exercise 1.5.4). However, one could imagine an algorithm watching out for the occurrence of many consecutive degenerate pivots in order to keep an eye out for the chance for a nondegenerate pivot, so as to avoid cycling. A better method is simply to use LS, according to the following theorem.

<small>degenerate pivot/ tableau/ basis/ extreme point/ LOP/ polyhedron</small>

Theorem 2.8.5 *The LS Simplex Algorithm halts on every LOP in standard max form.* ◇

<small>Bland's Theorem</small>

2.9 The Fundamental Theorem

There are three types of people in this world: those who can count, and those who can't.

Now we are ready to state the Fundamental Theorem of Linear Optimization.

Theorem 2.9.1 *Let P be a LOP in standard max form. Then*

<small>Fundamental Theorem</small>

a. *P is either infeasible, unbounded, or it has a maximum;*

b. *if P has a feasible solution, then it has a basic feasible solution; and*

c. *if P has an optimal solution then it has a basic optimal solution.*

Workout 2.9.2

a. Prove Theorem 2.9.1a. [HINT: Consider all the outcomes of LS Simplex.]

b. Prove Theorem 2.9.1b. [HINT: Think about Tableaux.]

  c. Prove Theorem 2.9.1c. [HINT: Think about Tableaux.]

Problem 2.9.3

$$\begin{aligned}
\text{Max. } z &= 3x_1 + 7x_2 \\
\text{s.t.} \quad -2x_1 + x_2 &\leq 6 \\
-x_1 + 5x_2 &\leq 2 \\
x_1 - 2x_2 &\leq 9 \\
2x_1 - 4x_2 &\leq 9 \\
\& \quad x_1, x_2 &\geq 0
\end{aligned}$$

Workout 2.9.4 *Consider Problem 2.9.3. Use Theorem 2.9.1 to prove that it has an optimum (i.e., do not solve it).*

2.10 Exercises

Practice

2.10.1 *Draw the feasible region of the LOP in Exercise 1.5.2. Label each boundary line by the index of the variable whose value is zero on that line.*

2.10.2 *Repeat Exercise 2.10.1 N times, each time with a different LOP of your own design, having 2 variables and 3–5 constraints.*

2.10.3 *Draw the feasible region of the LOP in Exercise 1.5.3. Label each boundary plane by the index of the variable whose value is zero on that line.*

2.10.4 *Repeat Exercise 2.10.3 N times, each time with a different LOP of your own design, having 3 variables and 2–4 constraints.*

2.10.5 *Circle (or otherwise locate) the pivot entry in each of the following tableaux, according to our LS Implementation of Simplex.*

  a.
$$\left[\begin{array}{ccc|cccc|c}
3 & 4 & 0 & -1 & -2 & 0 & 0 & 0 \\
0 & 2 & 0 & 0 & -5 & 3 & 0 & 16 \\
0 & -1 & 3 & -3 & 1 & 0 & 0 & 0 \\
\hline
0 & 6 & 0 & -1 & -4 & 0 & 3 & 0
\end{array}\right]$$

  b.
$$\left[\begin{array}{ccc|cccc|c}
-6 & 0 & 5 & -1 & 0 & 0 & 2 & 0 & -1 \\
1 & 0 & 0 & 8 & 5 & 0 & -3 & 0 & 7 \\
0 & 0 & 0 & -1 & 0 & 5 & -4 & 0 & -3 \\
2 & 5 & 0 & -5 & 0 & 0 & -6 & 0 & -3 \\
\hline
-11 & 0 & 0 & 16 & 0 & 0 & 13 & 5 & 43
\end{array}\right]$$

c.
$$\begin{bmatrix} 0 & 0 & 2 & | & 0 & 8 & 5 & 0 & 0 & | & 4 \\ 8 & 0 & -5 & | & 2 & 0 & 0 & 0 & 0 & | & 0 \\ 0 & 8 & 3 & | & 6 & 0 & -1 & 0 & 0 & | & 6 \\ 0 & 0 & 0 & | & 1 & 0 & 2 & 8 & 0 & | & 1 \\ \hline 0 & 0 & -1 & | & -7 & 0 & 4 & 0 & 8 & | & -27 \end{bmatrix}$$

2.10.6 *Solve the problem formulated in Problem 1.1.3.*

2.10.7 *Solve Problem 2.1.6.*

2.10.8 *Solve Problem 2.1.8.*

2.10.9 *Solve the problem formulated in Exercise 1.5.18.*

2.10.10 *Solve the problem formulated in Exercise 1.5.19.*

2.10.11 *Solve the problem formulated in Exercise 1.5.20.*

2.10.12 *Compare the Auxiliary and Shortcut Methods on the following infeasible LOP.*

$$\begin{aligned}
\text{Max.} \quad z \; = \; & x_1 + x_2 + x_3 \\
\text{s.t.} \quad & x_1 + x_2 + x_3 \leq 2 \\
& -2x_2 - x_3 \leq 3 \\
& -3x_1 + x_2 - 2x_3 \leq -5 \\
& 2x_1 - 6x_2 \leq 0 \\
\& \quad & x_1, \; x_2, \; x_3 \geq 0
\end{aligned}$$

2.10.13 *Compare the Auxiliary and Shortcut Methods on N infeasible LOPs of your own design. [HINT: You may have more luck in creating them if you have more constraints than variables.]*

2.10.14 *Solve the following LOP (with the Shortcut Method, if necessary).*

$$\begin{aligned}
\text{Max.} \quad z \; = \; & x_1 + 7x_2 \\
\text{s.t.} \quad & 3x_1 + 2x_3 \leq 6 \\
& -2x_1 + 2x_2 + 4x_3 \leq 4 \\
& 9x_1 + 3x_2 + x_3 \leq 18 \\
\& \quad & x_1, \; x_2, \; x_3 \geq 0
\end{aligned}$$

2.10.15 Solve the following LOP (with the Shortcut Method, if necessary).

$$\text{Max. } z = x_1 + 6x_2 + 3x_3$$

$$\text{s.t. } \begin{aligned} x_1 + 2x_2 + 3x_3 &\leq 2 \\ 2x_1 - 2x_2 + 4x_3 &\leq 4 \end{aligned}$$

$$\& \quad x_1, x_2, x_3 \geq 0$$

Include a certificate of your result.

2.10.16 Solve the following LOP (with the Shortcut Method, if necessary).

$$\text{Max. } z = x_1 - 2x_2 + 5x_3$$

$$\text{s.t. } \begin{aligned} x_1 - 2x_2 &\leq 10 \\ -2x_1 - 2x_3 + 3x_4 &\leq -4 \\ -x_1 - x_2 - 4x_3 &\leq -1 \\ -2x_1 + 5x_2 + 8x_3 - 10x_4 &\leq -5 \\ 2x_1 + 2x_3 + x_4 &\leq 1 \end{aligned}$$

$$\& \quad x_1, x_2, x_3, x_4 \geq 0$$

Include a certificate of your result.

2.10.17 Solve the following LOP (with the Shortcut Method, if necessary).

$$\text{Max. } z = 112x_1 + 125x_2 + 103x_3$$

$$\text{s.t. } \begin{aligned} -13x_1 - 14x_3 &\leq -63 \\ -14x_1 + 15x_2 - 17x_3 &\leq -49 \\ 11x_1 - 18x_2 + 16x_3 &\leq -56 \\ 12x_1 + 20x_2 &\leq 98 \end{aligned}$$

$$\& \quad x_1, x_2, x_3 \geq 0$$

Include a certificate of your result.

2.10.18 Solve the following LOP (with the Shortcut Method, if necessary).

$$\text{Max. } z = 5x_1 + 7x_2$$

$$\text{s.t. } \begin{aligned} -2x_1 - x_2 &\leq -2 \\ -x_1 + 2x_2 &\leq 4 \\ x_1 - 2x_2 &\leq 1 \end{aligned}$$

$$\& \quad x_1, x_2 \geq 0$$

Include a certificate of your result.

2.10. Exercises

2.10.19 *Solve the following LOP (with the Shortcut Method, if necessary).*

$$
\begin{array}{rlrrrrrl}
\text{Max.} & z = & 12x_1 & + \ 3x_2 & + \ 12x_3 & - \ 6x_4 & & \\
\text{s.t.} & & x_1 & + \ x_2 & + \ 5x_3 & - \ 3x_4 & \leq & 2 \\
& & 7x_1 & & + \ 2x_3 & & \leq & 14 \\
& & 8x_1 & - \ x_2 & & - \ 2x_4 & \leq & 16 \\
& & & & x_3 & - \ 4x_4 & \leq & 5 \\
\& & & x_1,\ & x_2,\ & x_3,\ & x_4 & \geq & 0
\end{array}
$$

Include a certificate of your result.

2.10.20 *Consider the following LOP P.*

$$
\begin{array}{rlrrrrl}
\text{Max.} & z = & 3x_1 & + \ 3x_2 & + \ 3x_3 & & \\
\text{s.t.} & & -5x_1 & & + \ x_3 & \leq & -1 \\
& & 4x_1 & + \ 5x_2 & + \ 6x_3 & \leq & 14 \\
& & & & - \ 3x_3 & \leq & 1 \\
& & x_1 & - \ 7x_2 & - \ x_3 & \leq & -5 \\
\& & & x_1,\ & x_2,\ & x_3 & \geq & 0
\end{array}
$$

    a. *Write the sequence of pivot operations $i \mapsto j$ that solves P.*

    b. *Write β^*, π^*, x^*, and z^*.*

2.10.21 *For a given LOP tableau T, let $T(r, c)$ denote the entry in row r and column c of T. Let T be such a tableau and suppose that the rules of the Simplex Algorithm determine the pivot entry to be $T(i, j)$, and that the resulting tableau is T'. However, by accident you pivot on the entry $T(i', j')$, resulting in tableau T''.*

    a. *Find the pivot entry of T'' that returns you to tableau T.*

    b. *Explain how, and under what circumstances you can reach T' in a single pivot.*

2.10.22 *Repeat Workout 2.6.4 N times, with n variables and m constraints, for various values of $m > n$. Form a general conjecture based on your results.*

2.10.23 *Consider the following LOP.*

$$\begin{array}{rrrrrrrrr}
\text{Max.} \ z = & -2x_1 & -3x_2 & -x_3 & -2x_4 & -8x_5 & -3x_6 & & \\
\text{s.t.} & x_1 & -6x_2 & +15x_3 & +13x_4 & -3x_5 & +7x_6 & \leq & 8 \\
 & x_1 & +x_2 & +5x_3 & & & +17x_6 & \leq & 7 \\
 & x_1 & +2x_2 & +5x_3 & +3x_4 & -4x_5 & +27x_6 & \leq & 0 \\
 & -13x_1 & -66x_2 & & +3x_4 & +9x_5 & +37x_6 & \leq & 3 \\
 & -6x_1 & +6x_2 & & +3x_4 & & +47x_6 & \leq & 1 \\
 & & & 5x_3 & & & +57x_6 & \leq & 1 \\
 & & 6x_2 & +5x_3 & +3x_4 & -18x_5 & +67x_6 & \leq & 5 \\
 & x_1 & & & -23x_4 & +3x_5 & +77x_6 & \leq & 9 \\
\& & x_1, & x_2, & x_3, & x_4, & x_5, & x_6 & \geq & 0
\end{array}$$

Use Theorem 2.9.1 to prove that it has an optimum (i.e., do not solve it).

2.10.24 *Repeat the following 5-step exercise N times successfully (consider it a success if an optimal solution is found).*

> a. *Create an arbitrary primal linear problem P in standard form, using 2–6 variables and 2–6 constraints (use different size LOPs each time).*
>
> b. *Write the dual linear problem D to your problem P.*
>
> c. *Solve the primal P.*
>
> d. *Solve the dual D.*
>
> e. *Record the optimal solutions and tableaux of each.*

Describe the patterns you witness as precisely as you can. Try to write down a conjecture about the general case.

2.10.25 *Create a standard form LOP with 2 variables and 2 constraints so that its feasible region is a quadrilateral and the Simplex Algorithm visits all four of its extreme points.*

Challenges

2.10.26 *Solve Exercise 1.5.21(a).*

2.10.27 *Solve Exercise 1.5.21(b).*

2.10.28 *Solve Exercise 1.5.33.*

2.10.29 *Solve Exercise 1.5.39.*

2.10.30 *Solve Exercise 1.5.40.*

2.10. Exercises

2.10.31 Solve Exercise 1.5.42.

2.10.32 Let S be the feasible region of a LOP and let $\mathbf{x}^0$ be interior to some face F of S. Prove that if $\mathbf{x}^0$ is optimal then every point of F is optimal. [HINT: See Workout 2.1.3.]

2.10.33 Prove that a variable that leaves the basis during an iteration of Phase II of the Simplex Algorithm cannot reenter the basis during the next iteration.

2.10.34 List all the bases for each of the following LOPs; which are feasible and which are infeasible?

a.
$$\text{Max. } z = 4x_1 + 5x_2$$
$$\text{s.t. } \begin{aligned} 2x_1 + x_2 &\leq 6 \\ x_1 + x_2 &\leq 4 \\ x_1 + 2x_2 &\leq 6 \end{aligned}$$
$$\& \quad x_1, x_2 \geq 0$$

b.
$$\text{Max. } z = 3x_1 + 4x_2 + 5x_3$$
$$\text{s.t. } \begin{aligned} 2x_1 - 7x_2 + 4x_3 &\leq 6 \\ x_1 - 2x_2 - 2x_3 &\leq -4 \\ -5x_1 + x_2 + 2x_3 &\leq 5 \end{aligned}$$
$$\& \quad x_1, x_2, x_3 \geq 0$$

c.
$$\text{Max. } z = x_1 + x_2 + x_3$$
$$\text{s.t. } \begin{aligned} x_1 + x_3 &\leq 3 \\ 3x_2 - 4x_3 &\leq -1 \\ x_2 &\leq 2 \\ 2x_1 - x_2 + 3x_3 &\leq -4 \end{aligned}$$
$$\& \quad x_1, x_2, x_3 \geq 0$$

2.10.35

a. Given an $M \times N$ matrix $\mathbf{A}$, and integers $i \in [M]$, $j \in [N]$, and r, s, and $d > 0$, write an algorithm (in code or pseudocode) to perform the row operation $R_i = (rr_i - sr_j)/d$.

b. Given a tableau T, arising from a linear problem involving m constraints and n problem variables, write an algorithm (in code or pseudocode) to perform a pivot on $a_{i,j}$. [HINT: What entries of T play the roles of r, s, and d?]

2.10.36

a. Given an $M \times N$ matrix $\mathbf{A}$, and integers $i \in [M]$, $j \in [N]$, and r, s, and $d > 0$, write an algorithm (in code or pseudocode) to perform the row operation $R_i = (rr_i - sr_j)/d$.

b. Given a tableau T, arising from a linear problem involving m constraints and n problem variables, write an algorithm (in code or pseudocode) to perform a pivot on $a_{i,j}$. [HINT: What entries of T play the roles of r, s, and d?]

2.10.37 Justify the "fairly high probability" statement made in Workout 2.6.4(a).

2.10.38 Let $\mathbf{A}$ be a matrix and $\mathbf{B}$ be a matrix derived from $\mathbf{A}$ after some number of pivots. Use induction to prove that every row of $\mathbf{B}$ is a linear combination of the rows of $\mathbf{A}$.

2.10.39 Let P be a standard form maximization LOP with integer coefficients and right-hand sides. Let T be a tableau of P derived from the initial tableau by pivots as described in Workout 2.3.3. Let T' be the subsequent tableau derived from T by pivoting on an entry having absolute value d. Prove that the basic coefficient of T' is d.

2.10.40 Use Exercise 2.10.38 to prove your conjecture made in Exercise 2.10.22.

2.10.41 Let P be a standard form maximization LOP for which every entry of $\mathbf{A}$ and $\mathbf{b}$ is positive. Use Theorem 2.9.1 to prove that it has an optimum.

2.10.42 Let P be a standard form maximization LOP having nonnegative $\mathbf{b}$ and for which some nonnegative linear combination of its constraints yields a positive linear combination of its variables to be at most some constant. Use Theorem 2.9.1 to prove that it has an optimum.

2.10.43 Modify Theorem 2.9.1 to allow for maximization LOPs not in standard form. Prove your modification.

2.10.44 Write pseudocode for the Simplex Algorithm (Phase I & II for standard max form LOPs).

2.10.45 Create a standard form LOP with 3 variables and 3 constraints so that its feasible region has 8 extreme points and the Simplex Algorithm visits all of them.

2.10.46 Write an algorithm (in code or pseudocode) that lists every basis of an LOP with n variables and m constraints. Can you do so in such a way that the bases are listed in numerical order? Can you do so in such a way that the bases are listed so that consecutive bases differ by exactly two elements (such as by a pivot operation)?

Projects

2.10.47 Present the **Big M Method** for Phase I.

Big M Method

2.10.48 Present the **Fourier–Motzkin Elimination Algorithm** for Phase I.

Fourier–Motzkin Elimination Algorithm

2.10.49 Present a proof of **Bland's Theorem** 2.8.5.

2.10.50 Find an example that shows that the MN (or GI, or other) Implementation can cycle.

2.10.51 Present a proof that the **Perturbation Method** avoids cycling.

Perturbation Method

Chapter 3

Geometry

3.1 Extreme Points

We have seen how the Graphic Method works and how the Simplex Algorithm works. Now we would like to explore connections between the two. The goal of this section is to discover the relationship between basic feasible solutions and extreme points of the feasible region. Consider the following problem.

Problem 3.1.1

$$
\begin{aligned}
\text{Max.}\ z\ =\ &2x_1 + 11x_2 \\
\text{s.t.}\quad &x_1 + x_2 \geq 13 \\
&x_1 - x_2 \leq 5 \\
&-2x_1 + x_2 \leq 4 \\
&3x_1 + 4x_2 \leq 92 \\
&x_2 \leq 14 \\
\&\quad &x_1,\ x_2 \geq 0
\end{aligned}
$$

The drawing of its feasible region is shown in Figure 3.1. The extreme points $\mathbf{x}^{(p)}$ are labeled by the number p of Simplex pivots required to reach them, $\mathbf{x}^*$ being the extreme point of optimality. Its initial tableau is below.

Tableau 3.1.2

$$
\begin{bmatrix}
-1 & -1 & 1 & 0 & 0 & 0 & 0 & 0 & -13 \\
1 & -1 & 0 & 1 & 0 & 0 & 0 & 0 & 5 \\
-2 & 1 & 0 & 0 & 1 & 0 & 0 & 0 & 4 \\
3 & 4 & 0 & 0 & 0 & 1 & 0 & 0 & 92 \\
0 & 1 & 0 & 0 & 0 & 0 & 1 & 0 & 14 \\
-2 & -11 & 0 & 0 & 0 & 0 & 0 & 1 & 0
\end{bmatrix}
$$

G. H. Hurlbert, *Linear Optimization*, Undergraduate Texts in Mathematics,
DOI: 10.1007/978-0-387-79148-7_3, © Springer Science+Business Media LLC 2010

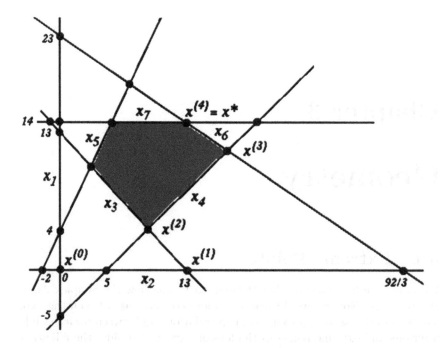

Figure 3.1: Feasible region for Problem 3.1.1, with variable and pivot labels

What are the values for x_1 and x_2 in the initial basic solution? Since they both begin as parameters, each is zero. We can think of the origin $\mathbf{x}^{(0)} = (0,0)^\mathsf{T}$ as being the intersection of the lines $x_1 = 0$ and $x_2 = 0$. Notice that each of the lines in Figure 3.1 has been labeled with an appropriate variable, so as to suggest that it is the line corresponding to that variable being zero. For example, the equality $x_1 + x_2 = 13$ implies that the slack variable $x_3 = 0$. So the line $x_1 + x_2 = 13$ is the same as the line $x_3 = 0$, illustrating why that line is labeled x_3.

The point $\mathbf{x}^{(1)} = (13,0)^\mathsf{T}$ is the intersection of the lines $x_2 = 0$ and $x_3 = 0$. Thus if we take x_2 and x_3 as parameters, this leaves x_1, x_4, x_5, x_6, and x_7 for the basis. That is, the point $(13,0)^\mathsf{T}$ corresponds to a basic solution (which incidentally arises from our first pivot operation).

Likewise, pivot once again and consider the resulting basic solution. It has parameters x_3 and x_4. Sure enough, the lines $x_3 = 0$ and $x_4 = 0$ intersect at $\mathbf{x}^{(2)} = (9,4)^\mathsf{T}$, just as the resulting tableau predicts. This particular tableau is feasible, so its corresponding basic solution is an extreme point of the feasible region. One more pivot takes us to the basis $\{1, 2, 3, 5, 7\}$. That is, parameters x_4 and x_6 are zero; these lines intersect at $\mathbf{x}^{(3)} = (16, 11)^\mathsf{T}$.

The point $X = (80, 204)^\mathsf{T}/11$ corresponds to the basis $\{1, 2, 3, 4, 7\}$. Since this point is on the infeasible side of the line $x_7 = 0$, it must be that $x_7 < 0$. Thus the pivot required to travel from $(16, 11)^\mathsf{T}$ to $(80/11, 204/11)^\mathsf{T}$ must be wrong according to our Simplex rules. Indeed, you can check that the pivot $4 \mapsto 5$ violates the **b**-ratio rule. The correct pivot of $4 \mapsto 7$ yields the extreme point $\mathbf{x}^{(4)} = (12, 14)^\mathsf{T}$, which is optimal.

Workout 3.1.3 *It seems clear that in two dimensions (i.e., two problem variables), there is a one-to-one correspondence between basic solutions and intersection points of two constraint lines. However, this is not quite true — why? What is true, and what is the general argument in $\mathbb{R}^n$? [HINT: Recall Workouts 2.1.5 and 2.8.1.]*

Surely, this is precisely what would please Descartes so. Knowing from geometry that the optimum must occur at a extreme point, it is quite satisfying to discover that the Simplex Algorithm spends its time jumping from one extreme point to another.

3.2 Convexity

Let's discuss the shapes of our feasible regions in greater detail. You might have noticed that the 2-dimensional problems considered up to this point all have had feasible regions sharing an interesting property. If you randomly pick any feasible point and your friend arbitrarily picks any other feasible point, then the straight line segment which joins these two points is contained entirely in the feasible region. (Check them and see that this is so!)

A region S is called **convex** if the line segment joining any two points of S is contained entirely in S. So the statement above can be rephrased to say that the feasible regions so far all have been convex. The question then arises, is every feasible region convex? We claim that the answer is yes.

convex region

Workout 3.2.1 *Consider the half-space R defined by $3x_1 - 7x_2 \leq -12$.*

  a. *Prove that R is convex.*

  b. *More generally, prove that the half-space defined by $\sum_{j=1}^n a_j x_j \leq b$ is convex.*

Workout 3.2.2 *Let S and T be two arbitrarily chosen convex regions.*

  a. *Prove by example that $S \cup T$ is not always convex.*

  b. *Prove that $S \cap T$ is always convex.*

  c. *Use induction to prove that the intersection of an arbitrary number (finite or infinite) of convex regions is convex.*

  d. *Prove that all polyhedra are convex.*

Theorem 3.2.3 *Let $\mathbf{A} \in \mathbb{R}^{m \times n}$, $\mathbf{b} \in \mathbb{R}^m$, and let $S = \{\mathbf{x} \in \mathbb{R}^n \mid \mathbf{Ax} \leq \mathbf{b}\}$. Then S is convex.*

Note that Theorem 3.2.3 implies that the feasible region of any LOP in standard form is convex. Indeed, the standard form $\{\mathbf{Ax} \leq \mathbf{b}, \mathbf{x} \geq \mathbf{0}\}$ is a special instance of the form $\{\mathbf{Ax} \leq \mathbf{b}\}$. In fact, the forms are equivalent since nonnegativity constraints can be considered as part of the problem constraints.

Workout 3.2.4

    a. *Use Workout 3.2.1 and 3.2.2 to prove Theorem 3.2.3.*

    b. *Does the same hold for LOPs not in standard form?*

Let $L = L(\mathbf{x}^1, \mathbf{x}^2)$ be the line segment that joins the two points $\mathbf{x}^1$ and $\mathbf{x}^2$ in $\mathbb{R}^n$ (here, the superscripts do not denote exponents, but rather merely first and second points). Can we describe L algebraically? If $\mathbf{x}^1 = (3,5)^\mathsf{T}$ and $\mathbf{x}^2 = (8,-1)^\mathsf{T}$ then the line that passes through both points is defined by the set of all points $(x_1, x_2)^\mathsf{T}$ which satisfy $6x_1 + 5x_2 = 43$. A parametric description can be given by those points of the form $(3,5)^\mathsf{T} + t(5,-6)^\mathsf{T}$, which can be written instead as $(1-t)(3,5)^\mathsf{T} + t(8,-1)^\mathsf{T}$. Then the substitutions $t_1 = 1-t$ and $t_2 = t$ yield the form $t_1\mathbf{x}^1 + t_2\mathbf{x}^2$, with $t_1 + t_2 = 1$. Notice that the point $\mathbf{x}^1$ corresponds to the value $t = 0$ ($(t_1, t_2) = (1, 0)$), and the point $\mathbf{x}^2$ corresponds to the value $t = 1$ ($(t_1, t_2) = (0, 1)$). So it seems that the points on the line segment between $\mathbf{x}^1$ and $\mathbf{x}^2$ must be defined as those which arise when both $t_1, t_2 \geq 0$ (for example, the midpoint $(5.5, 2)^\mathsf{T}$ has $(t_1, t_2) = (.5, .5)$). Therefore, we equivalently and algebraically can re-define a region (set) S to be convex if $t_1\mathbf{x}^1 + t_2\mathbf{x}^2 \in S$ whenever $\mathbf{x}^1, \mathbf{x}^2 \in S$, $t_1 + t_2 = 1$, and $t_1, t_2 \geq 0$. Of course, since $\mathbf{x}^1$ and $\mathbf{x}^2$ are linearly independent, we know from linear algebra that the region of all points of the form $t_1\mathbf{x}^1 + t_2\mathbf{x}^2$ is all of $\mathbb{R}^2$.

Workout 3.2.5 *With $\mathbf{x}^1$ and $\mathbf{x}^2$ as above, consider the four points $A = (2,6)^\mathsf{T}$, $B = (-2,11)^\mathsf{T}$, $C = (10,-1)^\mathsf{T}$ and $D = (\frac{25}{6}, \frac{18}{5})^\mathsf{T}$.*

    a. *For each of the points $(s_1, s_2)^\mathsf{T}$ above, find the values of t_1 and t_2 that yield*
$$t_1\mathbf{x}^1 + t_2\mathbf{x}^2 = (s_1, s_2)^\mathsf{T}.$$

    b. *Which of the above points lie on the line containing $\mathbf{x}^1$ and $\mathbf{x}^2$?*

    c. *Which of the above points lie between the rays from the origin through $\mathbf{x}^1$ and $\mathbf{x}^2$?*

    d. *Which of the above points lie on the line segment between $\mathbf{x}^1$ and $\mathbf{x}^2$?*

For a set $X = \{\mathbf{x}^1, \ldots, \mathbf{x}^k\}$ of k points in $\mathbb{R}^n$, we call the sum $\Sigma_{j=1}^k t_j \mathbf{x}^j$ a **linear combination** of X. If it is the case that $\Sigma_{j=1}^k t_j = 1$, then we call the sum an **affine combination** of X. If instead it is the case that each $t_j \geq 0$, then we call the sum a **conic combination** of X. Finally, if the linear combination is both affine and conic, then we call it a **convex combination** of X. (Note that for a LOP in standard form, the dual variables are used to form conic combinations of its constraints.)

3.2. Convexity

Workout 3.2.6 *Suppose that α and β are each convex combinations of X, and that γ is a convex combination of α and β. Prove that γ is a convex combination of X.*

Before proceeding, let's get more comfortable with these definitions. If we can think of pictures in $\mathbb{R}^3$ for a moment, imagine three points $\mathbf{x}^1$, $\mathbf{x}^2$, and $\mathbf{x}^3$ which are linearly independent. The points $(1,0,0)^\mathsf{T}$, $(0,1,0)^\mathsf{T}$, and $(0,0,1)^\mathsf{T}$ will do fine, but you can think of another three if you wish. (In fact, these points are completely general, as a change of basis matrix converts any other example to this one.) The region of all linear combinations of these three points consists of all of $\mathbb{R}^3$. The region of all affine combinations consists of the plane that contains all three points. (A 2-dimensional linear space is a plane going through the origin. A 2-dimensional affine space is a plane shifted way from the origin, like this one.) The region of all conic combinations consists of all the nonnegative points of $\mathbb{R}^3$. The region of all convex combinations consists of the points on the affine plane which lie on the triangle whose extreme points are $\mathbf{x}^1$, $\mathbf{x}^2$, and $\mathbf{x}^3$: the intersection of the affine and conic spaces. Let's test that what we have said here is true.

Workout 3.2.7 *Choose three points $\mathbf{x}^1, \mathbf{x}^2, \mathbf{x}^3 \in \mathbb{R}^3$. Let P be the plane containing $\mathbf{x}^1, \mathbf{x}^2$ and $\mathbf{x}^3$, and let T be the triangle having corners $\mathbf{x}^1, \mathbf{x}^2$ and $\mathbf{x}^3$. (Since every three noncolinear points determine a plane, for the purposes of this workout it may be simplest to first pick an equation $\sum_{j=1}^{3} a_j x_j = c$ and then pick $\mathbf{x}^1, \mathbf{x}^2$ and $\mathbf{x}^3$ satisfying it — what some call 'reverse engineering'.)*

  a. Pick a point γ that you know is on P but not T.
   (i) Verify that $t_1 + t_2 + t_3 = 1$.
   (ii) Verify that one of the t_js is negative.
  b. Pick a point γ that you know is on both P and T.
   (i) Verify that $t_1 + t_2 + t_3 = 1$.
   (ii) Verify that each $t_j \geq 0$.
  c. Pick a point τ that you know is not on P and verify that $t_1 + t_2 + t_3 \neq 1$.

Workout 3.2.8 *Draw the affine, conic and convex spaces determined by $\mathbf{x}^1 = (1,5,6)^\mathsf{T}$, $\mathbf{x}^2 = (7,1,5)^\mathsf{T}$ and $\mathbf{x}^3 = (6,7,1)^\mathsf{T}$.*

Why is it that all the points on the interior of the triangle T formed by $\mathbf{x}^1$, $\mathbf{x}^2$, and $\mathbf{x}^3$ are convex combinations of the $\mathbf{x}^i$s? Suppose $\gamma \in T$ and consider the line through γ and $\mathbf{x}^3$. That line intersects the line segment which joins $\mathbf{x}^1$ and $\mathbf{x}^2$ at some point which we can call δ. We know that δ is a convex combination of $\mathbf{x}^1$ and $\mathbf{x}^2$, and that γ is a convex combination of δ and $\mathbf{x}^3$. So indeed, by Workout 3.2.6, γ is a convex combination of $\mathbf{x}^1$, $\mathbf{x}^2$, and $\mathbf{x}^3$.

convex hull/span Define the **convex hull** of a set $X \subseteq \mathbb{R}^n$, denoted vhull(X), to be the smallest convex region containing X. Also define the **convex span** of X, denoted vspan(X), to be the set of all convex combinations of finitely many points in X.

Workout 3.2.9 *Use the definition of* vhull *and Workout 3.2.6 to prove that, for all* $X \subseteq \mathbb{R}^n$, *every* $\gamma \in$ vhull(X) *is in* vspan(X); *that is,* vhull$(X) \subseteq$ vspan(X). *(Note that X may be finite or infinite.)*

An obvious and interesting question to pursue is whether or not vhull(X) = vspan(X); in other words, whether or not vspan$(X) \subseteq$ vhull(X) also holds. To this end, define vspan$_k(X)$ to be the set of all convex combinations involving exactly k of the points in X. More precisely, $\mathbf{x} \in$ vspan$_k(X)$ if and only if there exist $X' = \{\mathbf{x}^1, \ldots, \mathbf{x}^k\} \subseteq X$ such that $\mathbf{x} \in$ vspan(X'). In particular, vspan$_1(X) = X$ and $\cup_{k=1}^{\infty}$ vspan$_k(X) =$ vspan(X).

Workout 3.2.10 *Let C be any convex region containing the set $X \subseteq \mathbb{R}^n$. Use induction on k to prove that* vspan$_k(X) \subseteq C$ *for all $k \geq 1$. [HINT: The technique is not too different from that used preceding Workout 3.2.9.]*

Theorem 3.2.11 *For every set $X \subseteq \mathbb{R}^n$ we have* vhull$(X) =$ vspan(X).

Workout 3.2.12 *Use Workouts 3.2.9 and 3.2.10 to prove Theorem 3.2.11.*

3.3 小试牛刀

Workout 3.3.1 *Consider the point set* $X = \{(5,14)^\mathsf{T}, (9,4)^\mathsf{T}, (16,11)^\mathsf{T}, (12,14)^\mathsf{T}, (7,8)^\mathsf{T}, (3,10)^\mathsf{T}\} \subset \mathbb{R}^2$.

   a. *Draw* vhull(X).

   b. *Which points of X can be thrown away without altering* vhull(X)?

   c. *For each such point γ, find the smallest k for which $\gamma \in$* vspan$_k(X)$.

   d. *Can you partition* vhull(X) *into t triangles for*

   (i) $t = 2$?
   (ii) $t = 3$?
   (iii) $t = 4$?
   (iv) $t = 5$?

   e. *Draw* vspan$_k(X)$ *for each of the values*

   (i) $k = 1$.
   (ii) $k = 2$.
   (iii) $k = 3$.
   (iv) $k = 4$.

f. Which of the $\text{vspan}_k(X)$ above are convex?

g. Find the inequalities that define the half-spaces whose intersection equals $\text{vhull}(X)$.

If $X = \{\mathbf{x}^1, \ldots, \mathbf{x}^t\}$ is a finite set of points in $\mathbb{R}^n$ then, for $L = \max\{|x_i^j| \mid 1 \leq j \leq t, 1 \leq i \leq n\}$, we have $X \subseteq B$, where B is the 'box' polytope $\{\mathbf{x} \in \mathbb{R}^n \mid -L \leq x_i \leq L, 1 \leq i \leq n\}$. However, it would be nice to find a smaller polytope containing X, one that fits its shape better, such as $\text{vhull}(X)$. But is $\text{vhull}(X)$ a polytope?

Workout 3.3.2 *Let X be a finite set of points in $\mathbb{R}^n$ for which $\text{vhull}(X)$ has dimension n. Use the points in X and your method from Workout 3.3.1(g) to define a polyhedron P that contains X. [HINT: For a set $S \subset X$, say that S qualifies if $\text{vhull}(S)$ has dimension $n-1$. What must be done with qualifying sets?]*

The problem with the construction in general is that all the points could lie on some hyperplane h. Then we would have $h_S = h$ for all qualifying S, making $P = h$. Of course, such a P is a polyhedron containing X, but it wouldn't be a polytope and, more importantly, it wouldn't equal $\text{vhull}(X)$, which is really what we're after in this sequence of workouts. Worse, if $\text{vhull}(X)$ has dimension less than $n-1$, then no S qualifies, meaning P is empty, not containing X.

3.4 Carathéodory's Theorem

Note that $\text{vspan}_{k-1}(X) \subseteq \text{vspan}_k(X)$ for every $X \subseteq \mathbb{R}^n$, because including an extra point of X with coefficient zero does not alter the convex combination requirements. Thus we have that $\text{vhull}(X) = \cup_{k=1}^{\infty} \text{vspan}_k(X) = \lim_{k \to \infty} \text{vspan}_k(X)$. It is natural then to ask if there is some finite k for which $\text{vhull}(X) = \text{vspan}_k(X)$. For example, the value $k = |X|$ suffices in the case that X is finite. But what about the case for infinite X? In either case, what is the smallest such k? Does the dimension n matter? In 1907, Carathéodory answered these questions in the following theorem.

Theorem 3.4.1 *Let $X \subseteq \mathbb{R}^n$. Then $\text{vhull}(X) = \text{vspan}_{n+1}(X)$.*

Carathéodory's Theorem

The result is, of course, best possible. For example, let X be any n linearly independent points with entirely nonnegative coordinates, plus the origin. Then the average of those $n+1$ points is in $\text{vspan}_{n+1}(X)$ but not $\text{vspan}_n(X)$. The proof is not too tricky, and uses the ideas from Workout 3.3.1c, as well as the Fundamental Theorem 2.9.1.

Proof. We already know that $\text{vspan}_{n+1}(X) \subseteq \text{vhull}(X)$. Therefore we need only show that $\text{vhull}(X) \subseteq \text{vspan}_{n+1}(X)$. Let $\gamma \in \text{vhull}(X)$. Then for some k and some $\mathbf{x}^1, \ldots, \mathbf{x}^k \in X$ we can write

$$\gamma = \sum_{j=1}^{k} t_j \mathbf{x}^j \quad \text{with} \quad \sum_{j=1}^{k} t_j = 1 \qquad (3.1)$$

and each $t_j \geq 0$. If we write the coordinates as $\mathbf{x}^j = (x_1^j, \ldots, x_n^j)^\mathsf{T}$ and $\boldsymbol{\gamma} = (\gamma_1, \ldots, \gamma_n)^\mathsf{T}$, then we may rewrite system 3.1 as

$$\sum_{j=1}^k x_i^j t_j = \gamma_i \quad (1 \leq i \leq n) \quad \text{with} \quad \sum_{j=1}^k t_j = 1 \,. \qquad (3.2)$$

Now (3.2) is a system of $n+1$ equations in k variables t_j that has a nonnegative solution. Thinking of this system as the constraints of a LOP (with arbitrary objective function) allows us to use the Fundamental Theorem 2.9.1 (actually, the more general version of the theorem for LOPs not in standard maximization form — see Exercise 2.10.43) to conclude that it has a basic feasible solution. Such a solution has at most $n+1$ nonnegative values because that is (an upper bound on) the size of its basis. Those points $\mathbf{x}^j$ for which $t_j = 0$ therefore can be thrown out, leaving a linear combination using at most $n+1$ points of X. Thus $\boldsymbol{\gamma} \in \text{vspan}_{n+1}(X)$. ◇

It is useful to consider Carathéodory's result in contrast to what we know about linear algebra. The maximum number of linearly independent vectors in $\mathbb{R}^n$ is n, and every maximal set of such vectors has this size. However, there is no limit to the number of convex independent vectors in $\mathbb{R}^n$ — the uncountably many vectors of length 1 are one example. Moreover, just as a **linear basis** for $\text{lspan}(X)^1$, for some $X \subseteq \mathbb{R}^n$, is defined to be a set of linearly independent vectors of maximum size, a **convex basis** for $\text{vspan}(X)$ is defined to be a set of **convex independent** (no one vector is a convex combination of the others) vectors of maximum size. While for each X there are infinitely many linear bases for $\text{lspan}(X)$, there is a unique convex basis for $\text{vspan}(X)$. From these perspectives, linear and convex algebra/geometry are quite different.

But Carathéodory's Theorem says that they share an interesting similarity: for every $X \subseteq \mathbb{R}^n$, each $\mathbf{v} \in \text{lspan}(X)$ can be written as a linear combination of at most n of the vectors in X, while each $\mathbf{v} \in \text{vspan}(X)$ can be written as a convex combination of at most $n+1$ of the vectors in X. Still, if X is a linear basis, then there is exactly one way to write $\mathbf{v}$ as such a linear combination, while if X is a convex basis, then there are (albeit finitely) many ways to write $\mathbf{v}$ as such a convex combination when $|X| > n+1$ (and just one if $|X| = n+1$).

We will come to see (and we already have seen in the Fundamental Theorem 2.9.1) convex bases as central to the Simplex Algorithm (we are already using the term *basic* solution), as well as at the heart the theory of LO and its applications.

---

[1] The artist formerly known as span— see its definition at the beginning of the Exercises.

3.5 Exercises

Practice

For many of the following exercises, the following definitions will be useful. Define the **linear** (resp. **affine** or **conic**) **span** of X, denoted lspan(X) (resp. aspan(X) or nspan(X)), to be the set of all linear (resp. affine or conic) combinations of finitely many points in X.

linear/
affine/
conic span,
lspan,
aspan,
nspan

3.5.1 Let $X = \{\mathbf{x}^1, \mathbf{x}^2, \mathbf{x}^3, \mathbf{x}^4\}$, where $\mathbf{x}^1 = (-3, -2)^\mathsf{T}$, $\mathbf{x}^2 = (1, 9)^\mathsf{T}$, $\mathbf{x}^3 = (7, -1)^\mathsf{T}$ and $\mathbf{x}^4 = (15, 8)^\mathsf{T}$. Let $\boldsymbol{\gamma}^1 = (10, 6)^\mathsf{T}$, $\boldsymbol{\gamma}^2 = (11, 5)^\mathsf{T}$ and $\boldsymbol{\gamma}^3 = (12, 4)^\mathsf{T}$. For each i determine if $\boldsymbol{\gamma}^i \in \text{vhull}(X)$. If so, show that $\boldsymbol{\gamma}^i \in \text{vspan}_3(X)$, as noted by Theorem 3.4.1; if not, prove not. [HINT: Pretend the convexity equations are the constraints of an arbitrary LOP and prove its infeasibility, following Workout 2.6.3.] Are any of the $\boldsymbol{\gamma}^i \in \text{vspan}_2(X)$?

3.5.2 Let $X = \{\mathbf{x}^1, \mathbf{x}^2, \mathbf{x}^3, \mathbf{x}^4, \mathbf{x}^5\}$, where $\mathbf{x}^1 = (6, -6, 3, 0, -3)^\mathsf{T}$, $\mathbf{x}^2 = (-1, -6, 0, 5, 10)^\mathsf{T}$, $\mathbf{x}^3 = (0, 10, 1, 4, 1)^\mathsf{T}$, $\mathbf{x}^4 = (3, 22, 2, -3, -16)^\mathsf{T}$ and $\mathbf{x}^5 = (-3, 6, 0, 9, 12)^\mathsf{T}$. Let $\boldsymbol{\gamma}^1 = (1, 10, 2, 7, 4)^\mathsf{T}$, $\boldsymbol{\gamma}^2 = (0, 6, 1, 5, 4)^\mathsf{T}$, $\boldsymbol{\gamma}^3 = (10, 12, 9, -5, 0)^\mathsf{T}$ and $\boldsymbol{\gamma}^4 = (-9, 22, -2, 13, 16)^\mathsf{T}$. For each i answer (with proof) the following questions.

   a. Is $\boldsymbol{\gamma}^i$ a linear combination of X?
   b. Is $\boldsymbol{\gamma}^i$ an affine combination of X?
   c. Is $\boldsymbol{\gamma}^i$ a conic combination of X?
   d. Is $\boldsymbol{\gamma}^i$ a convex combination of X?

For those $\boldsymbol{\gamma}^i$ that are in vhull(X), find the smallest t for which $\boldsymbol{\gamma}^i \in \text{vspan}_t(X)$.

3.5.3 Let $X = \{(1, 4)^\mathsf{T}, (2, 7)^\mathsf{T}, (8, 3)^\mathsf{T}\}$ and $\mathbf{w} = (12, 17)^\mathsf{T}$. Prove that $\mathbf{w} \in \text{aspan}(X) \cap \text{nspan}(X) - \text{vspan}(X)$.

3.5.4 Let $X = \{(-1, 3, 2)^\mathsf{T}, (1, 0, -3)^\mathsf{T}, (3, -4, 1)^\mathsf{T}, (1, 1, 3)^\mathsf{T}\}$ and $\mathbf{w} = (52, 37, 43)^\mathsf{T}$. Prove that $\mathbf{w} \in \text{aspan}(X) \cap \text{nspan}(X) - \text{vspan}(X)$.

3.5.5 For each of the regions R below, prove or disprove that R is convex.

   a. $R = \{\mathbf{x} \in \mathbb{R}^n \mid \sum_{i=1}^n x_i \geq 2\}$ for some $n \geq 1$.
   b. $R = \{\mathbf{x} \in \mathbb{R}^n \mid \sum_{i=1}^n x_i^2 \geq 2^2\}$ for some $n \geq 1$. (Here, the exponents actually mean squares, rather than indices.)
   c. $R = \{\mathbf{x} \in \mathbb{R}^n \mid x_n > \sum_{i=1}^{n-1} x_i\}$ for some $n \geq 1$.
   d. $R = \{\mathbf{x} \in \mathbb{R}^n \mid x_n^2 > \sum_{i=1}^{n-1} x_i\}$ for some $n \geq 1$. (Here, the exponent actually means a square, rather than an index.)

3.5.6 *State and prove an analogue of Workout 3.2.6 for linear combinations.*

3.5.7 *State and prove an analogue of Workout 3.2.6 for affine combinations.*

3.5.8 *State and prove an analogue of Workout 3.2.6 for conic combinations.*

3.5.9 *Find all extreme points of* vhull(X), *where* $X = \{(-5,4), (6,-9), (-1,1), (8,3), (-3,-5), (5,-2), (2,6)\}$.

3.5.10 *Find all extreme points of* $S = \{\mathbf{x} \in \mathbb{R}^3 \mid x_1 + 3x_2 + 8x_3 \leq 54, 7x_1 + x_2 + 3x_3 \leq 58, 3x_1 + 9x_2 + x_3 \leq 50, 7x_1 + 9x_2 + 8x_3 \geq 91, 8x_2 + 7x_2 + 9x_3 \leq 109\}$.

linear region/hull **3.5.11** *A* **linear region** L *is one having the property that* $r\mathbf{u} + s\mathbf{v} \in L$ *for all* $\mathbf{u}, \mathbf{v} \in L$ *and* $r, s \in \mathbb{R}$. *Define the* **linear hull** *of a set* $X \subseteq \mathbb{R}^n$, *denoted* lhull(X), *to be the smallest linear region containing* X. *Prove that for every set* $X \subseteq \mathbb{R}^n$ *we have* lhull$(X) =$ lspan(X).

affine region/hull **3.5.12** *An* **affine region** A *is one having the property that* $r\mathbf{u} + s\mathbf{v} \in A$ *for all* $\mathbf{u}, \mathbf{v} \in A$ *and* $r, s \in \mathbb{R}$ *for which* $r + s = 1$. *Define the* **affine hull** *of a set* $X \subseteq \mathbb{R}^n$, *denoted* ahull(X), *to be the smallest affine region containing* X. *Prove that for every set* $X \subseteq \mathbb{R}^n$ *we have* ahull$(X) =$ aspan(X).

conic region/hull **3.5.13** *A* **conic region** C *is one having the property that* $r\mathbf{u} + s\mathbf{v} \in C$ *for all* $\mathbf{u}, \mathbf{v} \in C$ *and* $r, s \geq 0$. *Define the* **conic hull** *of a set* $X \subseteq \mathbb{R}^n$, *denoted* nhull(X), *to be the smallest conic region containing* X.

3.5.14 *For any set* $X \subseteq \mathbb{R}^n$ *prove that* lhull$(X) =$ ahull$(X \cup \{\mathbf{0}\})$.

3.5.15 *For any set* $X \subseteq \mathbb{R}^n$ *prove that* vhull$(X \cup \{\mathbf{0}\}) \subseteq$ nhull(X).

3.5.16 *Let X be a finite set of points in $\mathbb{R}^n$. Modify your construction from Workout 3.3.2 to define a polyhedron P that contains X. [HINT: Modify the definition of qualifying.]*

3.5.17 *Prove that the polyhedron P constructed in Exercise 3.5.16 is a polytope.*

3.5.18 *Given the polyhedron P constructed in Workout 3.3.2, use induction on $|X|$ to prove that $P \subseteq$ vhull(X). [HINT: Use the idea from the paragraph preceding Workout 3.2.9.]*

3.5.19 *Let X be a finite set of points in $\mathbb{R}^n$. Prove that* vhull(X) *is a polytope.*

3.5. Exercises

3.5.20 *For a finite set $X \subset \mathbb{R}^2$, call a region an* **X-triangle** *if it is the convex hull of exactly three points of X. Let $X = \{(103, 69), (72, 101), (33, 99), (7, 64), (27, 15), (64, 7), (103, 42)\}$ and $\mathbf{y} = (55, 55)$. How many X-triangles contain $\mathbf{y}$? Which are they?*

3.5.21 *For a finite set $X \subset \mathbb{R}^3$, call a region an* **X-tetrahedron** *if it is the convex hull of exactly four points of X. Let $X = \{(655, 372, 503), (263, 174, 631), (218, 655, 264), (667, 223, 325), (347, 348, 692), (378, 579, 96), (433, 283, 23)\}$ and $\mathbf{y} = (357, 396, 307)$. How many X-tetrahedron contain $\mathbf{y}$? Which are they?*

3.5.22 *Let $X = \{(2,4)^\mathsf{T}, (11,6)^\mathsf{T}, (5,13)^\mathsf{T}\}$. Find some $\mathbf{w} \in \mathbb{R}^2$ such that $\mathbf{w} \in \mathrm{aspan}(X) \cap \mathrm{nspan}(X) - \mathrm{vspan}(X)$. [HINT: See Exercise 3.5.3 and draw the relevant region.]*

3.5.23 *Let $X = \{(2,5,6)^\mathsf{T}, (13,0,8)^\mathsf{T}, (1,17,4)^\mathsf{T}, (11,1,12)^\mathsf{T}\}$. Find some $\mathbf{w} \in \mathbb{R}^3$ such that $\mathbf{w} \in \mathrm{aspan}(X) \cap \mathrm{nspan}(X) - \mathrm{vspan}(X)$. [HINT: See Exercise 3.5.4. and draw the relevant region.]*

3.5.24 *For two sets $X_1, X_2 \subseteq \mathbb{R}^n$, define the* **Minkowski sum** $X_1 + X_2 = \{\mathbf{x}^1 + \mathbf{x}^2 \mid \mathbf{x}^i \in X_i\}$. *Prove that if each X_i is convex then so is $X_1 + X_2$.*

3.5.25 *Let $\mathbf{x}$ be a point of a polyhedron P. Prove that $\mathbf{x}$ is an extreme point of P if and only if the region $P - \{\mathbf{x}\}$ is convex.*

3.5.26 *Let $S = \{\mathbf{x} \in \mathbb{R}^2 \mid 5x_1 + 16x_2 \leq 80, 12x_1 + x_2 \geq 10, 15x_1 - 11x_2 \leq 9\}$, $Z = \{\mathbf{x} \in \mathbb{Z}^2 \mid 5x_1 + 16x_2 \leq 80, 12x_1 + x_2 \geq 10, 15x_1 - 11x_2 \leq 9\}$, $V = \mathrm{vspan}(Z)$, and X be the set of extreme points of V.*

  a. *Find Z.*
  b. *Find X.*
  c. *Write V as a system of constraints.*

3.5.27 *Find a family of three convex regions $\{F_1, F_2, F_3\}$ in $\mathbb{R}^2$ such that every pair $\{i, j\}$ satisfies $F_i \cap F_j \neq \emptyset$, but that $\cap_{i=1}^3 F_i = \emptyset$.*

Challenges

3.5.28 *Let $\mathbf{v}^1 = (4,1)^\mathsf{T}$, $\mathbf{v}^2 = (1,3)^\mathsf{T}$, $\mathbf{v}^3 = (3,4)^\mathsf{T}$, and $\mathbf{J}_2 = (1,1)^\mathsf{T}$.*

  a. *Draw $\mathrm{vhull}\{\mathbf{v}_1, \mathbf{v}_2, \mathbf{v}_3\}$, along with the line through the origin generated by $\mathbf{J}_2$.*
  b. *For each $\mathbf{x}$ below, find the maximum x_0 that satisfies $\mathbf{v}_1 x_1 + \mathbf{v}_2 x_2 + \mathbf{v}_3 x_3 \geq x_0 \mathbf{J}_2$. Add the results to your diagram from part a.*
   (i) $\mathbf{x} = (2,3,1)^\mathsf{T}/6$.
   (ii) $\mathbf{x} = (0,0,1)^\mathsf{T}$.

(iii) $\mathbf{x} = (1,0,3)^\mathsf{T}/4$.

c. Use your diagram from part a to explain one way of solving the following LOP.

$$\begin{array}{rlrcrcrcl}
\text{Max.} & x & = & x_0 & & & & & \\
\text{s.t.} & & & & x_1 & + & x_2 & + & x_3 & = & 1 \\
& & & x_0 & - & 4x_1 & - & x_2 & - & 3x_3 & \geq & 0 \\
& & & x_0 & - & x_1 & - & 3x_2 & - & 4x_3 & \geq & 0 \\
\& & & & & x_1 & , & x_2 & , & x_3 & \geq & 0
\end{array}$$

3.5.29 Repeat Exercise 3.5.28 with $\mathbf{v}^1 = (1,2)^\mathsf{T}$, $\mathbf{v}^2 = (-1,3)^\mathsf{T}$, and $\mathbf{v}^3 = (-2,-1)^\mathsf{T}$.

3.5.30 Repeat Exercise 3.5.26 with $S = \{\mathbf{x} \in \mathbb{R}^3 \mid x_1 + 3x_2 + 8x_3 \leq 54, 7x_1 + x_2 + 3x_3 \leq 58, 3x_1 + 9x_2 + x_3 \leq 50, 7x_1 + 9x_2 + 8x_3 \geq 91\}$.

3.5.31 Modify the proof of Theorem 3.4.1 to prove that if $X = \{\mathbf{x}^1, \ldots, \mathbf{x}^k\} \subset \mathbb{R}^n$ and $\boldsymbol{\gamma} \in \text{vhull}(X)$ then $\boldsymbol{\gamma} \in \text{vspan}_{k+1}(X)$.

3.5.32 Given $X = \{\mathbf{x}^1, \ldots, \mathbf{x}^k\} \subset \mathbb{R}^n$ find the smallest function $t = t(X)$ such that $\text{vspan}_t(X)$ is convex. Prove your result.

3.5.33 Let P be a polytope in $\mathbb{R}^n$ with extreme points X. Prove that $P = \text{vhull}(X)$. *[NOTE: This is the converse to Workout 3.5.19.]*

3.5.34 State and prove a linear analog of Carathéodory's Theorem 3.4.1.

3.5.35 State and prove an affine analog of Carathéodory's Theorem 3.4.1.

3.5.36 State and prove a conic analog of Carathéodory's Theorem 3.4.1.

3.5.37 Let $X = \{\mathbf{x}^1, \ldots, \dot{\mathbf{x}}^k\}$ be linearly but not affinely dependent. Prove that there is some $\mathbf{t} \in \mathbb{R}^k$ such that $\sum_{i=1}^k t_i \mathbf{x}^i = \mathbf{0}$ and $\sum_{i=1}^k t_i = 1$.

3.5.38 Let X be linearly but not affinely dependent. Use Exercise 3.5.37 to prove that $\text{aspan}(X) = \text{lspan}(X)$.

3.5.39 Let X be linearly but not affinely dependent. Use Exercise 3.5.38 to prove that there is some $\mathbf{w} \in \text{aspan}(X) \cap \text{nspan}(X) - \text{vspan}(X)$. *[HINT: See Exercises 3.5.22 and 3.5.23.]*

3.5.40 Let P be a polytope in $\mathbb{R}^n$ and $\mathbf{x} \in \mathbb{R}^n$. Suppose that $\{\mathbf{v}^1, \ldots, \mathbf{v}^n\}$ is linearly independent and define $P_i = \text{proj}_{\mathbf{W}_i} P$ and $\mathbf{x}^i = \text{proj}_{\mathbf{W}_i} \mathbf{x}$, where $\mathbf{W}_i = (\mathbf{v}^i)^\perp$ denotes the vector space orthogonal to $\mathbf{v}^i$. Prove or disprove the following.

a. If $\mathbf{x}^i$ is interior to P_i for all i then $\mathbf{x}$ is interior to P.
b. If $\mathbf{x}^i$ is extremal in P_i for all i then $\mathbf{x}$ is extremal in P.
c. If $\mathbf{x}$ is interior to P then $\mathbf{x}^i$ is interior to P_i for all i.
d. If $\mathbf{x}$ is extremal in P then $\mathbf{x}^i$ is extremal in P_i for all i.

3.5. Exercises

3.5.41 Let $X \subset \mathbb{R}^2$ be a set of m points in **convex position**; that is, no point $\mathbf{x} \in X$ is in the convex hull of $X - \{\mathbf{x}\}$. (For example, placing X around a circle suffices.) For any $\mathbf{y} \in \text{vhull}(X)$ find a formula for the number of X-triangles that contain it.

convex position

3.5.42 Use Exercise 3.5.41 to find the maximum, among all $X \subseteq \mathbb{R}^2$ of size $|X| = m$ and $\mathbf{y} \in \text{vhull}(X)$, of the number of X-triangles containing $\mathbf{y}$.

3.5.43 For a finite set $X \subset \mathbb{R}^n$, call a region an X-**simplex** if it is the convex hull of exactly $n+1$ points of X. Write an algorithm (or MAPLE code) that takes as input the set X and a point $\mathbf{y} \in \mathbb{R}^n$ and outputs the number of X-simplices that contain $\mathbf{y}$.

X-simplex

3.5.44 Prove that, given any 5 points in $\mathbb{R}^2$, some 4 of them form the corners of a convex quadrilateral.

3.5.45 Find a family of four convex regions $\{F_1, F_2, F_3, F_4\}$ in $\mathbb{R}^3$ such that every triple $\{i, j, k\}$ satisfies $F_i \cap F_j \cap F_k \neq \emptyset$, but that $\cap_{i=1}^4 F_i = \emptyset$.

3.5.46 Generalize both Theorem 3.4.1 and Exercise 3.5.31 by finding the best possible function $t = t(X)$ so that the following statement is true. If $X = \{\mathbf{x}^1, \ldots, \mathbf{x}^k\} \subset \mathbb{R}^n$ and $\gamma \in \text{vhull}(X)$ then $\gamma \in \text{vspan}_t(X)$. (The intention here is that $t \leq \min\{k+1, n+1\}$ holds for all X while, for some X, $t < \min\{k+1, n+1\}$.) Prove your result. [HINT: Consider Exercises 3.5.2 and 3.5.32.]

3.5.47 Let $X \subset \mathbb{R}^n$ be a set of m points in affine position; that is, no point $\mathbf{x} \in X$ is in the affine hull of $X - \{\mathbf{x}\}$. For any $\mathbf{y} \in \text{vhull}(X)$ find a formula for the number of X-triangles that contain it.

affine position

3.5.48 Use Exercise 3.5.47 to find the maximum, among all $X \subseteq \mathbb{R}^n$ of size $|X| = m$ and $\mathbf{y} \in \text{vhull}(X)$, of the number of X-simplices containing $\mathbf{y}$.

3.5.49 Generalize Exercises 3.5.27 and 3.5.45 by finding a family of $n+1$ convex regions $\{F_1, \ldots, F_{n+1}\}$ in $\mathbb{R}^n$ such that every set $H \subset \{1, \ldots, n+1\}$ of size n satisfies $\cap_{h \in H} F_h \neq \emptyset$, but that $\cap_{i=1}^{n+1} F_i = \emptyset$.

3.5.50 Write an algorithm that solves the general problem of Exercise 3.5.26.

3.5.51 Suppose that $\{F_1, \ldots, F_4\}$ is a family of four convex regions in $\mathbb{R}^2$ such that every triple $\{i, j, k\}$ satisfies $F_i \cap F_j \cap F_k \neq \emptyset$. Prove that $\cap_{i=1}^4 F_i \neq \emptyset$.

Projects

Graham's scan

3.5.52 *Present* **Graham's scan** *for computing the extremal points of* vhull(X) *for* $X \subseteq \mathbb{R}^2$.

(Perfect) Matching polytope

3.5.53 *Present the* **(Perfect) Matching polytope** *of a graph.*

3.5.54 *Find algorithms to solve problems like Exercises 3.5.26 and 3.5.30 in* $\mathbb{R}^n$.

Polyhedral Verification Problem

3.5.55 *Present the* **Polyhedral Verification Problem**.

Chapter 4
The Duality Theorem

4.1 Primal-Dual Relationship

Consider the following LOP.

Problem 4.1.1

$$\begin{array}{rlrcrcrcr}
\text{Max.} & z & = & 22x_1 & + & 31x_2 & + & 29x_3 & \\
\text{s.t.} & & & x_1 & + & 4x_2 & + & 6x_3 & \leq 73 \\
& & & 5x_1 & - & 2x_2 & + & 3x_3 & \leq 68 \\
\& & & & x_1 & , & x_2 & , & x_3 & \geq 0
\end{array}$$

As you are invited to confirm, the Simplex Algorithm produces the following final tableau.

Tableau 4.1.2

$$\left[\begin{array}{ccc|ccc|c} 0 & 22 & 27 & 5 & -1 & 0 & 297 \\ 22 & 0 & 24 & 2 & 4 & 0 & 418 \\ \hline 0 & 0 & 727 & 199 & 57 & 22 & 18403 \end{array}\right]$$

Tableau 4.1.2 shows the optimum solution $\mathbf{x}^* = (418, 297, 0 \mid 0, 0)^\mathsf{T}/22$ with corresponding optimum value $z^* = 18403/22$. Likewise, consider the dual LOP below.

Problem 4.1.3

$$\begin{array}{rlrcr}
\text{Min.} & w & = & 73y_1 & + & 68y_2 \\
\text{s.t.} & & & y_1 & + & 5y_2 & \geq 22 \\
& & & 4y_1 & - & 2y_2 & \geq 31 \\
& & & 6y_1 & + & 3y_2 & \geq 29 \\
\& & & & y_1 & , & y_2 & \geq 0
\end{array}$$

Problem 4.1.3 has the following final tableau.

Tableau 4.1.4

$$\left[\begin{array}{cc|ccc|c} 0 & 22 & -4 & 1 & 0 & 0 & 57 \\ 22 & 0 & -2 & -5 & 0 & 0 & 199 \\ 0 & 0 & -24 & -27 & 22 & 0 & 727 \\ \hline 0 & 0 & 418 & 297 & 0 & 22 & -18403 \end{array} \right]$$

Tableau 4.1.4 shows the optimum solution $y^* = (199, 57 \mid 0, 0, 727)/22$ with corresponding optimum value $w^* = 18403/22$.

Curiously, this data shows certain repetitions of values. It looks like the values of $\mathbf{x}^*$ show up in the final dual objective row, but switched around a little. Likewise, the values of $\mathbf{y}^*$ appear in the final primal objective row, with a similar swap of some sort. To be more precise, the pattern seems to be that the *problem* values of $\mathbf{x}^*$ are the final coefficients of the dual *slack* variables, while the *slack* values of $\mathbf{x}^*$ are the final coefficients of the dual *problem* variables.

Problem 4.1.5

$$\begin{array}{rrrrrrrrr} \text{Max.} \quad z = & -22x_1 & -18x_2 & -27x_3 & -23x_4 & +16x_5 & -12x_6 & & \\ \text{s.t.} & 4x_1 & +x_2 & -3x_3 & & +2x_5 & +7x_6 & \leq & 211 \\ & 6x_1 & & +2x_3 & +5x_4 & -x_5 & +8x_6 & \leq & 189 \\ & -5x_1 & +4x_2 & -2x_3 & & -x_5 & -7x_6 & \leq & -106 \\ & 3x_1 & +9x_2 & & -2x_4 & +x_5 & +4x_6 & \leq & 175 \\ \& & x_1, & x_2, & x_3, & x_4, & x_5, & x_6 & \geq & 0 \end{array}$$

Workout 4.1.6 *Consider Problem 4.1.5.*

  a. *Use the Simplex Algorithm to solve it.*

  b. *Without solving the dual linear problem, use the final primal tableau to find the optimal dual variable values $\mathbf{y}^*$ (including slacks).*

  c. *Verify that $\mathbf{y}^*$ is dual feasible and optimal.*

It will help to articulate this perceived pattern notationally. We return to the general descriptions of primals and duals below.

Problem 4.1.7

$$\text{Max.} \quad z = \sum_{j=1}^{n} c_j x_j \qquad (1)$$

$$\text{s.t.} \quad \sum_{j=1}^{n} a_{ij} x_j \leq b_i \quad (1 \leq i \leq m) \qquad (2)$$

$$\& \qquad x_j \geq 0 \quad (1 \leq j \leq n) \qquad (3)$$

4.1. Primal-Dual Relationship

Problem 4.1.8

$$\text{Min. } w = \sum_{i=1}^{m} b_i y_i \tag{4}$$

$$\text{s.t.} \quad \sum_{i=1}^{m} a_{ij} y_i \geq c_j \quad (1 \leq j \leq n) \tag{5}$$

$$\& \quad y_i \geq 0 \quad (1 \leq i \leq m) \tag{6}$$

In order to describe the pattern we will need to look at the optimal primal objective row. Just as with all optimal values, let's use c_k^* for the final coefficient of x_k, as shown.

$$\left[\begin{array}{ccccc|cccc|c} c_1^* & \cdots & c_j^* & \cdots & c_n^* & c_{n+1}^* & \cdots & c_{n+i}^* & \cdots & c_{n+m}^* & 1 & z^* \end{array} \right]$$

Notice that there is a 1 written instead of the more general d for the coefficient of z. Why did we do that?

It seems as though the pattern we have witnessed is given below.

$$\left[\begin{array}{ccccc|cccc|c} y_{m+1}^* & \cdots & y_{m+j}^* & \cdots & y_{m+n}^* & y_1^* & \cdots & y_i^* & \cdots & y_m^* & 1 & z^* \end{array} \right]$$

That is,

$$y_i^* = c_{n+i}^* \quad (1 \leq i \leq m) \quad \text{and} \quad y_{m+j}^* = c_j^* \quad (1 \leq j \leq n). \tag{7}$$

Furthermore, we have also noticed time and again that, when the primal has an optimal solution, then so does its dual — in fact, with the same optimal value. Quite possibly, we could take advantage of the detailed pattern above to verify such a phenomenon in general. From *(1)* the value in question is

$$z^* = \sum_{j=1}^{n} c_j x_j^*. \tag{8}$$

From the Weak Duality Theorem (Inequality 1.4.7) we know that if the dual problem is feasible then its optimum is at least this value z^*. In fact we get equality.

Theorem 4.1.9 *If a linear problem P has an optimum z^* then its dual linear problem D has an optimum w^*; moreover, $z^* = w^*$.* — Strong Duality Theorem

Proof. Because we already know that every feasible z and w satisfy $z \leq w$, we only need to find a feasible w for which $w = z$. For this we can turn to the y_i^*s defined in *(7)*, and show that they satisfy inequalities *(5)* and *(6)* as well as the equality

$$z^* = \sum_{i=1}^{m} b_i y_i^*. \tag{9}$$

Workout 4.1.10 *Show that the y_i^*s defined in (7) satisfy (6).*

One of the things we can do is write out the optimal objective row, solving for z. With the substitutions from (7) we have

$$z = z^* - \sum_{j=1}^{n} y_{m+j}^* x_j - \sum_{i=1}^{m} y_i^* x_{n+i}. \qquad (10)$$

Workout 4.1.11 *Use equation (1) and the definition of the slack variables x_{n+i} to derive from equation (10) the equality*

$$\sum_{j=1}^{n} c_j x_j = \left(z^* - \sum_{i=1}^{m} b_i y_i^*\right) + \sum_{j=1}^{n} \left(\left(\sum_{i=1}^{m} a_{ij} y_i^*\right) - y_{m+j}^*\right) x_j. \qquad (11)$$

Interestingly, since these are equations that hold for any values of the x_js, we may experiment with various choices. For example, if each $x_j = 1$ we obtain

$$\sum_{j=1}^{n} c_j = \left(z^* - \sum_{i=1}^{m} b_i y_i^*\right) + \sum_{j=1}^{n} \left(\sum_{i=1}^{m} a_{ij} y_i^* - y_{m+j}^*\right).$$

Unfortunately, that experiment tells us nothing to help us show that (5) or (9) hold.

Workout 4.1.12 *What choice of values for the x_js, plugged into (11), show immediately that (9) holds?*

Now that (9) holds, we see that (11) reduces to

$$\sum_{j=1}^{n} c_j x_j = \sum_{j=1}^{n} \left(\sum_{i=1}^{m} a_{ij} y_i^* - y_{m+j}^*\right) x_j. \qquad (12)$$

Of course, we can try similar experiments on equation (12).

Workout 4.1.13 *What choice of values for the x_js, plugged into (12), shows that*

$$c_1 = \sum_{i=1}^{m} a_{i1} y_i^* - y_{m+1}^* ?$$

Workout 4.1.14 *Do for any c_k what you did for c_1 and use your results to show that (5) holds.*

Now that (5), (6) and (9) have been verified for the y_i^*s, the Duality Theorem has been proved. ◇

4.2. Complementary Slackness Conditions

Workout 4.1.15 *Suppose P is a linear problem with 4 variables and 7 constraints.*

  a. *If x_3^* is in the basis, what does that say about some optimal objective coefficient?*

    (i) *In turn, what does that say about some optimal dual variable value?*

    (ii) *In particular, what does that say about some optimal dual constraint?*

  b. *If x_8^* is in the basis, what does that say about some optimal objective coefficient?*

    (i) *In turn, what does that say about some optimal dual variable value?*

    (ii) *Also, what does that say about some optimal primal constraint?*

4.2 Complementary Slackness Conditions

Let us return to Problem 4.1.5 and its dual Problem 4.2.1.

Problem 4.2.1

$$
\begin{array}{rrrrrrl}
\text{Min.} \; w = & 211y_1 + & 189y_2 & - & 106y_3 & + & 175y_4 \\
\text{s.t.} & 4y_1 + & 6y_2 & - & 5y_3 & + & 3y_4 \geq -22 \\
& y_1 & & + & 4y_3 & + & 9y_4 \geq -18 \\
& -3y_1 + & 2y_2 & - & 2y_3 & & \geq -27 \\
& & 5y_2 & & & - & 2y_4 \geq -23 \\
& 2y_1 - & y_2 & - & y_3 & + & y_4 \geq 16 \\
& 7y_1 + & 8y_2 & - & 7y_3 & + & 4y_4 \geq -12 \\
\& & y_1, & y_2, & & y_3, & & y_4 \geq 0
\end{array}
$$

As shown in the corresponding primal optimal Tableau 4.2.2 below, the optimal primal value of $z^* = 11813/7$ occurs at $\mathbf{x}^* = (0, 0, 1, 0, 740, 0)^\mathsf{T}/7$, and the optimal dual value of $w^* = 11813/7$ occurs at $\mathbf{y}^* = (59, 0, 6, 0)^\mathsf{T}/7$.

Tableau 4.2.2

$$
\left[\begin{array}{ccccc|ccccc|c}
23 & -10 & 0 & 0 & 7 & 35 & 2 & 0 & -3 & 0 & 0 & 740 \\
53 & 8 & 0 & 35 & 0 & 77 & 4 & 7 & 1 & 0 & 0 & 2061 \\
6 & -9 & 7 & 0 & 0 & 7 & -1 & 0 & -2 & 0 & 0 & 1 \\
-2 & 73 & 0 & -14 & 0 & -7 & -2 & 0 & 3 & 7 & 0 & 485 \\
\hline
360 & 209 & 0 & 161 & 0 & 455 & 59 & 0 & 6 & 0 & 7 & 11813
\end{array}\right]
$$

It is interesting to look back on the original constraints of Problem 4.1.5, evaluated at $\mathbf{x}^*$. Notice the results below.

$$
\begin{array}{rcrcrcrcrcrcr}
4x_1^* & + & x_2^* & - & 3x_3^* & & & + & 2x_5^* & + & 7x_6^* & = & 211 \\
6x_1^* & & & + & 2x_3^* & + & 5x_4^* & - & x_5^* & + & 8x_6^* & < & 189 \\
-5x_1^* & + & 4x_2^* & - & 2x_3^* & & & - & x_5^* & - & 7x_6^* & = & -106 \\
3x_1^* & + & 9x_2^* & & & - & 2x_4^* & + & x_5^* & + & 4x_6^* & < & 175 \\
\end{array}
$$

$$
\begin{array}{ccccccc}
x_1^* & , & x_2^* & , & x_4^* & , & x_6^* & = & 0 \\
& & & x_3^* & , & x_5^* & & > & 0 \\
\end{array}
$$

Likewise we observe below the results of plugging in $\mathbf{y}^*$ into the original constraints of Problem 4.2.1.

$$
\begin{array}{rcrcrcrcr}
4y_1^* & + & 6y_2^* & - & 5y_3^* & + & 3y_4^* & > & -22 \\
y_1^* & & & + & 4y_3^* & + & 9y_4^* & > & -18 \\
-3y_1^* & + & 2y_2^* & - & 2y_3^* & & & = & -27 \\
& & 5y_2^* & & & - & 2y_4^* & > & -23 \\
2y_1^* & - & y_2^* & - & y_3^* & + & y_4^* & = & 16 \\
7y_1^* & + & 8y_2^* & - & 7y_3^* & + & 4y_4^* & > & -12 \\
\end{array}
$$

$$
\begin{array}{ccccc}
& & y_2^* & , & y_4^* & = & 0 \\
y_1^* & , & & y_3^* & & > & 0 \\
\end{array}
$$

Notice the pattern that emerges. We know that the constraints of a LOP pair up with the variables of its dual. In this case it seems that every (constraint, nonnegative variable) pair is satisfied with equality for one of them. For example, the pair $(y_1^* + 4y_3^* + 9y_4^* \geq -18,\ x_2^* \geq 0)$ has $x_2^* = 0$. Why might this be so? For an answer let's take a closer look at the inequalities found in the proof of the Weak Duality Theorem (Inequality 1.4.7). In this case they look like the following.

Inequality 4.2.3

$$
\begin{aligned}
z &= -22x_1 - 18x_2 - 27x_3 - 23x_4 + 16x_5 - 12x_6 \\
&\leq (4y_1 + 6y_2 - 5y_3 + 3y_4)x_1 - 18x_2 - 27x_3 - 23x_4 + 16x_5 - 12x_6 \\
&\leq (4y_1 + 6y_2 - 5y_3 + 3y_4)x_1 + (y_1 + 4y_3 + 9y_4)x_2 - 27x_3 - 23x_4 \\
&\quad + 16x_5 - 12x_6 \\
&\leq \cdots \\
&\leq 211y_1 + 189y_2 + (-5x_1 + 4x_2 - 2x_3 - x_5 - 7x_6)y_3 \\
&\quad + (3x_1 + 9x_2 - 2x_4 + x_5 + 4x_6)y_4 \\
&\leq 211y_1 + 189y_2 - 106y_3 + (3x_1 + 9x_2 - 2x_4 + x_5 + 4x_6)y_4 \\
&\leq 211y_1 + 189y_2 - 106y_3 + 175y_4 \\
&= w
\end{aligned}
$$

Observe that when $\mathbf{x}^*$ and $\mathbf{y}^*$ are plugged into Inequality 4.2.3, every inequality becomes an equality (check this!). The reason for this is that $z^* = w^*$, so that no strict inequality can be allowed along the way. Now consider the following instance of one of the equalities that results from this analysis of Inequality 4.2.3.

Equality 4.2.4

$$(4y_1^* + 6y_2^* - 5y_3^* + 3y_4^*)x_1^* + (y_1^* + 4y_3^* + 9y_4^*)x_2^* + (-3y_1^* + 2y_2^* - 2y_3^*)x_3^*$$
$$-23x_4^* + 16x_5^* - 12x_6^*$$
$$= (4y_1^* + 6y_2^* - 5y_3^* + 3y_4^*)x_1^* + (y_1^* + 4y_3^* + 9y_4^*)x_2^* + (-3y_1^* + 2y_2^* - 2y_3^*)x_3^*$$
$$+(5y_2^* - 2y_4^*)x_4^* + 16x_5^* - 12x_6^*$$

Now Equality 4.2.4 holds if and only if

$$-23x_4^* = (5y_2^* - 2y_4^*)x_4^*,$$

which holds if and only if

$$x_4^* = 0 \quad \text{or} \quad -23 = 5y_2^* - 2y_4^*.$$

Workout 4.2.5 *Prove that every primal-dual optimal pair $\mathbf{x}^*$ and $\mathbf{y}^*$ satisfies at least one of $-5x_1^* + 4x_2^* - 2x_3^* - x_5^* - 7x_6^* = -106$ or $y_3^* = 0$.*

Consider the LOP P in Problem 4.1.7 and its dual D in Problem 4.1.8.

Theorem 4.2.6 *Let $\mathbf{x}'$ be P-feasible and $\mathbf{y}'$ be D-feasible. Then $\mathbf{x}'$ and $\mathbf{y}'$ are a primal-dual optimal pair if and only if both* — Complementary Slackness Theorem

$$x_j' = 0 \quad \text{or} \quad \sum_{i=1}^{m} a_{i,j} y_i' = c_j \quad \text{for all} \ 1 \le j \le n \quad (4.1)$$

and

$$\sum_{j=1}^{n} a_{i,j} x_j' = b_i \quad \text{or} \quad y_i' = 0 \quad \text{for all} \ 1 \le i \le m. \quad (4.2)$$

Workout 4.2.7 *Prove Theorem 4.2.6.*

4.3 Jizoezi, Jizoezi, Jizoezi

Workout 4.3.1 *Verify the Complementary Slackness Theorem 4.2.6 on Problem 4.1.1.*

Problem 4.3.2

$$\text{Max. } z = 2x_1 + 4x_2 + 5x_3 + 8x_4$$

s.t.
$$\begin{array}{rcrcrcrcr}
x_1 & & & + & x_3 & + & 4x_4 & \leq & 5 \\
-x_1 & + & 4x_2 & + & 2x_3 & - & 3x_4 & \leq & 22 \\
3x_1 & + & x_2 & + & x_3 & + & x_4 & \leq & 8
\end{array}$$

$$x_1, \; x_2, \; x_3, \; x_4 \geq 0$$

Workout 4.3.3 *Verify the Complementary Slackness Theorem 4.2.6 on Problem 4.3.2.*

Moral 4.3.4 *It is possible that both parts of conditions (4.1) or (4.2) are satisfied.*

Consider the following LOP.

Problem 4.3.5

$$\text{Max. } z = 13x_1 + 20x_2 + 17x_3$$

s.t.
$$\begin{array}{rcrcrcr}
8x_1 & + & 7x_2 & + & 9x_3 & \leq & 455 \\
5x_1 & - & x_2 & + & 6x_3 & \leq & 190 \\
4x_1 & + & 8x_2 & - & x_3 & \leq & 205 \\
-x_1 & + & 2x_2 & + & 3x_3 & \leq & 80 \\
-3x_1 & - & 5x_2 & - & 4x_3 & \leq & -200
\end{array}$$

$$\& \quad x_1, \; x_2, \; x_3 \geq 0$$

Workout 4.3.6 *Use Theorem 4.2.6 to decide whether or not either of the following pairs $(\mathbf{x}', \mathbf{y}')$ are primal-dual optimal for Problem 4.3.5.*

a. $\mathbf{x}' = (0, 2300, 2205)^\mathsf{T}/79$, $\mathbf{y}' = (156, 0, 61, 0, 0)^\mathsf{T}/79$.

b. $\mathbf{x}' = (4600, 2600, 2300)^\mathsf{T}/180$, $\mathbf{y}' = (0, 753, 546, 0, 0)^\mathsf{T}/180$.

c. $\mathbf{x}' = (4895, 5280, 5445)^\mathsf{T}/275$, $\mathbf{y}' = (390, 0, 230, 465, 0)^\mathsf{T}/275$.

4.4 Finding Optimal Certificates

Let us use the notation $(+, 0, * \mid +, +, 0, +, 0)^\mathsf{T}$ to describe any feasible solution to a LOP with 3 variables and 5 constraints for which coordinates labeled 0 have value 0, and coordinates labeled + (resp. *) have positive (resp. nonnegative) value. Now the Complementary Slackness Conditions 4.1 and 4.2 can be rewritten as

$$x'_j = 0 \quad \text{or} \quad y'_{m+j} = 0 \quad \text{for all } 1 \leq j \leq n \qquad (4.3)$$

and

$$x'_{n+i} = 0 \quad \text{or} \quad y'_i = 0 \quad \text{for all } 1 \leq i \leq m. \qquad (4.4)$$

4.4. Finding Optimal Certificates

Thus, if $\mathbf{x}'$ is of the form $(+,0,0,+ \mid 0,+,0)^\mathsf{T}$ then in order for it to be primal optimal $\mathbf{y}'$ must be of the form $(*, 0, * \mid 0, *, *, 0)^\mathsf{T}$.

Workout 4.4.1 *Let $\mathbf{x}^*$ be of the form $(0,0,+,+,0 \mid +,+,+,0,+,0,+)^\mathsf{T}$. Find the form of $\mathbf{y}^*$.*

Now imagine you work for Varyim Portint Co. Two weeks ago you witnessed your division boss make a report to the company president in which he claimed a particular $\mathbf{x}'$ was optimal. The president wanted proof, and because your boss didn't have the relevant multipliers on hand (and didn't know that they could be read off of the final primal tableau), he left to go compute the dual, which took three hours because the LOP was large. But the president needed the proof within the hour, and so fired your boss upon his return.

Last week, your boss's replacement found herself in the same situation, and knowing she had the final tableau on her office desk (but not knowing the Complementary Slackness Theorem), she left to get the optimal multipliers. When she returned after 20 minutes, she also learned that she was fired because the president was on the phone with investors who needed the proof within 5 minutes.

Now you are the division boss, and accidentally find yourself in the same situation. However, when you recover the multipliers on your laptop within minutes, the president rewards you with a red cape with the letters "LO" on the back, and commissions you to rid $\mathbb{R}^n$ of evil. Here's how you must have done it.

Suppose $\mathbf{x}' = (418, 297, 0)^\mathsf{T}/22$ is a proposed optimal solution for Problem 4.1.1. Because it has the form $(+, +, 0 \mid 0, 0)$, its supposed optimal (to Problem 4.1.3) partner $\mathbf{y}'$ must have the form $(*, * \mid 0, 0, *)$. That is, it must be a nonnegative solution to the system of equations

$$\begin{pmatrix} y_1 & + & 5y_2 & & & = & 22 \\ 4y_1 & - & 2y_2 & & & = & 31 \\ 6y_1 & + & 3y_2 & - & y_5 & = & 29 \end{pmatrix}.$$

Since the unique solution $\mathbf{y}' = (199, 57 \mid 0, 0, 727)^\mathsf{T}/22$ to the above system is nonnegative, $\mathbf{x}'$ is indeed primal-optimal (and $\mathbf{y}'$ is dual-optimal).

This technique can be recorded in the following theorem, which follows easily from the Complementary Slackness Theorem 4.2.6.

Theorem 4.4.2 *Let $\mathbf{x}'$ be P-feasible. Then $\mathbf{x}'$ is primal optimal if and only if there is a D-feasible $\mathbf{y}'$ such that both*

$$x'_j > 0 \quad \text{implies} \quad \sum_{i=1}^m a_{i,j} y'_i = c_j \quad \text{for all } 1 \le j \le n \quad (4.5)$$

and

$$\sum_{j=1}^n a_{i,j} x'_j < b_i \quad \text{implies} \quad y'_i = 0 \quad \text{for all } 1 \le i \le m \quad (4.6)$$

Problem 4.4.3

$$\text{Max. } z = 5x_1 + 8x_2 + 15x_3 + 20x_4 + 13x_5$$

$$\text{s.t.} \quad \begin{array}{rcrcrcrcrcr}
x_1 & + & 2x_2 & + & 3x_3 & + & 4x_4 & + & 5x_5 & \leq & 17 \\
-x_1 & & & + & 3x_3 & & & - & x_5 & \leq & 3 \\
3x_1 & + & 2x_2 & + & x_3 & - & x_4 & & & \leq & 15
\end{array}$$

$$\& \quad x_1, \; x_2, \; x_3, \; x_4, \; x_5 \geq 0$$

Workout 4.4.4 *Consider Problem 4.4.3.*

a. Use Theorem 4.4.2 to show that $\mathbf{x}' = (2, 0, 1, 3, 0)^\mathsf{T}$ is optimal.

b. Use Theorem 4.4.2 to show that $\mathbf{x}' = (0, 7, 1, 0, 0)^\mathsf{T}$ is not optimal.

4.5 Exercises

Practice

4.5.1 *Verify the Duality Theorem 4.1.9 on each of the following LOPs.*

a.

$$\text{Max. } z = -211x_1 - 189x_2 - 106x_3 - 175x_4$$

$$\text{s.t.} \quad \begin{array}{rcrcrcrcr}
4x_1 & +6x_2 & -5x_3 & +3x_4 & \leq & 22 \\
x_1 & & +4x_3 & +9x_4 & \leq & 18 \\
-3x_1 & +2x_2 & -2x_3 & & \leq & 27 \\
& 5x_2 & & -2x_4 & \leq & 23 \\
2x_1 & -x_2 & -x_3 & +x_4 & \leq & 16 \\
7x_1 & +8x_2 & -7x_3 & +4x_4 & \leq & 12
\end{array}$$

$$\& \quad x_1, \; x_2, \; x_3, \; x_4 \geq 0$$

b.

$$\text{Max. } z = x_1 + 2x_2$$

$$\text{s.t.} \quad \begin{array}{rcrcr}
17x_1 & + & 21x_2 & \leq & 51 \\
x_1 & - & 4x_2 & \leq & 12 \\
3x_1 & + & 6x_2 & \leq & 14
\end{array}$$

$$\& \quad x_1, \; x_2 \geq 0$$

4.5. Exercises

c.

$$
\begin{aligned}
\text{Max.} \quad z = \quad & 2x_1 - 40x_2 \qquad\qquad - 42x_4 \\
\text{s.t.} \quad & 2x_1 - 4x_2 + x_3 - 7x_4 \leq -31 \\
& x_1 + 6x_2 - 5x_3 + 8x_4 \leq 44 \\
& 3x_1 - 2x_2 - x_3 \qquad\quad \leq 6 \\
\& \quad & x_1, \; x_2, \; x_3, \; x_4 \geq 0
\end{aligned}
$$

4.5.2 *Verify the Duality Theorem 4.1.9 on N LOPs of your own making.*

4.5.3 *Let S be the following system of inequalities.*

$$
\begin{aligned}
-3x_1 - 4x_2 + 5x_3 &\leq -221 \\
x_1 - 6x_2 + 2x_3 &\leq -253 \\
5x_1 + 4x_2 - 9x_3 &\leq 268 \\
-8x_1 + 3x_2 + x_3 &\leq -173 \\
x_1, \; x_2, \; x_3 &\geq 0
\end{aligned}
$$

Find a corresponding standard maximization form LOP P so that S is solvable if and only if P is optimal. Implicit in the statement that P is optimal is the assumption that P must be feasible. Try to respond with the kind of answer that suggests a general method.

4.5.4 *Let P be the following LOP.*

$$
\begin{aligned}
\text{Max.} \quad z = \quad & 318x_1 + 301x_2 - 279x_3 - 313x_4 \\
\text{s.t.} \quad & 51x_1 - 42x_2 - 46x_3 + 36x_4 \leq -12 \\
& -39x_1 + 56x_2 + 41x_3 - 58x_4 \leq 15 \\
& 45x_1 - 37x_2 + 48x_3 - 50x_4 \leq -17 \\
\& \quad & x_1, \; x_2, \; x_3, \; x_4 \geq 0
\end{aligned}
$$

Find a corresponding system of linear inequalities S so that S is solvable if and only if P is optimal. Try to respond with the kind of answer that suggests a general method.

4.5.5 *Verify the Complementary Slackness Theorem 4.2.6 on each of the LOPs in Exercise 4.5.1.*

4.5.6 *Verify the Complementary Slackness Theorem 4.2.6 on each of the LOPs in Exercise 4.5.2.*

4.5.7 Consider the LOP Max. $z = \mathbf{c}^\mathsf{T}\mathbf{x}$ s.t. $\mathbf{Ax} \leq \mathbf{b}$, $\mathbf{x} \geq \mathbf{0}$, with

$$\mathbf{c} = \begin{pmatrix} 21 \\ 23 \\ -27 \\ 25 \\ -29 \\ 24 \end{pmatrix}, \quad A = \begin{pmatrix} -1 & 4 & 2 & 0 & 6 & -3 \\ 3 & -5 & 1 & 1 & 0 & 7 \\ 8 & -6 & 0 & 3 & -1 & -4 \\ 4 & -9 & 2 & 0 & 6 & -7 \\ 5 & 0 & -1 & 1 & 8 & 5 \end{pmatrix},$$

$$\text{and} \quad \mathbf{b} = \begin{pmatrix} 220 \\ -250 \\ -236 \\ 281 \\ 264 \end{pmatrix}.$$

Let $\mathbf{x}' = (100, 410, 0, 0, 0, 0)^\mathsf{T}/7$. Use the Complementary Slackness Theorem to prove or disprove that $\mathbf{x}'$ is optimal.

4.5.8 Let $\mathbf{x}' = (21, 58, 29)^\mathsf{T}$ be a feasible solution to the LOP below.

$$\begin{array}{rrcrcrcr}
\text{Max. } z = & 3x_1 & + & 2x_2 & + & 2x_3 & & \\
\text{s.t.} & 2x_1 & + & 3x_2 & + & 5x_3 & \leq & 361 \\
& 2x_1 & + & 12x_2 & + & 3x_3 & \leq & 910 \\
& 4x_1 & + & x_2 & - & x_3 & \leq & 113 \\
& 10x_1 & + & 2x_2 & + & 5x_3 & \leq & 494 \\
\& & x_1 &, & x_2 &, & x_3 & \geq & 0
\end{array}$$

a. Is $\mathbf{x}'$ basic? Explain without solving.

b. Is $\mathbf{x}'$ optimal? Explain without solving.

4.5.9 Consider the LOP Max. $z = \mathbf{c}^\mathsf{T}\mathbf{x}$ s.t. $\mathbf{Ax} \leq \mathbf{b}$, $\mathbf{x} \geq \mathbf{0}$, with

$$A = \begin{pmatrix} 8 & -2 & 0 & 5 \\ 3 & 6 & -4 & 0 \\ 0 & 7 & 1 & -9 \end{pmatrix}.$$

Let $\mathbf{x}' = (0, 5, 7, 0)^\mathsf{T}$ and $\mathbf{y}' = (3, 0, 4)^\mathsf{T}$. Find $\mathbf{b}$ and $\mathbf{c}$ that make $\mathbf{x}'$ and $\mathbf{y}'$ an optimal pair.

4.5.10 Which of the following primal-dual forms could be optimal pairs? (Note that all decision and slack variables are present; only the lines that separate them are missing.)

a. $\mathbf{x}' = (0, +, 0, 0, +, 0)^\mathsf{T}$, $\mathbf{y}' = (0, +, +, 0, +, +)^\mathsf{T}$.

b. $\mathbf{x}' = (0, +, +, 0, +, 0, 0, +, 0)^\mathsf{T}$, $\mathbf{y}' = (+, 0, 0, +, +, 0, 0, +, 0)^\mathsf{T}$.

c. $\mathbf{x}' = (0, +, +, 0, 0, +, +, 0, 0, +, 0, 0, 0, +, +)^\mathsf{T}$,
$\mathbf{y}' = (0, 0, +, 0, +, +, +, 0, 0, +, 0, 0, +, 0, +)^\mathsf{T}$.

4.5. Exercises

4.5.11 Consider the following LOP.

$$\text{Max.} \quad z = 3x_1 + 4x_2 - 5x_3$$

$$\text{s.t.} \quad \begin{aligned} 2x_1 - 3x_2 - x_3 &\leq -10 \\ x_1 + x_2 + 5x_3 &\leq 10 \end{aligned}$$

$$\& \quad x_1, \; x_2, \; x_3 \geq 0$$

a. Use Theorem 4.4.2 to show that $\mathbf{x}' = (0, 10, 0)^\mathsf{T}$ is optimal.

b. Use Theorem 4.4.2 to show that $\mathbf{x}' = (2, 5, 0)^\mathsf{T}$ is not optimal.

4.5.12 Here is a more brute-force method of finding optimal solutions by using the Complementary Slackness Theorem 4.2.6. A generalization of it will come in handy in Chapter 9. Consider the LOP in Exercise 4.4.3.

a. List all P-feasible bases.

b. Write each of their dual-basis partners.

c. List which of those are D-feasible, and write their values. Those are the optimal pairs.

4.5.13

a. Suppose that $\sum_{j=1}^{n} a_{ij} x_j = b_i$. For what values of y_i does the inequality $(\sum_{j=1}^{n} a_{ij} x_j) y_i \leq b_i y_i$ hold?

b. Suppose that $(\sum_{i=1}^{m} a_{ij} y_i) x_j \leq c_j x_j$ for every value of x_j. What is the relationship between $\sum_{i=1}^{m} a_{ij} y_i$ and c_j?

Challenges

4.5.14 None of the following general LOPs are in standard form. For each of them, write its corresponding general dual LOP directly, *without* converting it to standard form. [HINT: Make sure that the Weak Duality Inequality 1.4.7 still holds; use Exercise 4.5.13.]

a.

$$\text{Max.} \quad z = 18x_1 + 31x_2 - 29x_3 + 13x_4$$

$$\text{s.t.} \quad \begin{aligned} 4x_1 + 2x_2 + 7x_3 - 3x_4 &\leq 112 \\ 5x_1 - 6x_2 + x_3 + 8x_4 &\leq 137 \end{aligned}$$

$$\& \quad x_1, \; x_2, \; x_4 \geq 0$$

b.

$$\begin{array}{rrrrrrl}
\text{Max. } z = & 201x_1 & + 198x_2 & - 229x_3 & - 193x_4 & & \\
\text{s.t.} & x_1 & + 4x_2 & - 6x_3 & - 8x_4 & \leq & 27 \\
& 7x_1 & + 5x_2 & & - 2x_4 & = & -32 \\
& 3x_1 & & - 9x_3 & - x_4 & \leq & 25 \\
\& & x_1, & x_2, & x_3, & x_4 & \geq & 0
\end{array}$$

c.

$$\begin{array}{rrrrl}
\text{Min. } w = & 2y_1 & + 3y_2 & & \\
\text{s.t.} & -35y_1 & + 42y_2 & = & 723 \\
& 54y_1 & + 37y_2 & \geq & 712 \\
& 47y_1 & + 41y_2 & \geq & 736 \\
\& & y_1, & y_2 & \geq & 0
\end{array}$$

d.

$$\begin{array}{rrrrrl}
\text{Min. } w = & 337y_1 & - 402y_2 & + 385y_3 & & \\
\text{s.t.} & 36y_1 & - 90y_2 & + 35y_3 & \geq & -9 \\
& 39y_1 & + 48y_2 & - 89y_3 & \geq & 7 \\
& -81y_1 & + 42y_2 & + 43y_3 & \geq & 8 \\
\& & y_1, & & y_3 & \geq & 0
\end{array}$$

e.

$$\begin{array}{rrrrl}
\text{Max. } z = & 8x_1 & - 5x_2 & & \\
\text{s.t.} & 18x_1 & - 13x_2 & \leq & 965 \\
& -25x_1 & + 20x_2 & = & -1150 \\
& 17x_1 & - 26x_2 & \geq & 808 \\
& 16x_1 & + 15x_2 & \leq & 896 \\
\& & x_1 & & \geq & 0 \\
& & x_2 & \leq & 0
\end{array}$$

(strongly) **4.5.15** Call a system of inequalities **redundant** if there are nonnegative
redundant multipliers (not all zero) that yield the inequality $0 \leq \mathbf{b}'$ for some $\mathbf{b}' \geq$
system $\mathbf{0}$; call the system **strongly redundant** if it is redundant with $\mathbf{b}' = \mathbf{0}$.
Suppose that P is an optimal LOP whose constraints are strongly redundant.
Prove that its dual LOP D has infinitely many optima. [HINT: Analyze
Exercise 1.5.10.]

4.5.16 Let S be the following system of inequalities.

$$\begin{array}{rll}
\sum_{j=1}^{n} a_{i,j} x_j \leq b_i & (1 \leq i \leq m) \\
x_j \geq 0 & (1 \leq j \leq n)
\end{array}$$

4.5. Exercises

Find a corresponding standard maximization form LOP P so that S is solvable if and only if P is optimal.

4.5.17 Let P be the following LOP.

$$\text{Max. } z = \sum_{j=1}^{n} c_j x_j$$

$$\text{s.t. } \sum_{j=1}^{n} a_{i,j} x_j \leq b_i \quad (1 \leq i \leq m)$$

$$\& \quad x_j \geq 0 \quad (1 \leq j \leq n)$$

Find a corresponding system of linear inequalities S so that S is solvable if and only if P is optimal.

4.5.18 Let P be the following LOP.

$$\text{Max. } z = 30x_1 - 40x_2 \qquad - 42x_4$$

$$\begin{array}{rcrcrcrcl}
\text{s.t.} & 2x_1 & - & 4x_2 & + & x_3 & - & 7x_4 & \leq b_1 \\
& x_1 & + & 6x_2 & - & 5x_3 & + & 8x_4 & \leq b_2 \\
& 3x_1 & - & 2x_2 & - & x_3 & & & \leq b_3
\end{array}$$

$$\& \quad x_1, \; x_2, \; x_3, \; x_4 \geq 0$$

a. Determine b_1, b_2, b_3 so that $\mathbf{x}^* = (2, 0, 0, 5)^\mathsf{T}$ and $\mathbf{y}^* = (6, 0, 6)^\mathsf{T}$ are primal and dual optimal, respectively. [HINT: Use complementary slackness.]

b. Show that $\mathbf{x}^{**} = (16, 0, 42, 15)^\mathsf{T}$ is also optimal with the same values for the b_i.

4.5.19 Pick an arbitrary $\mathbf{x}^*$ and $\mathbf{y}^*$ and create a LOP for which they are primal and dual optimal, respectively.

4.5.20 Find a primal-dual LOP pair for which some condition (4.1) or (4.2) is satisfied in both parts, as in Workout 4.3.3.

4.5.21 Find n and m so that $\mathbf{x}^* = (0, +, 0, +, 0, +)^\mathsf{T}$ and $\mathbf{y}^* = (+, 0, +, 0, +, 0)^\mathsf{T}$ is a basic optimal pair form.

Projects

4.5.22 Present the relationship between **Lagrange multipliers** and LO.

4.5.23 Present **Karmarkar's Algorithm**.

4.5.24 Present the **Ellipsoid Algorithm**.

Lagrange multipliers
Karmarkar's Algorithm
Ellipsoid Algorithm

Chapter 5
Matrix Environment

5.1 Format and Dictionaries

We return to the matrix LO format of (1.2) in Section 1.4. Consider Problem 5.1.1, below, having initial Tableau 5.1.2.

Problem 5.1.1

$$
\begin{aligned}
\text{Max.} \quad z = 46x_1 &+ 15x_2 + 12x_3 \\
\text{s.t.} \quad 7x_1 + x_2 + 3x_3 &\geq 23 \\
2x_1 + 6x_2 + 8x_3 &\geq 14 \\
4x_1 + 5x_2 + x_3 &\leq 87 \\
9x_1 + 4x_2 + 3x_3 &\leq 112 \\
\& \quad x_1, \; x_2, \; x_3 &\geq 0
\end{aligned}
$$

Tableau 5.1.2

$$
\left[\begin{array}{ccc|ccccc|c}
-7 & -1 & -3 & 1 & 0 & 0 & 0 & 0 & -23 \\
-2 & -6 & -8 & 0 & 1 & 0 & 0 & 0 & -14 \\
4 & 5 & 1 & 0 & 0 & 1 & 0 & 0 & 87 \\
9 & 4 & 3 & 0 & 0 & 0 & 1 & 0 & 112 \\
\hline
-46 & -15 & -12 & 0 & 0 & 0 & 0 & 1 & 0
\end{array}\right]
$$

Once in standard form, Problem 5.1.1 can be written in matrix form as Max. $z = \mathbf{c}^\mathsf{T}\mathbf{x}$ s.t. $\mathbf{A}\mathbf{x} \leq \mathbf{b}$ *and* $\mathbf{x} \geq \mathbf{0}$, where

$$
\mathbf{c} = \begin{pmatrix} 46 \\ 15 \\ 12 \end{pmatrix}, \quad \mathbf{A} = \begin{pmatrix} -7 & -1 & -3 \\ -2 & -6 & -8 \\ 4 & 5 & 1 \\ 9 & 4 & 3 \end{pmatrix}, \quad \mathbf{b} = \begin{pmatrix} -23 \\ -14 \\ 87 \\ 112 \end{pmatrix}.
$$

Recall that the dual LOP is Min. $w = \mathbf{b}^\mathsf{T}\mathbf{y}$ s.t. $\mathbf{A}^\mathsf{T}\mathbf{y} \geq \mathbf{c}$ *and* $\mathbf{y} \geq \mathbf{0}$, and that weak duality is shown by $z = \mathbf{c}^\mathsf{T}\mathbf{x} \leq \mathbf{y}^\mathsf{T}\mathbf{A}\mathbf{x} \leq \mathbf{y}^\mathsf{T}\mathbf{b} = w^\mathsf{T} = w$ for feasible $\mathbf{x}$ and $\mathbf{y}$.

G. H. Hurlbert, *Linear Optimization*, Undergraduate Texts in Mathematics,
DOI: 10.1007/978-0-387-79148-7_5, © Springer Science+Business Media LLC 2010

Tableau 5.1.2 can also be represented in matrix form, but we will need finer notation. Since the initial parameters are $\pi = \{1, 2, 3\}$, let's use the notation $\mathbf{x}_\pi = (x_1, x_2, x_3)^\mathsf{T}$ and $\mathbf{x}_\beta = (x_4, x_5, x_6, x_7)^\mathsf{T}$, with $\mathbf{x} = (x_1, \ldots, x_7)^\mathsf{T}$. Then we have

$$\text{Max. } z = \mathbf{c}^\mathsf{T} \mathbf{x}_\pi \quad \text{s.t. } \mathbf{A}\mathbf{x}_\pi + \mathbf{I}\mathbf{x}_\beta = \mathbf{b} \ \& \ \mathbf{x} \geq \mathbf{0} . \tag{5.1}$$

$\mathbf{x}_\beta, \mathbf{x}_\pi$ in the left margin.

In fact, since the basis changes with each pivot operation, we would be better served with more general notation.

Given basis $\beta = \{j_1, \ldots, j_m\}$ and parameter set $\pi = \{j_{m+1}, \ldots, j_{m+n}\}$ (written so that $j_k < j_{k+1}$ for all k), let $\mathbf{c}_\beta = (c_{j_1}, \ldots, c_{j_m})^\mathsf{T}$ and $\mathbf{c}_\pi = (c_{j_{m+1}}, \ldots, c_{j_{m+n}})^\mathsf{T}$. That is, the entries of $\mathbf{c}_\beta$ (resp., $\mathbf{c}_\beta$) are the original objective coefficients of the current basic variables (resp., parameters) — note that $c_k = 0$ for all $k > n$ — in least subscript order. For example, with $\beta = \{2, 3, 5, 7\}$ we have $\mathbf{c}_\beta = (15, 12, 0, 0)^\mathsf{T}$ and $\mathbf{c}_\pi = (46, 0, 0)^\mathsf{T}$.

$\mathbf{c}_\beta, \mathbf{c}_\pi$ in the left margin.

Furthermore, we let $\mathbf{B}$ be the matrix whose columns are the columns of $\mathbf{A}$ or $\mathbf{I}$ that correspond to the basic variables, while $\mathbf{\Pi}$ is the corresponding matrix for parameters. With $\beta = \{2, 3, 5, 7\}$ as above we then have

$\mathbf{B}, \mathbf{\Pi}$ in the left margin.

$$\mathbf{B} = \begin{pmatrix} -1 & -3 & 0 & 0 \\ -6 & -8 & 1 & 0 \\ 5 & 1 & 0 & 0 \\ 4 & 3 & 0 & 1 \end{pmatrix} \quad \text{and} \quad \mathbf{\Pi} = \begin{pmatrix} -7 & 1 & 0 \\ -2 & 0 & 0 \\ 4 & 0 & 1 \\ 9 & 0 & 0 \end{pmatrix}.$$

With the initial basis $\beta = \{4, 5, 6, 7\}$ we have $\mathbf{B} = \mathbf{I}$ and $\mathbf{\Pi} = \mathbf{A}$.

Using this notation (5.1) becomes

$$\text{Max. } z = \mathbf{c}_\beta^\mathsf{T} \mathbf{x}_\beta + \mathbf{c}_\pi^\mathsf{T} \mathbf{x}_\pi \quad \text{s.t. } \mathbf{B}\mathbf{x}_\beta + \mathbf{\Pi}\mathbf{x}_\pi = \mathbf{b} \ \& \ \mathbf{x} \geq \mathbf{0} . \tag{5.2}$$

In fact, we can use (5.2) to write a dictionary corresponding to the basis β as follows. From $\mathbf{B}\mathbf{x}_\beta + \mathbf{\Pi}\mathbf{x}_\pi = \mathbf{b}$ we have $\mathbf{x}_\beta = \mathbf{B}^{-1}(\mathbf{b} - \mathbf{\Pi}\mathbf{x}_\pi)$, or

$$\mathbf{x}_\beta = \mathbf{B}^{-1}\mathbf{b} - \mathbf{B}^{-1}\mathbf{\Pi}\mathbf{x}_\pi , \tag{5.3}$$

provided that $\mathbf{B}$ *is invertible!*[1] Then also (5.3) can be substituted into $z = \mathbf{c}_\beta^\mathsf{T} \mathbf{x}_\beta + \mathbf{c}_\pi^\mathsf{T} \mathbf{x}_\pi$ from (5.2) to obtain

$$z = \mathbf{c}_\beta^\mathsf{T}(\mathbf{B}^{-1}\mathbf{b} - \mathbf{B}^{-1}\mathbf{\Pi}\mathbf{x}_\pi) + \mathbf{c}_\pi^\mathsf{T} \mathbf{x}_\pi .$$

Separating out the constant and collecting terms yields

$$z = (\mathbf{c}_\beta^\mathsf{T}\mathbf{B}^{-1}\mathbf{b}) - (\mathbf{c}_\beta^\mathsf{T}\mathbf{B}^{-1}\mathbf{\Pi} - \mathbf{c}_\pi^\mathsf{T})\mathbf{x}_\pi . \tag{5.4}$$

Notice that evaluating (5.3) and (5.4) at $\mathbf{x}_\pi = \mathbf{0}$ (setting the parameters to zero!) produces the current basic solution $\mathbf{x}_\beta = \mathbf{B}^{-1}\mathbf{b}$ and current objective value $z = z_\beta = \mathbf{c}_\beta^\mathsf{T}\mathbf{B}^{-1}\mathbf{b}$.

---

[1] In standard form LOPs this is not an issue, since the original basis (all slacks) has $\mathbf{B} = \mathbf{I}$; however, in the general form LOPs of Chapter 6 it is very much so, since there is no full basis to begin with.

5.2. Simplex Phases and Advantages

Workout 5.1.3 *Let $\beta = \{2, 3, 5, 7\}$ for Problem 5.1.1.*

a. Compute $\mathbf{B}^{-1}$.

b. Compute $\mathbf{B}^{-1}\mathbf{b}$.

c. Compute $\mathbf{B}^{-1}\mathbf{\Pi}$.

d. Compute $\mathbf{c}_\beta^\mathsf{T}\mathbf{B}^{-1}\mathbf{b}$.

e. Compute $\mathbf{c}_\beta^\mathsf{T}\mathbf{B}^{-1}\mathbf{\Pi} - \mathbf{c}_\pi^\mathsf{T}$.

Because we maintain integers in our tableau by clearing the denominators (multiplying by the basic coefficient), we should do the same with matrices in order to make fair comparisons. By Cramer's Rule (see Appendix A) we can write $\mathbf{B}^{-1} = \mathbf{B}'/|\det(\mathbf{B})|$; if $\mathbf{B}$ is integral then so is $\mathbf{B}'$. If we let $d_\beta = |\det(\mathbf{B})|$ then we can rewrite (5.3) and (5.4) as

$$d_\beta \mathbf{x}_\beta = \mathbf{B}'\mathbf{b} - \mathbf{B}'\mathbf{\Pi}\mathbf{x}_\pi , \qquad (5.5)$$

and

$$d_\beta z = (\mathbf{c}_\beta^\mathsf{T}\mathbf{B}'\mathbf{b}) - (\mathbf{c}_\beta^\mathsf{T}\mathbf{B}'\mathbf{\Pi} - d_\beta \mathbf{c}_\pi^\mathsf{T})\mathbf{x}_\pi . \qquad (5.6)$$

Workout 5.1.4 *Let $\beta = \{2, 3, 5, 7\}$ for Problem 5.1.1.*

a. Compute d_β.

b. Compute $\mathbf{B}'$.

c. Compute $\mathbf{B}'\mathbf{b}$.

d. Compute $\mathbf{B}'\mathbf{\Pi}$.

e. Compute $\mathbf{c}_\beta^\mathsf{T}\mathbf{B}'\mathbf{b}$.

f. Compute $\mathbf{c}_\beta^\mathsf{T}\mathbf{B}'\mathbf{\Pi} - d_\beta \mathbf{c}_\pi^\mathsf{T}$.

Workout 5.1.5 *Starting from Tableau 5.1.2, and ignoring the Simplex rules, perform the necessary pivots to obtain the basis $\beta = \{2, 3, 5, 7\}$ (2 pivots should suffice). Compare the resulting tableau with the data from Workout 5.1.4 and describe the relationship between them.*

Workout 5.1.6 *Repeat Workouts 5.1.4 and 5.1.5 with N other bases of your own choosing.*

5.2 Simplex Phases and Advantages

There is no difference in the Simplex Algorithm rules for the Matrix Environment as opposed to for the Tableau Environment. At this point, the only difference is in where to find in the matrices the information we're used to finding in the tableaux. We will discuss at the end of this section why anyone might prefer matrices to tableaux.

As noted, the initial basis is $\beta^{(0)} = \{4, 5, 6, 7\}$, giving $\mathbf{B} = \mathbf{I}$ and $\mathbf{\Pi} = \mathbf{A}$; thus $\mathbf{x}_{\beta^{(0)}} = \mathbf{B}'\mathbf{b} = (-23, -14, 87, 112)^\mathsf{T}$. Since the first coordinate is most negative, Phase I kicks the first basic variable, x_4, out of $\beta^{(0)}$. Because in the Tableau Environment we next look for the first negative in the first row of Tableau 5.1.2, we should look for the first negative in row 1 of $\mathbf{A}$ — that is where one finds the coefficients of parameters. The -7 signals that the first parameter, x_1, then enters the basis: now $\beta^{(1)} = \{1, 5, 6, 7\}$.

Tableau 5.2.1

$$\begin{bmatrix} 7 & 1 & 3 & -1 & 0 & 0 & 0 & 23 \\ 0 & -40 & -50 & -2 & 7 & 0 & 0 & -52 \\ 0 & 31 & -5 & 4 & 0 & 7 & 0 & 517 \\ 0 & 19 & -6 & 9 & 0 & 0 & 7 & 577 \\ \hline 0 & -59 & 54 & -46 & 0 & 0 & 0 & 1058 \end{bmatrix}$$

The first tableau pivot gives rise to Tableau 5.2.1. Let's compare it to what we compute in matrix form, below. We have

$$\mathbf{B} = \begin{pmatrix} -7 & 0 & 0 & 0 \\ -2 & 1 & 0 & 0 \\ 4 & 0 & 1 & 0 \\ 9 & 0 & 0 & 1 \end{pmatrix} \quad \text{and} \quad \mathbf{\Pi} = \begin{pmatrix} -1 & -3 & 1 \\ -6 & -8 & 0 \\ 5 & 1 & 0 \\ 4 & 3 & 0 \end{pmatrix}.$$

Thus $d_\beta = 7$,

$$\mathbf{B}' = \begin{pmatrix} -1 & 0 & 0 & 0 \\ -2 & 7 & 0 & 0 \\ 4 & 0 & 7 & 0 \\ 9 & 0 & 0 & 7 \end{pmatrix}, \quad \mathbf{B}'\mathbf{\Pi} = \begin{pmatrix} 1 & 3 & -1 \\ -40 & -50 & -2 \\ 31 & -5 & 4 \\ 19 & -6 & 9 \end{pmatrix},$$

$$\text{and} \quad \mathbf{B}'\mathbf{b} = \begin{pmatrix} 23 \\ -52 \\ 517 \\ 577 \end{pmatrix},$$

showing an uncanny resemblance to Tableau 5.2.1. The usual rules apply — the second basic variable is replaced by the first parameter: $2 \mapsto 5$. The subsequent tableau and matrices corresponding to $\beta^{(2)} = \{1, 2, 6, 7\}$ follow.

Tableau 5.2.2

$$\begin{bmatrix} 40 & 0 & 10 & -6 & 1 & 0 & 0 & 124 \\ 0 & 40 & 50 & 2 & -7 & 0 & 0 & 52 \\ 0 & 0 & -250 & 14 & 31 & 40 & 0 & 2724 \\ 0 & 0 & -170 & 46 & 19 & 0 & 40 & 3156 \\ \hline 0 & 0 & 730 & -246 & -59 & 0 & 0 & 0 \end{bmatrix}$$

Also,

$$\mathbf{B} = \begin{pmatrix} -7 & -1 & 0 & 0 \\ -2 & -6 & 0 & 0 \\ 4 & 5 & 1 & 0 \\ 9 & 4 & 0 & 1 \end{pmatrix} \quad \text{and} \quad \mathbf{\Pi} = \begin{pmatrix} -3 & 1 & 0 \\ -8 & 0 & 1 \\ 1 & 0 & 0 \\ 3 & 0 & 0 \end{pmatrix}.$$

5.2. Simplex Phases and Advantages

Thus $d_\beta = 40$,

$$\mathbf{B'} = \begin{pmatrix} -6 & 1 & 0 & 0 \\ 2 & -7 & 0 & 0 \\ 14 & 31 & 40 & 0 \\ 46 & 19 & 0 & 40 \end{pmatrix}, \quad \mathbf{B'\Pi} = \begin{pmatrix} 10 & -6 & 1 \\ 50 & 2 & -7 \\ -250 & 14 & 31 \\ -170 & 46 & 19 \end{pmatrix},$$

and $\mathbf{B'b} = \begin{pmatrix} 124 \\ 52 \\ 2724 \\ 3156 \end{pmatrix}$.

Again, the resemblance to Tableau 5.2.2 is unmistakable.

By the looks of $\mathbf{B'b}$, $\beta^{(2)}$ is feasible, and so the objective row is the place to look next. Equation (5.6) can be rewritten as

$$(\mathbf{c}_\beta^T \mathbf{B'\Pi} - d_\beta \mathbf{c}_\pi^T)\mathbf{x}_\pi + d_\beta z = (\mathbf{c}_\beta^T \mathbf{B'b}) \tag{5.7}$$

so as to look more like the objective row. The parameter coefficients are thus

$$\mathbf{c}_\beta^T \mathbf{B'\Pi} - d_\beta \mathbf{c}_\pi^T = (730, -246, -59)^T,$$

and so the second parameter, x_4, enters the basis.

Now we need to look at the coefficients of x_4 in $\mathbf{B'\Pi}$, namely $(-6, 2, 14, 46)^T$, as denominators for their partner terms in $\mathbf{B'b}$. The ratios

$$\text{``}\mathbf{B'b}/\mathbf{B'\Pi}_{x_4}\text{''} = \begin{pmatrix} -124/6 \\ 52/2 \\ 2724/14 \\ 3156/46 \end{pmatrix}$$

show that second term, $52/2$, is the smallest nonnegative. Hence the second basic variable leaves: $4 \mapsto 2$. The subsequent tableau and matrices corresponding to $\beta^{(3)} = \{1, 4, 6, 7\}$ follow.

Tableau 5.2.3

$$\begin{bmatrix} 2 & 6 & 8 & 0 & -1 & 0 & 0 & 0 & 14 \\ 0 & 40 & 50 & 2 & -7 & 0 & 0 & 0 & 52 \\ 0 & -14 & -30 & 0 & 4 & 2 & 0 & 0 & 118 \\ 0 & -46 & -66 & 0 & 9 & 0 & 2 & 0 & 98 \\ \hline 0 & 246 & 344 & 0 & -46 & 0 & 0 & 2 & 644 \end{bmatrix}$$

Also,

$$\mathbf{B} = \begin{pmatrix} -7 & 1 & 0 & 0 \\ -2 & 0 & 0 & 0 \\ 4 & 0 & 1 & 0 \\ 9 & 0 & 0 & 1 \end{pmatrix} \quad \text{and} \quad \mathbf{\Pi} = \begin{pmatrix} -1 & -3 & 0 \\ -6 & -8 & 1 \\ 5 & 1 & 0 \\ 4 & 3 & 0 \end{pmatrix}.$$

Thus $d_\beta = 2$,

$$\mathbf{B}' = \begin{pmatrix} 0 & -1 & 0 & 0 \\ 2 & -7 & 0 & 0 \\ 0 & 4 & 2 & 0 \\ 0 & 9 & 0 & 2 \end{pmatrix}, \quad \mathbf{B}'\mathbf{\Pi} = \begin{pmatrix} 6 & 8 & -1 \\ 40 & 50 & -7 \\ -14 & -30 & 4 \\ -460 & -66 & 9 \end{pmatrix},$$

$$\text{and} \quad \mathbf{B}'\mathbf{b} = \begin{pmatrix} 14 \\ 52 \\ 118 \\ 98 \end{pmatrix}.$$

As usual, the columns of $\mathbf{B}$ are those of Tableau 5.1.2 that correspond to $\beta^{(3)}$ — technically, column k of $\mathbf{B}$ equals column $\beta_k^{(3)}$ of Tableau 5.1.2. Because $\beta^{(3)}$ is feasible, we next compute the parameter coefficients

$$\mathbf{c}_\beta^\mathsf{T} \mathbf{B}'\mathbf{\Pi} - d_\beta \mathbf{c}_\pi^\mathsf{T} = (246, 344, -46),$$

which signal that the third parameter, x_5, enters the basis.

Now we need to look at the coefficients of x_5 in $\mathbf{B}'\mathbf{\Pi}$, namely $(-1, -7, 4, 9)^\mathsf{T}$, as denominators for their partner terms in $\mathbf{B}'\mathbf{b}$. The ratios

$$\text{``}\mathbf{B}'\mathbf{b}/\mathbf{B}'\mathbf{\Pi}_{x_5}\text{''} = \begin{pmatrix} - \\ - \\ 118/4 \\ 98/9 \end{pmatrix}$$

show that fourth term, $98/9$, is the smallest nonnegative. Hence the fourth basic variable leaves: $5 \mapsto 7$. The subsequent tableau and matrices corresponding to $\beta^{(3)} = \{1, 4, 5, 6\}$ follow. (Look for a slight twist!)

Tableau 5.2.4

$$\left[\begin{array}{rrr|rrrrr|r} 9 & 4 & 3 & 0 & 0 & 0 & 1 & 0 & 112 \\ 0 & 19 & -6 & 9 & 0 & 0 & 7 & 0 & 577 \\ 0 & 29 & -3 & 0 & 0 & 9 & -4 & 0 & 335 \\ 0 & -46 & -66 & 0 & 9 & 0 & 2 & 0 & 98 \\ \hline 0 & 49 & 30 & 0 & 0 & 0 & 46 & 9 & 5152 \end{array}\right]$$

Also,

$$\mathbf{B} = \begin{pmatrix} -7 & 1 & 0 & 0 \\ -2 & 0 & 1 & 0 \\ 4 & 0 & 0 & 1 \\ 9 & 0 & 0 & 0 \end{pmatrix} \quad \text{and} \quad \mathbf{\Pi} = \begin{pmatrix} -1 & -3 & 0 \\ -6 & -8 & 0 \\ 5 & 1 & 0 \\ 4 & 3 & 1 \end{pmatrix}.$$

Thus $d_\beta = 9$,

$$\mathbf{B}' = \begin{pmatrix} 0 & 0 & 0 & 1 \\ 9 & 0 & 0 & 7 \\ 0 & 9 & 0 & 2 \\ 0 & 0 & 9 & -4 \end{pmatrix}, \quad \mathbf{B}'\mathbf{\Pi} = \begin{pmatrix} 4 & 3 & 1 \\ 19 & -6 & 7 \\ -46 & -66 & 2 \\ 29 & -3 & -4 \end{pmatrix},$$

5.2. Simplex Phases and Advantages

$$\text{and} \quad \mathbf{B'b} = \begin{pmatrix} 112 \\ 577 \\ 98 \\ 335 \end{pmatrix}.$$

Because $\beta^{(4)}$ is feasible, we next compute the parameter coefficients

$$\mathbf{c}_\beta^\mathsf{T} \mathbf{B'} \mathbf{\Pi} - d_\beta \mathbf{c}_\pi^\mathsf{T} = (49, 30, 46),$$

which signal that we are at the optimum. Hence $\mathbf{x}^* = (112, 0, 0 \mid 577, 98, 335)^\mathsf{T}/9$.

Workout 5.2.5 *What is the "slight twist" and why does it occur?*

In order to compute z^* we plug $\mathbf{x}^*$ into the original objective function, obtaining

$$z^* = \mathbf{c}^\mathsf{T} \mathbf{x}^* = \mathbf{c}_\beta^\mathsf{T} \mathbf{x}_\beta = \mathbf{c}_\beta^\mathsf{T} \mathbf{B}^{-1} \mathbf{b} = \mathbf{c}_\beta^\mathsf{T} \mathbf{B'b}/d = 5152/9.$$

Workout 5.2.6

a. *Calculate $\mathbf{c}_\beta^\mathsf{T} \mathbf{B'}$.*

b. *Relate your result to Tableau 5.2.4.*

c. *Explain your findings. [HINT: Write the dual of Problem 5.1.1 in matrix form.]*

Workout 5.2.7

a. *Count the number of arithmetic operations performed in calculating Tableau 5.2.3 from Tableau 5.2.2.*

b. *Count the number of arithmetic operations performed in calculating $\mathbf{B'}$, $\mathbf{B'b}$, $\mathbf{B'\Pi}$ and $\mathbf{c}_\beta^\mathsf{T} \mathbf{B'\Pi} - d_\beta \mathbf{c}_\pi^\mathsf{T}$.*

c. *Count the number of arithmetic operations performed in a single pivot operation in the Tableau Environment, supposing the given LOP has m inequalities and n variables.*

d. *Count the number of arithmetic operations performed in a single pivot operation in the Matrix Environment, supposing the given LOP has m inequalities and n variables.*

e. *Which would you consider to be the faster implementation on large LOPs, Tableau or Matrix?*

Let's get one thing out in the open. Applied LOPs can be huge. While hundreds of variables and constraints might seem large (and were considered large fifty years ago), recently solved problems have contained *millions* of variables and constraints! Thus we should keep in mind that subtle improvements on a small scale can have tremendous implications on a large scale.

No one will argue that the Matrix Environment is simpler to visualize than the Tableau Environment. The organization of the data in tableaux is far simpler to digest. So why would anyone prefer matrices? The answer lies in the fact that we use computers for solving large LOPs and, lacking eyes, the computer doesn't share our sentiments in that regard, emotional as they can be. For computers, the relevant issues boil down to time and space. (We'll ignore energy, or you might think this is Physics.)

unstable tableau   Actually, there is one other issue — correctness. As it turns out, tableaux are what numerical analysts call **unstable**, which means essentially that, because of how they're stored and manipulated in a computer, errors can crop into the computations. Because of this, one realizes that if a mistake is made during a given tableau iteration then every tableau thereafter is no longer working with the original constraints. The error may, in fact, propogate to every row, meaning that the polyhedron defined by the new constraints may have no relation to the original polyhedron. How disastrous! On the other hand, if a mistake is made during a given matrix iteration then the worst that can happen is that a poor pivot decision is made. The reason is that every new set of pivot calculations *uses the original problem information*. Thus the original polyhedron remains intact. Moreover, because of the ways in which matrices can be stored,[2] the inverse of one **B** can be used to compute the inverse of its subsequent **B** quickly (since only one basic variable is exchanged).

With regard to speed we first remind the reader that the Matrix and Tableau Environments, assuming error-free calculations, make identical pivot decisions. Therefore, each requires the same number of pivots to solve a particular LOP. Hence the only possibility for a difference in speed occurs during each individual iteration. Here there can be a marked difference in practice. At first glance, a tableau pivot operation requires roughly $m(m+n)$ operations, while a matrix pivot operation requires about m^3 operations just to invert B, never mind perform the remaining calculations. However, knowing the inverse of one basis helps in computing the inverse of the next basis (since only one column is different), and so it can be shown that both methods perform on the same order of magnitude in general. At second glance, the matrix method can be shown to be far faster in practice.

sparse matrix/ tableau   The reason for this is that practical problems tend to be **sparse**, meaning that most of the entries are zero. (We will encounter many sparse problems in Chapter 10.) It has even been reported that, on practical problems, regardless of the size of a problem, only a constant number per row, maybe as small as 10 (ten!), might be nonzero. In hindsight this tendency isn't so surprising, considering that variables often come in groups — type of employee, hours of operation, materials of construction, location of office, etc. With this in mind, we note that there are especially fast methods for

---

[2]Various *triangular* factorizations of **B**, into upper and lower triangular *eta* matrices and permutation matrices, have been used over the years — we will not discuss them here but instead refer the reader to any numerical analysis text (such as G. Golub and C. F. Van Loan, *Matrix Computations* (3rd ed.), Johns Hopkins University Press, Baltimore, 1996) that covers matrix inversion.

inverting sparse matrices, methods that take on the order of *linear* time. On problems as large as just a thousand variables and constraints, this improvement is very noticeable indeed.

The final consideration of space becomes apparent on sparse problems, since zero entries do not need to be stored in clever implementations of matrix Simplex. On the other hand, after a number of Tableau Environment pivots, any measure of sparseness is lost, and so every entry needs to be stored. Again, linear (in m) storage space in the matrix case compares quite favorably with quadratic storage space in the tableau case, especially on large problems.

We will see in Chapter 10 methods that resemble the Matrix Environment in always using original problem data.

5.3 קװת תרגול

Here is an infeasible problem.

Problem 5.3.1

$$\begin{array}{rrrrrrr}
\text{Max.} \ z = & 63x_1 & + & 84x_2 & + & 51x_3 & \\
\text{s.t.} & 3x_1 & + & x_2 & & & \leq 15 \\
& -x_1 & & & - & 9x_3 & \leq -12 \\
& -2x_1 & + & 7x_2 & - & 3x_3 & \leq 23 \\
& x_1 & - & 4x_2 & - & 5x_3 & \leq -22 \\
& -9x_1 & + & 6x_2 & + & 8x_3 & \leq -14 \\
\& & x_1 & , & x_2 & , & x_3 & \geq 0
\end{array}$$

Workout 5.3.2

    a. Use the Matrix Environment to solve Problem 5.3.1.

    b. Where in $\mathbf{B}'$ do you find a certificate of infeasibility? (Recall Workout 2.6.3.)

    c. Explain your result from part b.

Here is an unbounded problem.

Problem 5.3.3

$$\begin{array}{rrrrrrr}
\text{Max.} \ z = & -22x_1 & +10x_2 & -29x_3 & +16x_4 & +25x_5 & \\
\text{s.t.} & x_1 & +3x_2 & -5x_3 & +7x_4 & -4x_5 & \leq -43 \\
& -6x_1 & & -2x_3 & -5x_4 & +9x_5 & \leq -48 \\
& 2x_1 & +8x_2 & & -8x_4 & +5x_5 & \leq -62 \\
\& & x_1, & x_2, & x_3, & x_4, & x_5 & \geq 0
\end{array}$$

Workout 5.3.4

    a. Use the Matrix Environment to solve Problem 5.3.3.

    b. Where in your final set of matrices do you find a certificate of unboundedness? (Recall Workout 2.7.1.)

    c. Explain your result from part b.

Here is an optimal problem.

Problem 5.3.5

$$\begin{array}{rl} \text{Min.} & w = 7y_1 + 4y_2 + 2y_3 + y_4 + 8y_5 + 3y_6 + 9y_7 + 5y_8 \\ \text{s.t.} & -y_1 - y_2 + y_4 + y_7 \leq 13 \\ & y_1 - y_3 \geq 11 \\ & y_3 - y_4 - y_5 + y_6 \leq -17 \\ & y_2 - y_6 + y_8 \leq 16 \\ & y_5 - y_7 - y_8 \geq -19 \\ \& & y_1, y_2, y_3, y_4, y_5, y_6, y_7, y_8 \geq 0 \end{array}$$

Workout 5.3.6 *Which LOP would you rather solve, Problem 5.3.5 or its dual? Why?*

Workout 5.3.7

    a. Use the Matrix Environment to solve Problem 5.3.5. (Maximize $u = -w$ rather than solve its dual.)

    b. Find a certificate of optimality.

    c. Did you notice anything especially interesting about one of the repeated computations in this problem?

5.4 Basic Coefficients

As noticed in the examples of Section 5.1, in particular Equations (5.5) and (5.6), it seems as if the basic coefficient for a tableau with basis β equals $d_\beta = |\det(\mathbf{B}_\beta)|$. Of course, while the coefficient d_β does clear all denominators, it is possible that some smaller coefficient does the same. It is our objective now to prove that this is not the case, and also to discuss some of its ramifications.

Theorem 5.4.1 *Let d_T be the basic coefficient for the tableau T having basis β from a standard form LOP. Then $d_T = d_\beta$.*

Proof. We proceed by induction. Clearly the theorem is true at the start. Now suppose that d_T is the basic coefficient for the tableau T having basis β, and that $d_T = d_\beta$. Let T_β be the submatrix of T having columns corresponding to β (and not including the objective row). Then $\det(T_\beta) = \rho \det(\mathbf{B}_\beta)$ for some ρ determined by the sequence of row operations that transformed $\mathbf{B}_\beta$ to T_β.

5.4. Basic Coefficients

Workout 5.4.2 *Use the induction hypothesis and the structure of T_β to show that $|\rho| = d_T^{m-1}$, where m is the number of rows of T.*

Likewise, let T' be the subsequent tableau derived from T by pivoting from β to β', and let $T'_{\beta'}$ be the analogous submatrix of T'. As above, we have $d_{T'}^m = |\det(T'_{\beta'})| = |\rho' \det(\mathbf{B}_{\beta'})|$ for the analogous ρ', whose value is yet to be determined.

Now suppose that a is the pivot entry that transforms T to T'.

Workout 5.4.3 *Use the form of the row operations of the transformation to prove that $|\rho'| = d_{T'}^{m-1}$. [HINT: Write ρ' in terms of ρ and use Workout 5.4.2.]*

Workout 5.4.4 *Use Workout 5.4.3 to finish the proof of the theorem.*

⋄

Workout 5.4.5 *Prove Theorem 5.4.1 for general form LOPs.*

What is the moral of Theorem 5.4.1? One answer lies in Problem 5.3.5. If at any stage we find $d_\beta = 1$ then the corresponding basic solution and objective value are both integer-valued. In particular, if $d_{\beta^*} = 1$ then the optimal LO solution also solves the corresponding ILO. As you may have discovered in Workout 5.3.7, *every* basic determinant has absolute value 1. For such a problem it is guaranteed that its optimal LO solution also solves the corresponding ILO. We say a matrix is **totally unimodular** (TU) if every square submatrix has determinant 0, 1 or −1. Thus we discover that, while general ILOPs are significantly more time consuming (NP-complete — see Appendix C) than LOPs in general, for the special class of problems having a TU constraint matrix the two problems are identical. Before tossing this class aside as an extreme anomaly, know that the very important subclass of network problems that we will encounter in Chapter 10 are all of this variety. In fact, almost all problems having a TU constraint matrix arise from networks.

totally unimodular matrix

Workout 5.4.6 *Consider Problem 5.3.5.*

  a. *Write its 5×13 constraint matrix $\mathbf{A}$ without first converting to standard form. That is, add or subtract slack variables as necessary.*
  b. *Write every one of its $\binom{13}{2}\binom{5}{2} = 780$ 2×2 submatrices and compute its determinant. Well, okay, write and compute N of them.*
  c. *Argue that every 2×2 submatrix of $\mathbf{A}$ has determinant 0, 1 or −1.*
  d. *Suppose that every 3×3 submatrix of $\mathbf{A}$ has determinant 0, 1 or −1, and use that supposition to prove that the same holds for every 4×4 submatrix. [HINT: Consider the three cases according to the number of nonzero entries of a column of the submatrix.]*

e. Use the ideas from part d to prove that $\mathbf{A}$ is TU.

Let us now make the integrality arguments above more formal. We say a nondegenerate polyhedron is **integral** if each of its extreme points is integral. Given a TU matrix $\mathbf{A}$ and integral $\mathbf{b}$, let $Q_\mathbf{b}$ be the polyhedron of solutions to $S_\mathbf{b} = \{\mathbf{Ax} \leq \mathbf{b}, \mathbf{x} \geq \mathbf{0}\}$. As mentioned above, every basic solution of $S_\mathbf{b}$ is integral, and hence $Q_\mathbf{b}$ is integral. Hence, if $\mathbf{A}$ is TU then $Q_\mathbf{b}$ is integral for every integral $\mathbf{b}$. In 1956 Hoffman and Kruskal proved that the converse is also true. The proof of their result is beyond the scope of this book.

Theorem 5.4.7 *The matrix $\mathbf{A}$ is TU if and only if the polyhedron $Q_\mathbf{b}$ is integral for every integral vector $\mathbf{b}$.* ◇

Workout 5.4.8 *Consider the polytope $Q_\mathbf{b}$ for*

$$\mathbf{A} = \begin{pmatrix} -1 & -1 \\ -1 & 1 \\ 1 & 1 \end{pmatrix} \quad \text{and} \quad \mathbf{b} = \begin{pmatrix} -1 \\ 2 \\ 4 \end{pmatrix}.$$

a. Show that $\mathbf{A}$ is not TU.

b. Show that $Q_\mathbf{b}$ is integral.

c. Why does this example not contradict Theorem 5.4.7?

5.5 Exercises

Practice

5.5.1 *Consider the LOP* Max. $z = \mathbf{c}^T\mathbf{x}$ s.t. $\mathbf{Ax} \leq \mathbf{b}$, $\mathbf{x} \geq \mathbf{0}$, *with*

$$\mathbf{c} = \begin{pmatrix} -13 \\ 11 \\ 17 \\ -16 \\ -19 \end{pmatrix}, \ A = \begin{pmatrix} 1 & 1 & 0 & 0 & 0 \\ 1 & 0 & 0 & -1 & 0 \\ 0 & -1 & -1 & 0 & 0 \\ -1 & 0 & 1 & 0 & 0 \\ 0 & 0 & 1 & 0 & 1 \\ 0 & 0 & -1 & 1 & 0 \\ -1 & 0 & 0 & 0 & -1 \\ 0 & 0 & 0 & -1 & -1 \end{pmatrix} \text{ and } \mathbf{b} = \begin{pmatrix} 7 \\ 4 \\ -2 \\ -1 \\ 8 \\ 3 \\ 9 \\ 5 \end{pmatrix}.$$

Solve it using the Matrix Environment.

5.5.2 *Use the Matrix Environment to solve the following LOPs. Provide certificates of your results.*

a. Problem 1.1.3

b. Problem 1.1.5

5.5. Exercises

    c. Problem 1.2.4

    d. Problem 1.2.5

    e. Problem 1.3.1

    f. Problem 1.4.2

    g. Problem 1.4.6

5.5.3 Use the Matrix Environment to solve the following LOPs. Provide certificates of your results.

    a. Exercise 1.5.1(a)

    b. Exercise 1.5.1(b)

    c. Exercise 1.5.1(c)

    d. Exercise 1.5.2

    e. Exercise 1.5.3

    f. Exercise 1.5.4

    g. Exercise 1.5.7

    h. Exercise 1.5.8

    i. Exercise 1.5.15

    j. Exercise 1.5.5

    k. Exercise 1.5.6

    l. Exercise 1.5.10

    m. Exercise 1.5.11

    n. Exercise 1.5.12

    o. Exercise 1.5.18

    p. Exercise 1.5.19

    q. Exercise 1.5.20

    r. Exercise 1.5.21(a)

    s. Exercise 1.5.21(b)

    t. Exercise 1.5.33

    u. Exercise 1.5.39

    v. Exercise 1.5.40

    w. Exercise 1.5.42

5.5.4 Use the Matrix Environment to solve the following LOPs. Provide certificates of your results.

    a. Problem 2.1.6

    b. Problem 2.1.8

    c. Problem 2.3.1

    d. Problem 2.4.1

    e. Problem 2.4.3

    f. Problem 2.5.1

    g. Problem 2.5.2

    h. Problem 2.5.5

    i. Workout 2.5.6(a)

    j. Problem 2.6.1

    k. Workout 2.6.4(a)

    l. Problem 2.7.1

    m. Problem 2.7.3

    n. Problem 2.8.3

    o. Problem 2.9.3

5.5.5 *Use the Matrix Environment to solve the following LOPs. Provide certificates of your results.*

    a. Exercise 2.10.12

    b. Exercise 2.10.13

    c. Exercise 2.10.14

    d. Exercise 2.10.15

    e. Exercise 2.10.16

    f. Exercise 2.10.17

    g. Exercise 2.10.18

    h. Exercise 2.10.19

    i. Exercise 2.10.22

    j. Exercise 2.10.23

    k. Exercise 2.10.24(a)

    l. Exercise 2.10.34(a)

    m. Exercise 2.10.34(b)

    n. Exercise 2.10.34(c)

5.5.6 *Use the Matrix Environment to solve the following LOPs. Provide certificates of your results.*

    a. Problem 3.1.1

    b. Problem 4.1.1

    c. Problem 4.1.3

    d. Problem 4.1.5

    e. Problem 4.2.1

5.5. Exercises

    f. Problem 4.3.2

    g. Problem 4.3.5

    h. Problem 4.4.3

5.5.7 *Use the Matrix Environment to solve the following LOPs. Provide certificates of your results.*

    a. Exercise 4.5.1(a)

    b. Exercise 4.5.1(b)

    c. Exercise 4.5.1(c)

    d. Exercise 4.5.2

    e. Exercise 4.5.3

    f. Exercise 4.5.4

    g. Exercise 4.5.11

    h. Exercise 4.5.14(a)

    i. Exercise 4.5.14(b)

    j. Exercise 4.5.14(c)

    k. Exercise 4.5.14(d)

    l. Exercise 4.5.14(e)

    m. Exercise 4.5.20

5.5.8 *Use the Matrix Environment to solve the following LOPs. Provide certificates of your results.*

    a. Problem 5.1.1

    b. Problem 5.3.1

    c. Problem 5.3.3

    d. Problem 5.3.5

5.5.9 *Use the Matrix Environment to solve the following LOPs. Provide certificates of your results.*

    a. Problem 6.1.1

    b. Problem 6.1.3

    c. Problem 6.1.4

    d. Workout 6.1.5

    e. Problem 6.3.3

    f. Problem 6.4.7

    g. Problem 6.4.8

    h. Problem 6.4.13

    i. Exercise 6.5.8

5.5.10 *Use the Matrix Environment to solve the following LOPs. Provide certificates of your results.*

   a. *Problem 7.1.4*
   b. *Exercise 7.5.1*
   c. *Exercise 7.5.2*
   d. *Exercise 7.5.3*

5.5.11 *Prove that the following matrix* $\mathbf{A}$ *is TU.*

$$\mathbf{A} = \begin{pmatrix} 1 & 1 & 1 & 1 & 1 \\ 1 & 1 & 1 & 0 & 0 \\ 1 & 0 & 1 & 1 & 0 \\ 1 & 0 & 0 & 1 & 1 \\ 1 & 1 & 0 & 0 & 1 \end{pmatrix}$$

5.5.12 *Prove that the following matrix* $\mathbf{A}$ *is TU.*

$$\mathbf{A} = \begin{pmatrix} 1 & -1 & 0 & 0 & -1 \\ -1 & 1 & -1 & 0 & 0 \\ 0 & -1 & 1 & -1 & 0 \\ 0 & 0 & -1 & 1 & -1 \\ -1 & 0 & 0 & -1 & 1 \end{pmatrix}$$

5.5.13 *Prove that the following matrix* $\mathbf{A}$ *is TU.*

$$\mathbf{A} = \begin{pmatrix} 1 & 1 & 0 & 0 & 0 \\ 0 & 0 & 1 & 0 & 0 \\ 0 & 0 & 0 & 1 & 1 \\ 1 & 0 & 1 & 1 & 0 \\ 0 & 1 & 0 & 0 & 1 \end{pmatrix}$$

5.5.14 *Prove that the following matrix* $\mathbf{A}$ *is TU.*

$$\mathbf{A} = \begin{pmatrix} 1 & 0 & 0 & 1 & 0 \\ 1 & 0 & 1 & 1 & 0 \\ 1 & 1 & 1 & 1 & 0 \\ 1 & 1 & 0 & 0 & 1 \\ 0 & 1 & 0 & 0 & 1 \end{pmatrix}$$

5.5.15 *Prove that the matrix A in Exercise 5.5.1 is TU.*

Challenges

5.5.16 *Let* $\beta = \beta^*$ *be the optimal basis for Problem 1.1.3 (see Exercise 5.5.2(a)). Keeping* γ *fixed, compare the final data with the data derived from replacing* $\mathbf{c} = (10, 10, 15, 15)$ *by*

5.5. Exercises

    a. $\mathbf{c}' = (11, 10, 15, 15)$,

    b. $\mathbf{c}' = (12, 10, 15, 15)$, and

    c. $\mathbf{c}' = (13, 10, 15, 15)$.

Describe and explain the pattern you see.

5.5.17 *Let $\beta = \beta^*$ be the optimal basis for Problem 1.1.3 (see Exercise 5.5.2(a)). Keeping β fixed, compare the final data with the data derived from replacing $\mathbf{b} = (400, 250, 370, 490)^\mathsf{T}$ by*

    a. $\mathbf{b}' = (400, 250, 121, 490)^\mathsf{T}$,

    b. $\mathbf{b}' = (400, 250, 120, 490)^\mathsf{T}$, and

    c. $\mathbf{b}' = (400, 250, 119, 490)^\mathsf{T}$.

Describe and explain the pattern you see.

5.5.18 *Write pseudocode that takes $\mathbf{A}$, $\mathbf{b}$, $\mathbf{c}$ and β as input and outputs the pivot operation $i \mapsto j$.*

5.5.19 *Write pseudocode for the Simplex Algorithm in the Matrix Environment.*

5.5.20 *Let $\mathbf{A}$ be a TU matrix and let $\mathbf{A}'$ be constructed from $\mathbf{A}$ by a pivot operation. Prove that $\mathbf{A}'$ is TU.*

5.5.21 *Let $\mathbf{A}$ be a $\{0, \pm 1\}$-matrix such that each column has at most one 1 and at most one -1. Prove that $\mathbf{A}$ is TU. [HINT: See Workout 5.4.6.]*

5.5.22 *Let $\mathbf{A} = (a_{i,j})$ be a $\{0, 1\}$-matrix having distinct columns with the property that every column has exactly two 1s, and such that there is some k such that, whenever $a_{i,j} = a_{i',j} = 1$ with $i < i'$, we have $i \leq k < i'$. Prove that $\mathbf{A}$ is TU. [For example, see Exercise 5.5.13.]*

5.5.23 *Let $\mathbf{A} = (a_{i,j})$ be as in Workout 5.4.6. Prove that, for every set J of columns there is some subset $J_1 \subseteq J$ (with $J_2 = J - J_1$) so that every row i satisfies*

$$\left| \sum_{j \in J_1} a_{i,j} - \sum_{j \in J_2} a_{i,j} \right| \leq 1 \ .$$

5.5.24 *A $\{0, 1\}$-matrix $\mathbf{A}$ is called* **interval** *if the 1s are found consecutively in every column. Prove that interval matrices are TU. [HINT: See Workout 5.5.14.]*   interval matrix

Projects

5.5.25 *Present Bixby's 2002 paper,* Solving real-world linear programs.

5.5.26 *Present matrix decomposition ideas regarding sparse matrix inversion.*   Cholesky factorization

5.5.27 *Present the* **Cholesky factorization** *of $\mathbf{A}\mathbf{A}^\mathsf{T}$ and the* **Log Barrier Method** *for LO.*   Log Barrier Method

Chapter 6

General Form

6.1 Nonstandard Duals

Consider the following general LOP. We refer to it as general because it contains an equality constraint, as well as the free variable x_2.

Problem 6.1.1

$$
\begin{array}{rrrrrrrcr}
\text{Max.} & z = & 36x_1 & - & 31x_2 & + & 37x_3 & & \\
\text{s.t.} & & 7x_1 & + & 6x_2 & + & 2x_3 & \leq & 419 \\
& & -5x_1 & + & 3x_2 & + & 8x_3 & \leq & 528 \\
& & 3x_1 & - & x_2 & & & = & 272 \\
& & & - & 9x_2 & - & 4x_3 & \leq & 168 \\
\& & & x_1 & , & & & x_3 & \geq & 0
\end{array}
$$

Instead of putting Problem 6.1.1 in standard form by creating a larger LOP with more constraints and variables, our aim is to modify the Simplex Algorithm to handle LOPs in **general form** like this. First, let's see what its dual LOP must look like. For this we need to recall how dual problems were first developed in Chapter 1, with the Weak Duality Inequality 1.4.7 in mind. First, we write down a proof of the inequality in this case, and then decide which components of the dual are necessary for each step of the proof to hold.

general form

G. H. Hurlbert, *Linear Optimization*, Undergraduate Texts in Mathematics, DOI: 10.1007/978-0-387-79148-7_6, © Springer Science+Business Media LLC 2010

Inequality 6.1.2

$$
\begin{aligned}
z &= 36x_1 - 31x_2 + 37x_3 &\text{(6.1)}\\
&\leq (7y_1 - 5y_2 + 3y_3)x_1 - 31x_2 + 37x_3 &\text{(6.2)}\\
&\leq (7y_1 - 5y_2 + 3y_3)x_1 + (6y_1 + 3y_2 - y_3 - 9y_4)x_2 + 37x_3 &\text{(6.3)}\\
&\leq (7y_1 - 5y_2 + 3y_3)x_1 + (6y_1 + 3y_2 - y_3 - 9y_4)x_2 \\
&\quad + (2y_1 + 8y_2 - 4y_4)x_3 &\text{(6.4)}\\
&= (7x_1 + 6x_2 + 2x_3)y_1 + (-5x_1 + 3x_2 + 8x_3)y_2 \\
&\quad + (3x_1 - x_2)y_3 + (-9x_2 - 4x_3)y_4 &\text{(6.5)}\\
&\leq 419y_1 + (-5x_1 + 3x_2 + 8x_3)y_2 + (3x_1 - x_2)y_3 \\
&\quad + (-9x_2 - 4x_3)y_4 &\text{(6.6)}\\
&\leq 419y_1 + 528y_2 + (3x_1 - x_2)y_3 + (-9x_2 - 4x_3)y_4 &\text{(6.7)}\\
&\leq 419y_1 + 528y_2 + 272y_3 + (-9x_2 - 4x_3)y_4 &\text{(6.8)}\\
&\leq 419y_1 + 528y_2 + 272y_3 + 168y_4 &\text{(6.9)}\\
&= w\,. &\text{(6.10)}
\end{aligned}
$$

For inequality 6.2 to hold we need $7y_1 - 5y_2 + 3y_3 \geq 36$, as usual. In fact, every line is as usual except inequalities 6.3 and 6.8.

In the former case we need $-31x_2 \leq (6y_1 + 3y_2 - y_3 - 9y_4)x_2$. However, if $6y_1 + 3y_2 - y_3 - 9y_4 \neq -31$ then we risk having $-31x_2 > (6y_1 + 3y_2 - y_3 - 9y_4)x_2$ if x_2 has the appropriate sign — remember, it can be negative! Thus the dual must have $6y_1 + 3y_2 - y_3 - 9y_4 = -31$ (and thus equality holds after multiplying by any value of x_2).

In a similar vein, the latter case requires $(3x_1 - x_2)y_3 \leq 272y_3$, and since we have equality in the constraint $3x_1 - x_2 = 272$, equality holds after multiplying by any value of y_3. Therefore we can allow y_3 to be free in the dual LOP, below.

Problem 6.1.3

$$
\begin{aligned}
\text{Min. } w = 419y_1 + 528y_2 + 272y_3 + 168y_4 \\
\text{s.t.} \quad 7y_1 - 5y_2 + 3y_3 &\geq 36 \\
6y_1 + 3y_2 - y_3 - 9y_4 &= -31 \\
2y_1 + 8y_2 - 4y_4 &\geq 37 \\
\& \quad y_1,\ y_2, y_4 &\geq 0
\end{aligned}
$$

Try another example.

6.1. Nonstandard Duals

Problem 6.1.4

Max. $z = 735x_1 \quad +658x_2 \quad +863x_3 \quad +514x_4$

s.t.
$$
\begin{array}{rrrrcr}
341208x_1 & -402515x_2 & +358922x_3 & -287130x_4 & \leq & -256312 \\
610425x_1 & +383611x_2 & -491039x_3 & -523646x_4 & \leq & 328051 \\
283301x_1 & +566242x_2 & +453948x_3 & +370066x_4 & = & 429148 \\
547515x_1 & +811028x_2 & -220383x_3 & -414116x_4 & = & 269723 \\
670104x_1 & +355367x_2 & -252983x_3 & -592604x_4 & \leq & -215062
\end{array}
$$

& $\qquad\qquad x_2, \qquad\qquad\qquad x_4 \geq 0$

Workout 6.1.5 *Write the dual of Problem 6.1.4.*

Now suppose we are given a LOP in the following general form.

Problem 6.1.6

Max. $z = \sum_{j=1}^{n} c_j x_j$

s.t.
$$
\begin{array}{rcll}
\sum_{j=1}^{n} a_{ij} x_j & \leq & b_i & (i \in I_P) \\
\sum_{j=1}^{n} a_{ij} x_j & = & b_i & (i \in E_P)
\end{array}
$$

& $\qquad\qquad x_j \geq 0 \qquad (j \in R_P)$

Here, $I_P \subseteq \{1, \ldots, m\}$ identifies those (primal) constraints that use an *Inequality*, while $E_P = \{1, \ldots, m\} - I_P$ identifies those that use an *Equality*. For example, in Problem 6.1.1 we have $I = \{1, 2, 4\}$ and $E = \{3\}$. Likewise, $R_P \subseteq \{1, \ldots, n\}$ identifies those variables that are *Restricted* to be nonnegative, while $F_P = \{1, \ldots, n\} - R_P$ will denote those that are *Free*. We have $R = \{1, 3\}$ and $F = \{2\}$ in Problem 6.1.1. Define the analogous sets I_D, E_D, R_D, F_D for dual LOPs, and then notice their values for Problem 6.1.3.

Problem 6.1.7

Min. $w = \sum_{i=1}^{m} b_i y_i$

s.t.
$$
\begin{array}{rcll}
\sum_{i=1}^{m} a_{ij} y_i & \geq & c_j & (j \in I_D) \\
\sum_{i=1}^{m} a_{ij} y_i & = & c_j & (j \in E_D)
\end{array}
$$

& $\qquad\qquad y_i \geq 0 \qquad (i \in R_D)$

Workout 6.1.8 *Suppose that Problem 6.1.7 is the dual of Problem 6.1.6 (in other words the Weak Duality Inequality holds: every primal-feasible z and dual-feasible w satisfy $z \leq w$). Prove that*

    a. $I_D = R_P$,

    b. $E_D = F_P$,

    c. $R_D = I_P$, and

d. $F_D = E_P$.

There are occasions (see Chapter 10) in which problems arise even more generally, such as the following LOP.

Problem 6.1.9

$$
\begin{array}{rrrrrrl}
\text{Max.} \quad z = & 37x_1 & - & 51x_2 & - & 44x_3 & \\
\text{s.t.} & 4x_1 & - & x_2 & + & 3x_3 & \leq 146 \\
& -7x_1 & + & 8x_2 & + & 2x_3 & \geq 105 \\
& 9x_1 & + & 5x_2 & - & 6x_3 & = 89 \\
\& & x_1 \geq 0, & & & & x_3 \leq 0 &
\end{array}
$$

Workout 6.1.10 *Write the dual of LOP 6.1.9. Explain without converting, instead using the reasoning of multipliers.*

6.2 General Simplex and Phase 0

Let us return to an analysis of Problem 6.1.1. Below is its initial tableau (we have made note of the free variable x_2 as shown).

Tableau 6.2.1

$$
\left[
\begin{array}{rrr|rrrr|r}
 & \text{F} & & & & & & \\
7 & 6 & 2 & 1 & 0 & 0 & 0 & 419 \\
-5 & 3 & 8 & 0 & 1 & 0 & 0 & 528 \\
3 & -1 & 0 & 0 & 0 & 0 & 0 & 272 \\
0 & -9 & -4 & 0 & 0 & 1 & 0 & 168 \\
\hline
-36 & 31 & -37 & 0 & 0 & 0 & 1 & 0
\end{array}
\right]
$$

Notice that one cannot write down the dictionary for Tableau 6.2.1 because only $4, 5, 7 \in \beta$ so far — the third constraint, because it is an equality, delivers no basic variable (which is why we skipped over what would have been its basic variable, x_6, in the partial basis). Fear not, however: just pick one! In fact, since both phases of the Simplex Algorithm include mechanisms for keeping basic variables from becoming negative, why not throw in a basic variable that you don't care whether it is negative or not? That is, let's make x_2 basic in the third constraint; i.e., pivot on the -1, obtaining the next tableau.

Tableau 6.2.2

$$
\left[
\begin{array}{rrr|rrrr|r}
 & \text{F} & & & & & & \\
25 & 0 & 2 & 1 & 0 & 0 & 0 & 2051 \\
4 & 0 & 8 & 0 & 1 & 0 & 0 & 1344 \\
-3 & 1 & 0 & 0 & 0 & 0 & 0 & -272 \\
-27 & 0 & -4 & 0 & 0 & 1 & 0 & -2280 \\
\hline
57 & 0 & -37 & 0 & 0 & 0 & 1 & 8432
\end{array}
\right]
$$

6.2. General Simplex and Phase 0

In any LOP having equality constraints, the first part of what we will call **Phase 0** will be to *fill the basis*, choosing nonbasic variables (in least subscript order) to be basic in those constraints. As above, preference is given to free variables in these choices. In fact, after the basis is filled, any LOP having free variables may still have some that are nonbasic. It makes the same sense to put those in the basis, replacing basic restricted variables (both in least subscript order).

Workout 6.2.3 *Restate Problem 6.1.1 in standard form by solving for x_2 in constraint 3 and substituting its formula back into the other constraints and objective function. Compare your result with Tableau 6.2.2 and explain.*

Recall from Section 2.2 the notation $i \mapsto j$ that denotes the pivot operation which replaces the basic variable x_j by the parameter x_i. By $i \mapsto \emptyset$ we mean the Phase 0 pivot in which x_i replaces nothing because the current basis is not full.

Workout 6.2.4 *For each of the following instances of LOPs write the pending Phase 0 pivot $i \mapsto j$.*

a. $n = 6$, $m = 5$, $I = \{1, 2, 4\}$, $R = \{1, 3, 4\}$ and $\beta = \{2, 7, 8, 10\}$.

b. $n = 4$, $m = 9$, $I = \{1, 3, 4, 6, 7, 9\}$, $R = \{3, 4\}$ and $\beta = \{1, 2, 5, 7, 8, 10, 11\}$.

c. $n = 6$, $m = 7$, $I = \{1, 3, 4, 7\}$, $R = \{2, 3, 6\}$ and $\beta = \{1, 3, 5, 7, 9, 10, 13\}$.

d. $n = 3$, $m = 8$, $I = \{2, 5, 6, 7\}$, $R = \{\}$ and $\beta = \{1, 2, 3, 5, 8, 9, 10\}$.

After Phase 0 has been completed, we will assume that every free variable is in the basis, and that some basic variable is restricted (see Exercise 6.5.20). We now need to see the effect of free variables on Phases I and II. It is useful to think in terms of Workout 6.2.3, which suggests that a LOP with free variables is equivalent to a smaller LOP, derived from the first but without those free variables. In that smaller LOP all pivots are exchanges of restricted variables, and so then they should be in the original LOP. Let's see how this holds true.

In Phase I we first look for negative basic current values because they signal infeasibility. But now free variables can be ignored because their negativity is allowed. Thus Phase I will never kick a free variable out of a basis. In Phase II, once an incoming variable is found, the ratio calculations place upper bounds on its increasing value that are induced by preventing basic variables from becoming negative. But not caring that free variables go negative means that their ratios are irrelevant. Therefore Phase II will never kick out a free variable either. One can keep floating in the back of one's mind, then, the guideline of the general Simplex Algorithm that all free variables go in to the basis and never come out.

It's time to continue with Tableau 6.2.2 from Problem 6.1.1. Phase I works as usual in this case since -2280 is the most negative current basic value among restricted variables. Hence we pivot $1 \mapsto 7$ (it looks like 6 in the tableau, but we prefer to save x_6 for the "missing" slack variable).

Tableau 6.2.5

$$\left[\begin{array}{ccc|cccc|c} & & \text{F} & & & & & \\ 0 & 0 & -46 & 27 & 0 & 25 & 0 & -1623 \\ 0 & 0 & 200 & 0 & 27 & 4 & 0 & 27168 \\ 0 & 27 & 12 & 0 & 0 & -3 & 0 & -504 \\ 27 & 0 & 4 & 0 & 0 & -1 & 0 & 2280 \\ \hline 0 & 0 & -1227 & 0 & 0 & 57 & 27 & 97704 \end{array}\right]$$

The next pivot is likewise as usual: $3 \mapsto 4$.

Tableau 6.2.6

$$\left[\begin{array}{ccc|cccc|c} & & \text{F} & & & & & \\ 0 & 0 & 46 & -27 & 0 & -25 & 0 & 1623 \\ 0 & 0 & 0 & 200 & 46 & 192 & 0 & 34264 \\ 0 & 46 & 0 & 12 & 0 & 6 & 0 & -1580 \\ 46 & 0 & 0 & 4 & 0 & 2 & 0 & 3644 \\ \hline 0 & 0 & 0 & -1227 & 0 & -1039 & 46 & 240215 \end{array}\right]$$

Finally we arrive at Phase II — Tableau 6.2.6 is feasible because x_2 is free. At this stage, x_4 enters β, and while the x_2 **b**-ratio isn't considered because it is negative, it wouldn't be in any case because x_2 is free. Now we pivot $4 \mapsto 5$ to reach the optimal tableau, with $\mathbf{x}^* = (12864, -15808, 27168 \mid 34264, 0, 0)^\mathsf{T}/200$ and $z^* = 1958368/200$.

Tableau 6.2.7

$$\left[\begin{array}{ccc|cccc|c} & & \text{F} & & & & & \\ 0 & 0 & 200 & 0 & 27 & 4 & 0 & 27168 \\ 0 & 0 & 0 & 200 & 46 & 192 & 0 & 34264 \\ 0 & 200 & 0 & 0 & -12 & -24 & 0 & -15808 \\ 200 & 0 & 0 & 0 & -4 & -8 & 0 & 12864 \\ \hline 0 & 0 & 0 & 0 & 1227 & 604 & 200 & 1958368 \end{array}\right]$$

Workout 6.2.8 *Write an outline for all Phases of the general Simplex Algorithm.*

General Fundamental Theorem

Theorem 6.2.9 *Let P be a LOP in general form (Problem 6.1.6). Then*

  a. *P is either infeasible, unbounded, or it has a maximum;*

  b. *if P has a feasible solution then it has a basic feasible solution; and*

  c. *if P has an optimal solution then it has a basic optimal solution.*

Workout 6.2.10 *Prove Theorem 6.2.9.*

6.3 Plus de Pratique

Workout 6.3.1 *Use the general Simplex Algorithm to solve Problem 6.1.3 (the dual of Problem 6.1.1).*

Workout 6.3.2 *Use the general Simplex Algorithm to solve Problem 6.1.4 and its dual problem from Workout 6.1.5.*

Problem 6.3.3

$$
\begin{array}{rl}
\text{Max.} \ z = & x_1 + 2x_2 + 3x_3 + 4x_4 \\
\text{s.t.} & x_1 + x_2 + 2x_3 + x_4 \leq 10 \\
& 2x_1 - x_2 + x_3 + x_4 \leq 20 \\
& x_1 + 5x_4 \leq 30 \\
\& & x_1, \ x_4 \geq 0
\end{array}
$$

Workout 6.3.4 *Use the general Simplex Algorithm to solve Problem 6.3.3 and its dual.*

Workout 6.3.5 *Use Workouts 6.3.1, 6.3.2 and 6.3.4 to find a pattern in the optimal tableau/dual value relationship. [HINT: It will be useful to maintain the numbering of the slack variables from the standard case; that is, use slack variables x_4, x_5 and x_7 for Problem 6.1.1 — one can think of the nonexistent x_6 as existing with constant value 0.]*

6.4 General Duality and Slackness

Theorem 6.4.1 *If a general linear problem P has an optimum z^* then its dual linear problem D has an optimum w^*; moreover, $z^* = w^*$.* — General Duality Theorem

Proof. Suppose P has the form of Problem 6.1.6. Then D is Problem 6.1.7, satisfying the relations of Workout 6.1.8. We begin by putting Problem 6.1.6 into standard form as follows, using the transformations discussed in Section 1.4.

Problem 6.4.2

$$
\begin{array}{rl}
\text{Max.} \ z' = \sum_{j \in R_P} c_j x_j + \sum_{j \in F_P} c_j x_j^+ - \sum_{j \in F_P} c_j x_j^- & \\
\text{s.t.} \ \sum_{j \in R_P} a_{ij} x_j + \sum_{j \in F_P} a_{ij} x_j^+ - \sum_{j \in F_P} a_{ij} x_j^- \leq b_i & (i \in I_P) \\
\sum_{j \in R_P} a_{ij} x_j + \sum_{j \in F_P} a_{ij} x_j^+ - \sum_{j \in F_P} a_{ij} x_j^- \leq b_i & (i \in E_P) \\
-\sum_{j \in R_P} a_{ij} x_j - \sum_{j \in F_P} a_{ij} x_j^+ + \sum_{j \in F_P} a_{ij} x_j^- \leq -b_i & (i \in E_P) \\
\& \quad x_j, \ x_j^+, \ x_j^- \geq 0 & (\forall j)
\end{array}
$$

Workout 6.4.3 *Put Problem 6.1.7 into standard minimization form using similar transformations.*

Workout 6.4.4 *Write the dual of Problem 6.4.2.*

Workout 6.4.5 *Compare your results from Workouts 6.4.3 and 6.4.4.*

Workout 6.4.6 *Use Workout 6.4.5 and the Strong Duality Theorem 4.1.9 to finish the proof.*

⋄

Certificates of unboundedness and infeasibility (more on infeasibility in Chapter 7) are found as before, with slight modifications involving free variables and equality constraints. For example, $\mathbf{y} = (1, -3, 2)^\mathsf{T}$ certifies that Problem 6.4.7 below is infeasible, while $\mathbf{x} = (-2t, t, t)^\mathsf{T}$ as $t \to \infty$ certifies that Problem 6.4.8 below is unbounded.

Problem 6.4.7

$$\begin{aligned}
\text{Max.} \quad z &= 6x_1 + 3x_2 + 5x_3 \\
\text{s.t.} \quad 4x_1 + 4x_2 + 3x_3 &\leq 13 \\
2x_1 + 3x_2 + x_3 &= 8 \\
x_1 + 3x_2 + x_3 &\leq 3 \\
\& \quad x_2, \ x_3 &\geq 0
\end{aligned}$$

Problem 6.4.8

$$\begin{aligned}
\text{Max.} \quad z &= 6x_1 + 3x_2 + 5x_3 \\
\text{s.t.} \quad 4x_1 + 4x_2 + 3x_3 &\leq 13 \\
2x_1 + 3x_2 + x_3 &\leq 8 \\
x_1 + 3x_2 - x_3 &\leq 3 \\
\& \quad x_2, \ x_3 &\geq 0
\end{aligned}$$

How do we go about finding certificates of optimality? The optimal objective row of Tableau 6.2.7 of Problem 6.1.1 seems to yield $\mathbf{y}^* = (0, 1227, 604 \mid 0, 0, 0)^\mathsf{T}/200$ and $w^* = 1958368/200$. However, because x_2 is free, the second dual constraint is an equality, and consequently y_6 doesn't exist. (Here we are using the numbering hint from Workout 6.3.5.) Moreover, since the third primal constraint is an equality, x_6 doesn't exist, and so its optimal coefficient doesn't show up in the tableau to reveal the value of the free variable y_3. Therefore the optimal objective row correctly yields $\mathbf{y}^* = (0, 1227, y_3^*, 604 \mid 0, 0)^\mathsf{T}/200$. In order to compute y_3^* we simply solve for it from any of the dual constraints containing it: $y_3^* = (36 - 7y_1^* + 5y_2^* - y_5^*)/3$ or $y_3^* = 31 + 6y_1^* + 3y_2^* - 9y_4^*$. In an arbitrary general problem, the number of primal constraints is at least as large as the number of free dual variables, so there will always be enough constraints to solve for the "missing" optimal free values.

Let P be Problem 6.1.6 and D be its dual Problem 6.1.7.

Theorem 6.4.9 *Let x' be P-feasible and $\mathbf{y}'$ be D-feasible. Then $\mathbf{x}'$ and $\mathbf{y}'$ are a primal-dual optimal pair if and only if both*

General Complementary Slackness Theorem

$$x'_j = 0 \quad \text{or} \quad \sum_{i=1}^{m} a_{i,j} y'_i = c_j \quad \text{for all } j \in R_P \quad (6.11)$$

and

$$\sum_{j=1}^{n} a_{i,j} x'_j = b_i \quad \text{or} \quad y'_i = 0 \quad \text{for all } i \in I_P. \quad (6.12)$$

Workout 6.4.10 *Prove the General Complementary Slackness Theorem 6.4.9.*

Workout 6.4.11 *Generalize Theorem 4.4.2 to handle general LOPs.*

Workout 6.4.12 *Prove your theorem from Workout 6.4.11.*

Problem 6.4.13

$$
\begin{array}{rrrrrrrrrrl}
\text{Max. } z = & -10x_1 & - & 6x_2 & - & x_3 & + & 4x_4 & + & 3x_5 & \\
\text{s.t.} & x_1 & + & x_2 & + & 3x_3 & + & 4x_4 & + & 5x_5 & \leq 12 \\
& -x_1 & & & + & x_3 & - & x_4 & - & x_5 & \leq -5 \\
& 3x_1 & + & 2x_2 & + & x_3 & - & x_4 & & & \leq 0 \\
& 2x_1 & + & 4x_2 & & & - & 5x_4 & - & 3x_5 & = -25 \\
\& & x_1 & , & & & x_3 & , & x_4 & & & \geq 0
\end{array}
$$

Workout 6.4.14 *Consider Problem 6.4.13.*

   a. *Use your theorem from Workout 6.4.11 to show that $\mathbf{x}' = (2, -5, 0, 0, 3)^\mathsf{T}$ is optimal.*

   b. *Use your theorem from Workout 6.4.11 to show that $\mathbf{x}' = (0, 8, 0, 21, -16)^\mathsf{T}$ is not optimal.*

6.5 Exercises

Practice

6.5.1 *Use the general Simplex Algorithm to solve*

   a. *the primal LOP from Exercise 4.5.14(a).*

   b. *the dual LOP from Exercise 4.5.14(a).*

Include certificates for infeasibility, unboundedness and optimality.

6.5.2 *Use the general Simplex Algorithm to solve*

  a. *the primal LOP from Exercise 4.5.14(b).*
  b. *the dual LOP from Exercise 4.5.14(b).*

Include certificates for infeasibility, unboundedness and optimality.

6.5.3 *Use the general Simplex Algorithm to solve*

  a. *the primal LOP from Exercise 4.5.14(c).*
  b. *the dual LOP from Exercise 4.5.14(c).*

Include certificates for infeasibility, unboundedness and optimality.

6.5.4 *Use the general Simplex Algorithm to solve*

  a. *the primal LOP from Exercise 4.5.14(d).*
  b. *the dual LOP from Exercise 4.5.14(d).*

Include certificates for infeasibility, unboundedness and optimality.

6.5.5 *Use the general Simplex Algorithm to solve*

  a. *Problem 6.1.9.*
  b. *the dual of Problem 6.1.9.*

6.5.6 *Use the general Simplex Algorithm to solve the following LOP.*

$$\begin{aligned}
\text{Min.} \quad w &= y_1 \\
\text{s.t.} \quad y_2 + y_3 + y_4 &= 1 \\
y_1 + y_2 - y_3 + 2y_4 &\geq 0 \\
y_1 - y_2 + y_3 - y_4 &\geq 0 \\
y_1 - 2y_2 + y_3 - y_4 &\geq 0 \\
\& \quad y_2, y_3, y_4 &\geq 0
\end{aligned}$$

Include certificates for infeasibility, unboundedness and optimality.

6.5.7 *Consider the matrix A, below.*

$$\begin{pmatrix} 3 & 0 & -2 \\ -1 & 4 & 0 \\ 2 & -3 & 1 \\ 0 & -2 & 1 \end{pmatrix}$$

  a. *Find the maximum value x_0 such that $A\mathbf{x} \geq x_0 \mathbf{J}_4$ for all $\mathbf{x} \geq \mathbf{0}$ whose coordinates sum to 1.*
  b. *Find the minimum value y_0 such that $A^T\mathbf{y} \leq y_0 \mathbf{J}_3$ for all $\mathbf{y} \geq \mathbf{0}$ whose coordinates sum to 1.*

6.5. Exercises

c. Compare your results from the two parts above.

6.5.8 *Use the general Simplex Algorithm to solve n LOPs of your own devising, as well as their dual LOPs. Include certificates for infeasibility, unboundedness and optimality.*

6.5.9 *Verify Theorem 6.4.1 on the primal LOP from Exercise 4.5.14(b).*

6.5.10 *Verify Theorem 6.4.1 on the primal LOP from Exercise 4.5.14(c).*

6.5.11 *Verify Theorem 6.4.1 on Problem 6.1.1.*

6.5.12 *Verify Theorem 6.4.1 on Problem 6.1.4.*

6.5.13 *Verify Theorem 6.4.1 on any of the optimal LOPs from Exercise 6.5.8.*

6.5.14 *Verify Theorem 6.4.9 on the primal LOP from Exercise 4.5.14(b).*

6.5.15 *Verify Theorem 6.4.9 on the primal LOP from Exercise 4.5.14(c).*

6.5.16 *Verify Theorem 6.4.9 on Problem 6.1.1.*

6.5.17 *Verify Theorem 6.4.9 on Problem 6.1.4.*

6.5.18 *Verify Theorem 6.4.9 on any of the optimal LOPs from Exercise 6.5.8.*

6.5.19 *Consider the following LOP P.*

Max. $z = 212x_1 \quad -320x_2 \quad +273x_3 \quad -347x_4 \quad +295x_5$

s.t.
$$\begin{array}{rl}
-4x_1 \quad\quad\quad -2x_3 \quad\quad\quad +8x_5 & \leq -22 \\
2x_1 \quad +3x_2 \quad\quad\quad -x_4 \quad\quad & = 31 \\
-5x_2 \quad +3x_3 \quad\quad\quad -2x_5 & \leq 27 \\
-7x_1 \quad\quad\quad -8x_3 \quad +6x_4 \quad\quad & = -38 \\
-9x_3 \quad -2x_4 \quad +x_5 & \leq -40 \\
-x_2 \quad\quad\quad -3x_4 \quad -5x_5 & \leq 42
\end{array}$$

& $\quad x_1, \quad\quad\quad x_3, \quad x_4 \quad\quad\quad \geq 0$

a. Find $\mathbf{x}^*$ and write the Phase 0, I and II pivots that solve P.

b. Use the General Complementary Slackness Theorem to find the optimal certificate $\mathbf{y}^*$ [do *not* solve the dual LOP D!].

Challenges

6.5.20 *What can you say about a LOP with as many free variables as constraints?*

6.5.21 *Write the pseudocode (or code) for the general Simplex Algorithm.*

6.5.22 *Workout 6.2.3 suggests a method for comparing a general LOP to a smaller LOP having no free variables. Use this idea to give another proof of the General Strong Duality Theorem 6.4.1.*

Projects

6.5.23 *Present the Applegate-Cook-Dash-Espinoza paper,* Exact solutions to linear programming problems.

6.5.24 *Present the equivalence of our standard form for LO to the following form:* Min. x_n s.t. $\mathbf{Ax} = \mathbf{0}$, $\mathbf{J}_n^\mathsf{T}\mathbf{x} = 1$, & $\mathbf{x} \geq \mathbf{0}$, *where* $\mathbf{A}$ *satisfies* $\mathbf{AJ}_n = \mathbf{0}$.

Chapter 7

Unsolvable Systems

7.1 Infeasible Certificates

Recall the Fundamental Theorem of Linear Optimization (Theorem 2.9.1). In particular, it states that there are only three kinds of linear problems: infeasible, unbounded and optimal. Thus, if a LOP and its dual are both feasible then they are both optimal (by the Weak Duality Theorem — Inequality 1.4.7). Of course, the converse is far more obvious. Moreover, weak duality also implies that any primal-dual feasible pair that satisfies $z \geq w$ is an optimal pair. We record these observations in Theorem 7.1.3.

Problem 7.1.1

$$\text{Max.} \quad z = \sum_{j=1}^{n} c_j x_j$$

$$\text{s.t.} \quad \sum_{j=1}^{n} a_{i,j} x_j \leq b_i \quad (1 \leq i \leq m)$$

$$\& \quad x_j \geq 0 \quad (1 \leq j \leq n)$$

System 7.1.2

$$\sum_{j=1}^{n} c_j x_j \geq \sum_{i=1}^{m} b_i y_i$$

$$\sum_{j=1}^{n} a_{i,j} x_j \leq b_i \quad (1 \leq i \leq m)$$

$$x_j \geq 0 \quad (1 \leq j \leq n)$$

$$\sum_{i=1}^{m} a_{i,j} y_i \geq c_j \quad (1 \leq j \leq n)$$

$$y_i \geq 0 \quad (1 \leq i \leq m)$$

Theorem 7.1.3 *Problem 7.1.1 is optimal if and only if System 7.1.2 is solvable.* ◇

G. H. Hurlbert, *Linear Optimization*, Undergraduate Texts in Mathematics,
DOI: 10.1007/978-0-387-79148-7_7, © Springer Science+Business Media LLC 2010

Problem 7.1.4 *Consider the following LOP.*

$$\text{Max. } z = x_1 + x_2 + x_3 + x_4 + x_5$$

$$\begin{aligned}
\text{s.t.} \quad & x_2 + x_3 + x_4 && \leq 86 \\
& x_1 + x_2 && \leq 59 \\
& x_3 + x_4 + x_5 && \leq 81 \\
& x_2 + x_3 && \leq 68 \\
& x_4 + x_5 && \leq 62 \\
& x_1 + x_2 + x_3 + x_4 && \leq 97 \\
& x_1 && \leq 40 \\
& x_3 + x_4 && \leq 75 \\
\& \quad & x_1, x_2, x_3, x_4, x_5 \geq 0
\end{aligned}$$

Workout 7.1.5

a. Write the system S that corresponds to Problem 7.1.4.

b. Find a solution to S.

c. Use your S-solution to find optimal solutions to Problem 7.1.4 and its dual.

What many find interesting about Theorem 7.1.3 is that it "reduces" the study of optimizing linear functions over linear systems to that of solving linear systems. While the latter amounts to performing Phase I, do not be fooled by thinking that the latter process is faster than Simplex — indeed, it is slower (see Appendix C) since the number of constraints has at least doubled (compared to the smaller of m and n). On the other hand, there are fairly efficient implementations of the idea of working on both problems simultaneously, collectively known as Primal-Dual methods, conveniently. We will not discuss such methods in this text. However, Theorem 7.1.3 does illustrate the importance of understanding the (un)solvability of linear systems, and we devote the remainder of this chapter to this aim.

Workout 7.1.6 *Generalize Theorem 7.1.3 to handle general LOPs.*

Let us return to the situation of infeasibility. Consider the following problem.

Problem 7.1.7

$$\text{Max. } z = 216x_1 - 160x_2 + 254x_3 + 303x_4$$

$$\begin{aligned}
\text{s.t.} \quad & 2x_1 - 4x_2 + x_3 + 3x_4 \leq 22 \\
& x_1 + 3x_2 - 5x_3 + 8x_4 \leq -17 \\
& 6x_1 + 4x_2 + 7x_4 \leq 27 \\
& 9x_1 - 2x_2 - x_3 + 4x_4 \leq -16 \\
& -3x_1 + 7x_3 - 5x_4 \leq -32 \\
\& \quad & x_1, x_2, x_3, x_4 \geq 0
\end{aligned}$$

7.1. Infeasible Certificates

Note that four Simplex pivots produces the following tableau, suggesting that Problem 7.1.7 is infeasible.

Tableau 7.1.8

$$\begin{bmatrix} 0 & 0 & 1459 & 0 & 61 & 130 & 0 & 73 & 303 & 0 & -11732 \\ 0 & 0 & -806 & 0 & 0 & -150 & 61 & -103 & -237 & 0 & 13429 \\ 0 & 0 & -164 & 61 & 0 & -6 & 0 & -9 & -29 & 0 & 1174 \\ 0 & 61 & 292 & 0 & 0 & 33 & 0 & 19 & 68 & 0 & -3041 \\ 61 & 0 & 131 & 0 & 0 & 10 & 0 & 15 & 28 & 0 & -1306 \\ \hline 0 & 0 & -83610 & 0 & 0 & -4938 & 0 & -2527 & -13619 & 61 & 560186 \end{bmatrix}$$

A certificate that proves Problem 7.1.7 is infeasible is given by the multipliers $\mathbf{y} = (61, 130, 0, 73, 303)^\mathsf{T}$, found as the coefficients of the slack variables in the halting row of Tableau 7.1.8.

Theorem 7.1.9 *Suppose that Problem 7.1.1 halts in Phase I on a row that represents the equation $\sum_{j=1}^{n+m} a'_j x_j = b'$. Then the multipliers $y_i = a'_{n+i}$ ($1 \leq i \leq m$) certify that Problem 7.1.1 is infeasible.*

Proof. The proof proceeds in 3 steps.

Lemma 7.1.10 *Every row of a Simplex tableau is a linear combination of the rows of the original tableau.*

Workout 7.1.11 *Prove Lemma 7.1.10.*

Lemma 7.1.12 *Let R be a row of a Simplex tableau with entries a'_j ($1 \leq j \leq n+m$), with right-hand side b'. Suppose that the multipliers y_i ($1 \leq i \leq m$) of the rows of the original tableau produce row R. Then $y_i = a'_{n+i}$ for $1 \leq i \leq m$.*

Workout 7.1.13 *Prove Lemma 7.1.12.*

Lemma 7.1.14 *The multipliers defined in Lemma 7.1.12 produce a contradiction.*

Workout 7.1.15 *Prove Lemma 7.1.14.*

$\diamond$

Note that $(61, 130, 0, 73, 303)^\mathsf{T}/61$ is not dual-feasible. However, it does have a nice property — it can be used to find a certificate for dual unboundedness. Before explaining how, let's create a small example to study.

Workout 7.1.16

a. Draw a diagram of an unbounded LOP D with two variables.

b. Write its corresponding dual P.

c. Obtain a P-infeasible certificate $\mathbf{y}'$.

d. Obtain a D-unbounded certificate $\mathbf{y}(t) = \mathbf{y}^0 + t\overrightarrow{\mathbf{y}}$.

e. Compare $\mathbf{y}'$ and $\overrightarrow{\mathbf{y}}$.

f. Is it possible that $\mathbf{y}' = \overrightarrow{\mathbf{y}}$?

Workout 7.1.17 *Suppose that $\mathbf{y}'$ is a feasible solution to the dual D of Problem 7.1.7. Use $(61, 130, 0, 73, 303)^\mathsf{T}/61$ to find a certificate for dual unboundedness. Will your method always work in general?*

7.2 Inconsistency

Back at the Varyim Portint Co.,[1] one must remember to be prepared — recall from Section 4.4 how swiftly the president can act. Without the offending row in hand, how might one construct an infeasible certificate for the boss on the spot? Is there a method that is independent of the problem?

Because a LOP is infeasible precisely when its system of constraints is unsolvable, we hereafter consider systems of constraints only. Let us now consider the general System 7.2.1 below.

System 7.2.1

$$\sum_{j=1}^n a_{i,j} x_j \leq b_i \quad (i \in I)$$

$$\sum_{j=1}^n a_{i,j} x_j = b_i \quad (i \in E)$$

$$x_j \geq 0 \quad (j \in R)$$

Suppose that System 7.2.1 (S) is unsolvable, and let P be a LOP having S as its system of constraints. Then P is infeasible and the Simplex Algorithm will halt in Phase I. What this means, according to Lemma 7.1.10, is that some linear combination of the constraints of S produces an offending row in the halting Simplex tableau of P. What does such a row look like? Phase I means that the right-hand side is negative, and the halting condition means that there is no coefficient on which to pivot. For free variables, this means that their coefficients are zero, and for restricted variables, this means their coefficients are nonnegative. With this in mind we consider the following **alternative** System 7.2.2.

alternative system

System 7.2.2

$$\sum_{i=1}^m b_i y_i = -1$$

$$\sum_{i=1}^m a_{i,j} y_i \geq 0 \quad (j \in R)$$

$$\sum_{i=1}^m a_{i,j} y_i = 0 \quad (j \in F)$$

$$y_i \geq 0 \quad (i \in I)$$

---

[1] Page 81.

7.2. Inconsistency

Note that part of this system would be the system of constraints for the alternative of P if the objective function for P were identically zero. The added constraint seems to force the dual objective function to be -1. It seems natural, then, that both systems cannot be solvable, a line of reasoning we will explore. The following definition suggests a link between the two systems.

Definition 7.2.3 *We say that a system is **inconsistent** if its alternative system is solvable.*

inconsistent system

The above argument almost shows that an unsolvable system is inconsistent. Demanding that the right-hand side equals -1 instead of an arbitrary negative is not much of a stretch: once negative, the multipliers can be scaled to make it any negative number requested. The only thing missing from the discussion is the restriction on some of the y_is. Notice that the presence of equalities means that some of the coefficients expected to appear by Theorem 7.1.9 do not exist. However, those that do correspond to values of $i \in I$, which must be nonnegative in the offending row of the halting tableau.

Workout 7.2.4 *Consider Problem 6.4.7 (P), having constraint system S.*

  a. *Write the alternative system S'.*
  b. *Solve S'.*
  c. *Find the offending row that halts the Simplex Algorithm on P.*
  d. *Compare your answers to parts b and c.*

The result of the above discussion is the following theorem, often called the Theorem of the Alternative.

Theorem 7.2.5 *A system is unsolvable if and only if it is inconsistent.*

Theorem of the Alternative

Proof. It is clear that inconsistent implies unsolvable – a solution to the alternative system serves as a certificate. While the converse has been argued above, we offer a more direct inference using duality. Suppose that System 7.2.1 is unsolvable and consider the following LOP, where $s_i = \text{sign}(b_i)$.

Problem 7.2.6

$$\text{Max.} \quad z = \sum_{i=1}^{m} -x_{n+i}$$

$$\text{s.t.} \quad \sum_{j=1}^{n} a_{i,j} x_j + s_i x_{n+i} \leq b_i \quad (i \in I)$$

$$\sum_{j=1}^{n} a_{i,j} x_j + s_i x_{n+i} = b_i \quad (i \in E)$$

$$\& \quad x_j \geq 0 \quad (j \in R)$$

$$x_{n+i} \geq 0 \quad (1 \leq i \leq m)$$

Workout 7.2.7 *Use the General Fundamental Theorem 6.2.9 to prove that Problem 7.2.6 is optimal, and that its optimum is negative.*

Workout 7.2.8

a. *Write the dual of Problem 7.2.6.*

b. *Use the General Duality Theorem 6.4.1 to draw conclusions about its dual.*

c. *Use the optimal dual solution $\mathbf{y}^*$ to construct a solution to System 7.2.2.*

⋄

7.3 इसक अध्ययन करो

Consider the following systems.

System 7.3.1

$$\begin{aligned}
2x_1 + x_2 - 3x_3 &= 7 \\
-x_1 + 3x_2 + x_3 &= -3 \\
x_1 + x_3 &\leq 2 \\
2x_1 + 2x_2 + 3x_3 &\leq 5 \\
4x_1 - 2x_2 - 3x_3 &\leq 6 \\
x_2, x_3 &\geq 0
\end{aligned}$$

Workout 7.3.2 *Verify Theorem 7.2.5 on System 7.3.1.*

System 7.3.3

$$\begin{aligned}
3x_1 - x_2 + 2x_3 - 6x_4 &= 21 \\
-x_1 + 9x_2 - 4x_3 + 2x_4 &\leq -12 \\
x_1 + 4x_2 - 3x_3 - 5x_4 &= -18 \\
4x_1 + 3x_2 + 3x_3 - 5x_4 &\leq 24 \\
-2x_1 + 5x_2 - x_3 + 7x_4 &= 19 \\
x_1 - 9x_2 + 8x_3 + 4x_4 &\leq 14 \\
x_2, x_4 &\geq 0
\end{aligned}$$

Workout 7.3.4 *Find all basic solutions to the alternative of System 7.3.3.*

7.4 Unsolvable Subsystems

Observe that the system of constraints of Problem 7.1.7 is shown to be unsolvable because the aggregate of its individual constraints is contradictory. Moreover, as shown by the certificate $\mathbf{y} = (61, 130, 0, 73, 303)^\mathsf{T}$, only four (all but the third) of its constraints are necessary to derive a contradiction.

7.4. Unsolvable Subsystems

On closer inspection, Tableau 7.1.8 reveals (in row 4 or 5) that only three constraints are necessary (the 2nd, 4th and 5th). One might wonder in general how small an unsolvable subsystem can be guaranteed inside a given unsolvable system. We can answer this soon enough. First we look back to the General Fundamental Theorem 6.2.9.

Theorem 7.4.1 *Every solvable system of r linear constraints (not including nonnegativity constraints) has a solution with at most r nonzero variables.*

Workout 7.4.2 *Prove Theorem 7.4.1. [HINT: Apply the General Fundamental Theorem 6.2.9 to a LOP having the given system of constraints.]*

Now we can answer the question above.

Theorem 7.4.3 *If a system of linear constraints in n variables is unsolvable then it has a subsystem of at most $n+1$ constraints that is unsolvable.*

Workout 7.4.4 *Prove Theorem 7.4.3. [HINT: What is the value of r in the alternative system?]*

One should notice that Theorem 7.2.5 implies that exactly one of the Systems 7.2.1 and 7.2.2 is solvable. Other theorems of this sort include the following.

Theorem 7.4.5 *Exactly one of the following two systems S and T is solvable.* — Farkas's Theorem

$\underline{S}:$

$\sum_{j=1}^{n} a_{i,j} x_j \leq 0 \quad (1 \leq i \leq m)$

$\sum_{j=1}^{n} c_j x_j > 0$

$\underline{T}:$

$\sum_{i=1}^{m} a_{i,j} y_i = c_j \quad (1 \leq j \leq n)$

$y_i \geq 0 \quad (1 \leq i \leq m)$

Proof. We invent two LOPs based on S and T as follows. Let P have objective function $z = \sum_{j=1}^{n} c_j x_j$ with constraints $\sum_{j=1}^{n} a_{i,j} x_j \leq 0$ for $1 \leq i \leq m$. Let D have objective function $w = 0$ with constraints $\sum_{i=1}^{m} a_{i,j} y_i = c_j$ for $1 \leq j \leq n$ and $y_i \geq 0$ for $1 \leq i \leq m$. Then P and D are dual to each other.

Now P is feasible (at $\mathbf{x} = \mathbf{0}$), so it is either optimal or unbounded by the General Fundamental Theorem 6.2.9. If P is unbounded then some feasible $\mathbf{x}$ has $z(\mathbf{x}) > 0$ (so S is solvable) and D is infeasible (so T is unsolvable). If P is optimal then D is feasible (and so T is solvable) and, by the Weak Duality Theorem (Inequality 1.4.7), every feasible $\mathbf{x}$ and $\mathbf{y}$ satisfy $z(\mathbf{x}) \leq w(\mathbf{y}) = \mathbf{0}$ (so S is unsolvable). ◇

Workout 7.4.6 *Use Theorem 7.2.5 to prove Farkas's Theorem 7.4.5.*

7.5 Exercises

Practice

7.5.1 *Use Theorem 7.1.3 instead of the Simplex Algorithm to solve the following LOP.*

$$\text{Max. } z = 5x_1 + 4x_2 + 3x_3 + x_4 + 2x_5 + 6x_6$$

$$\text{s.t.} \quad \begin{aligned} x_1 + 2x_2 + 3x_3 + 4x_4 + 5x_5 + 6x_6 &\leq 7 \\ 6x_1 + 5x_2 + 4x_3 + 3x_4 + 2x_5 + x_6 &\leq 23 \end{aligned}$$

$$\& \quad x_1, \, x_2, \, x_3, \, x_4, \, x_5, \, x_6 \geq 0$$

7.5.2 *Use Theorem 7.1.3 instead of the Simplex Algorithm to solve the following LOP.*

$$\text{Max. } z = -2x_1 + 9x_2 + 5x_3$$

$$\text{s.t.} \quad \begin{aligned} 2x_1 \quad\quad - 4x_3 &\leq -7 \\ x_1 + 3x_2 \quad\quad &= 8 \\ -2x_2 + 6x_3 &\leq 9 \end{aligned}$$

$$\& \quad x_1, \, x_2, \, x_3 \geq 0$$

7.5.3 *Use Theorem 7.1.3 instead of the Simplex Algorithm to solve N LOPs of your own design.*

7.5.4 *Verify Theorems 7.2.5 and 7.4.3 on the following system.*

$$\begin{aligned} 2x_1 + 3x_2 &\leq 9 \\ 3x_1 + 3x_2 &\leq 6 \\ -4x_1 + x_2 &\leq 13 \\ -2x_1 + 4x_2 &\leq 2 \\ -x_1 - 2x_2 &\leq -4 \end{aligned}$$

$$x_1, \, x_2 \geq 0$$

7.5.5 *Verify Theorems 7.2.5 and 7.4.3 on the following system.*

$$\begin{aligned} x_1 + 4x_2 - x_3 &= 2 \\ -2x_1 - 3x_2 + x_3 &= 1 \\ -3x_1 - 2x_2 + x_3 &= 0 \\ 4x_1 + x_2 - x_3 &= -1 \\ 2x_1 - 3x_2 + 5x_3 &= 2 \end{aligned}$$

7.5.6 *Verify Theorems 7.2.5 and 7.4.3 on the following system.*

7.5. Exercises

$$\begin{array}{rcrcrcrcrcrcrcr}
3x_1 & - & 5x_2 & - & 4x_3 & & & + & 8x_5 & & & = & -114 \\
 & & x_2 & - & 9x_3 & + & 2x_4 & & & + & 6x_6 & \leq & 240 \\
-x_1 & & & & & & & + & 3x_5 & + & 4x_6 & = & 65 \\
2x_1 & - & 3x_2 & & & + & x_4 & - & 6x_5 & & & \leq & 188 \\
 & & & & 7x_3 & + & 7x_4 & & & + & 2x_6 & \leq & -97 \\
9x_1 & - & 6x_2 & + & 4x_3 & & & + & x_5 & - & x_6 & = & 451 \\
-6x_1 & & & & & - & 3x_4 & + & x_5 & & & = & -312 \\
8x_1 & - & 4x_2 & + & x_3 & & & & & & & \leq & 209 \\
 & & 2x_2 & + & 7x_3 & - & 5x_4 & & & + & x_6 & = & 383 \\
 & & x_2 & , & x_3 & , & & & & & x_6 & \geq & 0
\end{array}$$

7.5.7 Verify Theorems 7.2.5 and 7.4.3 on N systems of your own devising.

7.5.8 In the definition of the alternative System 7.2.2, the first constraint is $\sum_{i=1}^{m} b_i y_i = -1$. Suppose instead we require only $\sum_{i=1}^{m} b_i y_i < 0$ — then any solution to the resulting system still certifies the unsolvability of System 7.2.1. Prove that the region of all such resulting solutions is conic.

7.5.9 Let $X, Y \subset \mathbb{R}^n$ be finite sets of points. We say that X and Y are **separable** if there is some linear inequality satisfied by all the points of X and none of the points of Y. Prove that X and Y are not separable if and only if there are subsets $X' \subseteq X$ and $Y' \subseteq Y$, with $|X'| + |Y'| \leq n + 2$, that are not separable.

separable points

7.5.10 Let S be the system $\{\mathbf{Ax} \leq \mathbf{b}, \mathbf{x} \geq \mathbf{0}\}$. The **homogenization** of S is the system $T = \{\mathbf{Ax} - x_0 \mathbf{b} \leq \mathbf{0}, \mathbf{x} \geq \mathbf{0}, x_0 > 0\}$. Prove that T has a solution if and only if S has a solution.

homogenization

Challenges

7.5.11 Prove that exactly one of the following two systems S and T is solvable. [HINT: Use a trick similar to that used to prove the Theorem of the Alternative 7.2.5.]

$\underline{S}:$

$\sum_{j=1}^{n} a_{i,j} x_j = 0 \quad (1 \leq i \leq m)$

$x_j > 0 \quad (1 \leq j \leq n)$

$\underline{T}:$

$\sum_{i=1}^{m} a_{i,j} y_i \geq 0 \quad (1 \leq j \leq n)$

$\sum_{i=1}^{m} a_{i,j} y_i \neq 0 \quad (\text{some } j)$

7.5.12 Prove that exactly one of the following two systems S and T is solvable. [HINT: Use a trick similar to that used to prove the Theorem of the

Alternative 7.2.5.]

$\underline{S}$:

$\sum_{j=1}^{n} a_{i,j} x_j < 0 \quad (1 \leq i \leq m)$

$x_j \geq 0 \quad (1 \leq j \leq n)$

$\underline{T}$:

$\sum_{i=1}^{m} a_{i,j} y_i \geq 0 \quad (1 \leq j \leq n)$

$y_i \geq 0 \quad (1 \leq i \leq m)$

$y_i \neq 0 \quad (\text{some } i)$

7.5.13 Prove that exactly one of the following two systems S and T is solvable. [HINT: Use a trick similar to that used to prove the Theorem of the Alternative 7.2.5.]

$\underline{S}$:

$\sum_{j=1}^{n} a_{i,j} x_j < 0 \quad (1 \leq i \leq m)$

$\underline{T}$:

$\sum_{i=1}^{m} a_{i,j} y_i = 0 \quad (1 \leq j \leq n)$

$y_i \geq 0 \quad (1 \leq i \leq m)$

$y_i \neq 0 \quad (\text{some } i)$

7.5.14 Prove that $\mathbf{Ax} = \mathbf{0}$ has a unique solution if and only if $\mathbf{A}^T \mathbf{y} = \mathbf{c}$ is solvable for every $\mathbf{c}$.

7.5.15 Let S be the system

$$S: \sum_{j=1}^{n} a_{i,j} x_j \leq 0 \quad (1 \leq i \leq m), \quad x_j \neq 0 \quad (\text{some } j)$$

and, for given $\mathbf{c} = (c_1, \ldots, c_n)^T$, define the system $T(\mathbf{c})$ by

$$T(\mathbf{c}): \sum_{i=1}^{m} a_{i,j} y_i = c_j \quad (1 \leq j \leq n), \quad y_i \geq 0 \quad (1 \leq i \leq m).$$

Prove that S is unsolvable if and only if $T(\mathbf{c})$ is solvable for every $\mathbf{c}$.

7.5.16 Use Theorem 7.2.5 to give a heuristic argument for your answer to Exercise 2.10.37.

Projects

7.5.17 Investigate the probability that a random system of m inequalities in n variables is feasible, and that a random objective function over those feasible systems that are feasible is optimal.

Primal-Dual Method

7.5.18 Present the **Primal-Dual Method**.

Hermite normal form

7.5.19 Present the **Hermite normal form** of a matrix and its use in solving integral systems of equations.

Chapter 8

Geometry Revisited

8.1 Helly's Theorem

Let $\mathcal{F}$ be a finite family of convex regions in $\mathbb{R}^n$. We say that $\mathcal{F}$ is **k-intersecting** if every k sets of $\mathcal{F}$ intersect ($\forall \mathcal{G} \subseteq \mathcal{F} : |\mathcal{G}| = k \Longrightarrow \cap_{S \in \mathcal{G}} S \neq \emptyset$), and **full-intersecting** if the intersection of all the sets of $\mathcal{F}$ is nonempty. For example, Figure 8.1 shows a 2-intersecting family in $\mathbb{R}^2$ that is not 3-intersecting.

k-intersecting/full-intersecting family

Game 8.1.1 *Thelma and J. J. alternate (with Thelma first) drawing distinct convex regions in $\mathbb{R}^2$ under the rule that the resulting family of regions (after the first move) is 2-intersecting but (after the second move) is not full-intersecting. A player unable to draw such a region loses the game.*

○

For example, Figure 8.1 shows an ellipse, pentagon and line segment as the first three regions drawn in a particular start to Game 8.1.1.

Workout 8.1.2 *Who wins Game 8.1.1? [HINT: Thelma loses if she starts with a point.]*

Game 8.1.3 *Thelma and J. J. alternate (with Thelma first) drawing distinct convex regions in $\mathbb{R}^2$ under the rule that the resulting family of regions (after the first move) is 2-intersecting and (after the second move) 3-intersecting but (after the third move) not full-intersecting. A player unable to do so loses the game.*

Workout 8.1.4 *Who wins Game 8.1.3?*

Theorem 8.1.5 *Let $\mathcal{F}$ be a finite family of convex regions in $\mathbb{R}^n$. If $\mathcal{F}$ is $(n+1)$-intersecting then $\mathcal{F}$ is full-intersecting.*

Helly's Theorem

Workout 8.1.6 *Prove Helly's Theorem 8.1.5 for the case $n = 1$. [HINT: Induction on $|\mathcal{F}|$ is one approach.]*

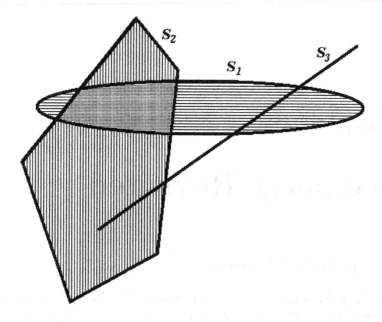

Figure 8.1: A 2-intersecting family of convex regions

Proof. The first part of the proof involves converting the arbitrary convex regions $C_i \in \mathcal{F}$ into polytopes $P_i \subseteq C_i$ such that $\mathcal{G} = \{P_1, \ldots, P_t\}$ is also $(n+1)$-intersecting. The strategy is to prove that if $\mathcal{G}$ is full-intersecting then so is $\mathcal{F}$. Also, it may be easier to prove the result for $\mathcal{G}$ because of the nice structure of its regions.

Workout 8.1.7 *Prove that if $\mathcal{G}$ is full-intersecting, with each $P_i \subseteq C_i$, then so is $\mathcal{F}$.*

Let $H \subseteq \{1, \ldots, t\}$ have size $n+1$. The hypothesis asserts that $\cap_{h \in H} C_h \neq \emptyset$, so choose $\mathbf{p}_H \in \cap_{h \in H} C_h$. Now do the same for every such $(n+1)$-subset of $\{1, \ldots, t\}$. Finally, define $X_h = \{\mathbf{p}_H \mid H \ni h\}$ and let $P_h = \text{vhull}(X_h)$. (Recall that P_h is a polytope by Workout 3.5.19.) For example, Figure 8.2 shows an instance in $\mathbb{R}^2$ with 4 regions.

Workout 8.1.8 *Prove that each $P_i \subseteq C_i$.*

The final task is to show that $\mathcal{G}$ is full-intersecting. For this we recall that every polytope is the intersection of a finite number of half-spaces. This means that for each i there exists a finite system S_i of inequalities whose solution region equals P_i. Now let S be the system of inequalities including every inequality from each system S_i.

Workout 8.1.9 *Let S' be any collection of $n+1$ of the inequalities of S. Prove that S' is solvable.*

Workout 8.1.10 *Use Theorem 7.4.3 to finish the proof.*

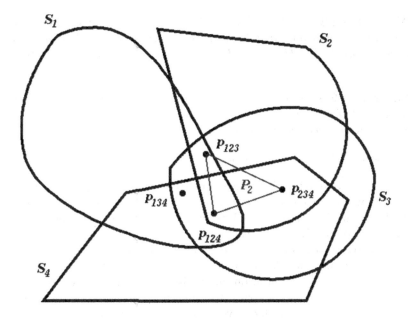

Figure 8.2: Constructing polytopes inside convex regions

⋄

Workout 8.1.11 *Finish the example from Figure 8.2 and find a point* **q** *common to all four regions.*

8.2 Permutation Matrices

Workout 8.2.1 *Let P be the polyhedron defined as the region of solutions of the following system.*

$$5x_1 - x_2 \leq 30$$
$$-x_1 + x_2 \leq 7$$
$$x_1 - 3x_2 \leq -8$$
$$x_2 \leq 10$$
$$-3x_1 - 4x_2 \leq -24$$

$$x_1, x_2 \geq 0$$

a. *Draw a diagram of P.*

b. *Find all extreme points of P.*

c. *Write and plot the sum of the first and third constraints.*

d. *For each extreme point* **x** *of P write an objective function that is maximized at* **x**.

Workout 8.2.2 Let $P \subseteq \mathbb{R}^2$ be a polyhedron. Prove that $\mathbf{x}^*$ is an extreme point of P if and only if it is the unique optimal solution to some linear problem having feasible region P. [HINT: What was the moral of Workout 8.2.1c?]

Theorem 8.2.3 Let $P \subseteq \mathbb{R}^n$ be a polyhedron. Prove that $\mathbf{x}^*$ is an extreme point of P if and only if it is the unique optimal solution to some linear problem having feasible region P.

Workout 8.2.4 Prove Theorem 8.2.3.

stochastic vector/ matrix

doubly stochastic/ permutation matrix

A vector is **stochastic** if it is nonnegative and its components sum to 1; in other words, its components could be used as the coefficients of a convex combination. A matrix is **stochastic** if each of its rows is stochastic, and **doubly stochastic** (DS) if each of its rows and columns is stochastic. A **permutation matrix** is a square $\{0, 1\}$-matrix with exactly one 1 per row and per column. The identity matrix is an example of a permutation matrix; indeed, every permutation matrix is a rearrangement of the columns (or rows) of an identity matrix. A permutation matrix is one type of doubly stochastic matrix; in fact, every integral doubly stochastic matrix is a permutation matrix.

Workout 8.2.5 Prove that every DS matrix is square. [HINT: Sum its entries.]

Workout 8.2.6 Prove that the convex combination of permutation matrices is DS.

Interestingly, the converse is a 1936 theorem of König, somehow attributed instead to the 1946 and 1953 work of Birkhoff and von Neumann, respectively.

Birkhoff–von Neumann Theorem

Theorem 8.2.7 Every DS matrix is a convex combination of permutation matrices.

Workout 8.2.8 Prove that every 2×2 doubly stochastic matrix can be written uniquely as a convex combination of permutation matrices.

Proof. The key idea is to think of an $n \times n$ matrix as a vector in $\mathbb{R}^{n^2}$. The strategy is to use the DS property to impose linear constraints on such vectors. If the extreme points of the polytope defined by the constraints correspond to permutation matrices (the bulk of the work in the proof) then the result follows by Exercise 3.5.33.

Workout 8.2.9 Let $\mathbf{X} = (x_{r,s})$ be an $n \times n$ DS matrix. Write the constraints of the system on $\{x_{r,s}\}$ that defines the DS property.

Workout 8.2.10 Prove that the polyhedron P defined in Workout 8.2.9 is a polytope.

8.2. Permutation Matrices

We now proceed to show that every extreme point of P is integral, by contrapositive. We will show that any nonintegral point of P is the center of some line segment residing inside P. It may be illustrative to follow how the following argument behaves on the DS matrix

$$\mathbf{x}' = \begin{pmatrix} .7 & .3 & 0 & 0 \\ 0 & 0 & .4 & .6 \\ 0 & .6 & 0 & .4 \\ .3 & .1 & .6 & 0 \end{pmatrix},$$

which is the center of the line segment with endpoints

$$\mathbf{x}'^{-}(.1) = \begin{pmatrix} .6 & .4 & 0 & 0 \\ 0 & 0 & .5 & .5 \\ 0 & .5 & 0 & .5 \\ .4 & .1 & .5 & 0 \end{pmatrix} \quad \text{and} \quad \mathbf{x}'^{+}(.1) = \begin{pmatrix} .8 & .2 & 0 & 0 \\ 0 & 0 & .3 & .7 \\ 0 & .7 & 0 & .3 \\ .2 & .1 & .7 & 0 \end{pmatrix},$$

for example.

Suppose that $\mathbf{x} \in P$ is not integral, and let $0 < x_{r_1, s_1} < 1$. Because of the row constraint $\sum_{s=1}^{n} x_{r_1, s} = 1$, there must be some s_2 such that $0 < x_{r_1, s_2} < 1$. Likewise, because of the column constraint $\sum_{r=1}^{n} x_{r, s_2} = 1$, there must be some r_2 such that $0 < x_{r_2, s_2} < 1$. This process can be iterated, and we will stop when some index (r, s) is repeated. Moreover, we will assume that we chose the iterated process having the shortest such sequence of indices. Then we know that the final index is the first repeated index, namely (r_1, s_1).

Workout 8.2.11 Prove that there is some $k > 1$ that satisfies $(r_k, s_k) = (r_1, s_1)$; that is, the length of the sequence is even. [HINT: Otherwise, can a shorter sequence be found?]

Now let $\epsilon_0 = \min\{r_j, 1 - r_j, s_j, 1 - s_j\}_{j=1}^{n}$. Then for any $0 < \epsilon < \epsilon_0$ define $\mathbf{x}^{+}(\epsilon)$ (resp. $\mathbf{x}^{-}(\epsilon)$) by decreasing (resp. increasing) the value of each x_{r_j, s_j} by ϵ, while increasing (resp. decreasing) the value of each $x_{r_j, s_{j+1}}$ by ϵ.

Workout 8.2.12 Prove that $\mathbf{x}^{+}(\epsilon), \mathbf{x}^{-}(\epsilon) \in P$.

Thus we have shown that the line segment $\overline{\mathbf{x}^{-}(\epsilon)\mathbf{x}^{+}(\epsilon)}$ lies entirely in P and has $\mathbf{x}$ as its center. Therefore, $\mathbf{x}$ is not extreme. Hence, every extreme point of P is integral, and so corresponds to a permutation matrix.

Workout 8.2.13 Finish the proof.

$\diamond$

We will see another proof of the Birkhoff–von Neumann Theorem 8.2.7 in Chapter 11 (Exercises 11.5.10 and 11.5.11), one that actually constructs a solution.

8.3 Pratique de Novo

Workout 8.3.1 *Let $\mathcal{F}$ be an $(n+1)$-intersecting family of convex regions in $\mathbb{R}^n$. In the proof of Helly's Theorem 8.1.5,*

   a. *how many points $\mathbf{p}_I$ are constructed?*

   b. *how many of those points $\mathbf{p}_I$ are in each P_i?*

Workout 8.3.2 *Let $X = \{(1,4,-2), (5,2,8), (-6,0,7), (9,-4,1)\}$ and let $P = \text{vhull}(X)$.*

   a. *Find the equation for each of the four facets of P.*

   b. *Write the system of inequalities that has P as its solution.*

   c. *For each point $\mathbf{x}^* \in X$ find an objective function that is maximized (over P) at $\mathbf{x}^*$, and that is different from the one generated by the the appropriate sum (as in Workout 8.2.1).*

Workout 8.3.3 *Write the following matrix as a convex combination of permutation matrices.*

$$\frac{1}{9}\begin{pmatrix} 6 & 3 & 0 \\ 0 & 4 & 5 \\ 3 & 2 & 4 \end{pmatrix}$$

8.4 Cones

(polyhedral) cone (polycone)

ray

Recall from Exercise 3.5.13 that a **cone** (conic region) C is a region having the property that $r\mathbf{u} + s\mathbf{v} \in C$ for all $\mathbf{u}, \mathbf{v} \in C$ and $r, s \geq 0$. For example, the positive quadrant of $\mathbb{R}^2$ is a cone. For points $\mathbf{u}$ and $\mathbf{v}$, define the **ray** $\overrightarrow{\mathbf{uv}} = \{\mathbf{u} + t(\mathbf{v} - \mathbf{u}) \mid t \geq 0\}$. Of course, every cone C has the property that $\overrightarrow{\mathbf{0v}} \in C$ for all $\mathbf{v} \in C$. A **polyhedral cone** (**polycone**) is the intersection of half-spaces, each of whose boundary contains the origin. Figures 8.3 and 8.4 show a generic cone and a polyhedral cone, respectively, in $\mathbb{R}^3$.

Workout 8.4.1 *Let S and T be two arbitrarily chosen cones.*

   a. *Prove by example that $S \cup T$ is not always conic.*

   b. *Prove that $S \cap T$ is always conic.*

   c. *Use induction to prove that the intersection of an arbitrary number (finite or infinite) of cones is conic.*

   d. *Prove that all polycones are conic.*

Note that a half-space whose boundary contains the origin is defined by an inequality of the form $\sum_{j=0}^{n} a_j x_j \leq 0$. Thus a polycone can be defined as set of $\mathbf{x}$ satisfying some system $\mathbf{Ax} \leq \mathbf{0}$. In this formulation it is easy to see that polycones are conic, since $\mathbf{A}(r\mathbf{u}+s\mathbf{v}) = r\mathbf{Au} + s\mathbf{Av} \leq \mathbf{0}$ whenever $\mathbf{Au} \leq \mathbf{0}$, $\mathbf{Av} \leq \mathbf{0}$ and $r, s \geq 0$.

8.4. Cones

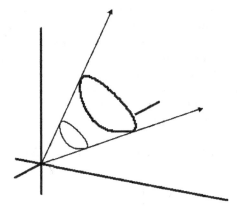

Figure 8.3: A generic cone in $\mathbb{R}^3$

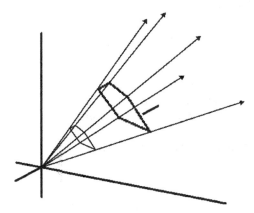

Figure 8.4: A polyhedral cone in $\mathbb{R}^3$

extreme ray of cone

A ray $R = \overrightarrow{\mathbf{0v}}$ is an **extreme ray** of the cone C if no line segment, with both its endpoints in $C - R$, intersects R. Of course, for R to be an extreme ray it is necessary that it lie on the boundary of C. In Figure 8.3 one can see that every boundary ray of the generic cone is extreme, while in Figure 8.4 only the five rays (shown) through the extreme points of the pentagons are extreme for the polycone. Naturally, these rays are on the intersections of the bounding planes. It is interesting to note that some polycones have no extreme rays; a cone defined by a single constraint is one such example. We will assume here that our polycones are not so degenerate — such cones contain a line, and hence have no extreme points (see Exercises 8.5.5, 8.5.11, and 8.5.30 — and leave the more general situation to the reader. In particular, the point $\mathbf{0}$ is extreme.

Workout 8.4.2 *Let C be the polycone defined by $\mathbf{Ax} \leq \mathbf{0}$, where*

$$\mathbf{A} = \begin{pmatrix} 3 & -8 \\ -7 & 4 \\ 2 & -9 \end{pmatrix}.$$

Find the extreme rays of C.

Lemma 8.4.3 *Suppose that the LOP P*

$$\text{Max.} \quad z = \mathbf{c}^\top \mathbf{x}$$

$$\text{s.t.} \quad \mathbf{Ax} \leq \mathbf{0}$$

is unbounded. Then the polycone C defined by $\mathbf{Ax} \leq \mathbf{0}$ has an extreme ray $\overrightarrow{\mathbf{0v}}$ such that $\mathbf{c}^\top \mathbf{v} > 0$.

Workout 8.4.4 *Prove Lemma 8.4.3. [HINT: What does the basis look like when Phase II halts?]*

Notice that the polycone in Figure 8.4 can be described as the conic hull of its extreme rays. The same is true in general; the following theorem for cones is the analogue of Workout 3.5.19 and Exercise 3.5.33 for polytopes.

Minkowski–Weyl Theorem

Theorem 8.4.5 *Every polycone C containing no line is the conic hull of some finite set of rays $\overrightarrow{\mathbf{0v}}$.*

Proof. We begin by rephrasing the statement as follows, using the result of Exercise 3.5.13. For every system $\mathbf{Ax} \leq \mathbf{0}$ there exist points $\mathbf{v}_1, \ldots, \mathbf{v}_m$ such that $\mathbf{Au} \leq \mathbf{0}$ if and only if $\mathbf{u} = \sum_{i=1}^m t_i \mathbf{v}_i$ for some $t_1, \ldots, t_m \geq 0$. We claim that the set of all extreme rays of C has this property.

Workout 8.4.6 *Show that if each $\overrightarrow{\mathbf{0v}}_i$ is an extreme ray of C then any conic combination of the $\mathbf{v}_i$s is in C.*

Now suppose that **u** is any point satisfying $\mathbf{Au} \leq \mathbf{0}$, and suppose that there is no solution to $\mathbf{u} = \sum_{i=1}^{m} t_i \mathbf{v}_i$ with each $t_i \geq 0$. By Farkas's Theorem 7.4.5, there must be some **c** such that $\mathbf{c}^\mathsf{T} \mathbf{v}_i \leq 0$ for all $1 \leq i \leq m$ and $\mathbf{c}^\mathsf{T} \mathbf{u} > 0$. Now consider the following LOP P.

$$\text{Max.} \quad z = \mathbf{c}^\mathsf{T} \mathbf{x}$$

$$\text{s.t.} \quad \mathbf{Ax} \leq \mathbf{0}$$

Workout 8.4.7 *Use Lemma 8.4.3 and the General Fundamental Theorem 6.2.9 to prove that P is optimal at $\mathbf{0}$.*

Now the fact that **u** is feasible and $\mathbf{c}^\mathsf{T} \mathbf{u} > 0$ implies that $z^* > 0$, contradicting the result of Workout 8.4.7. This contradiction proves that $\mathbf{u} \in \text{nspan}(\{\mathbf{v}_1, \ldots, \mathbf{v}_m\}) = \text{nhull}(\{\mathbf{v}_1, \ldots, \mathbf{v}_m\})$. ◊

The Minkowski–Weyl Theorem can be combined with Workout 3.5.19 and Exercise 3.5.33 to prove that every polyhedron is the Minkowski sum of a polytope and a polycone — see Exercises 8.5.19, 8.5.20 and 8.5.36.

It is worthwhile to note that this result is an example of a theme common to many areas of mathematics, that a structure (such as the set of solutions to a linear differential equation or recurrence relation) can be decomposed into its homogeneous (polycone) and particular (polytope) parts.

8.5 Exercises

Practice

8.5.1 Let $\mathbf{x}^1 = (-20, 2, 15)^\mathsf{T}$, $\mathbf{x}^2 = (9, -6, 23)^\mathsf{T}$, $\mathbf{x}^3 = (-4, 9, 3)^\mathsf{T}$, $\mathbf{x}^4 = (15, 11, 18)^\mathsf{T}$, $\mathbf{x}^5 = (-8, 16, -9)^\mathsf{T}$, $\mathbf{x}^6 = (1, -5, 8)^\mathsf{T}$, $\mathbf{x}^7 = (10, 3, -1)^\mathsf{T}$, $\mathbf{x}^8 = (-6, -4, -3)^\mathsf{T}$, $\mathbf{x}^9 = (24, -4, -13)^\mathsf{T}$, and $\mathbf{x}^{10} = (-7, -12, -17)^\mathsf{T}$, and define $X = \{\mathbf{x}^1, \ldots, \mathbf{x}^{10}\}$.

  a. Prove that $\mathbf{x}^3$ is not an extreme point of $\text{vhull}(X)$.

  b. Prove that $\mathbf{x}^4$ is an extreme point of $\text{vhull}(X)$.

  c. Find all extreme points of $\text{vhull}(X)$.

8.5.2 Let P be a polytope in $\mathbb{R}^3$ with k extreme points (and that is fully 3-dimensional). At most how many facets does P have?

8.5.3 Let P be a polytope in $\mathbb{R}^4$ with k extreme points (and that is full dimensional). At most how many facets does P have?

8.5.4 Let P be the polyhedron defined as the region of solutions of the following system.

$$\begin{array}{rcrcrcrcr}
-2x_1 & + & 5x_2 & + & 7x_3 & & & \leq & 782 \\
 & & 8x_2 & - & 4x_3 & + & 9x_4 & \leq & 829 \\
3x_1 & & & + & 6x_3 & - & 1x_4 & \leq & 765 \\
1x_1 & - & 7x_2 & & & - & 8x_4 & \leq & -94 \\
 & & 4x_2 & - & 3x_3 & - & 5x_4 & \leq & -28 \\
-6x_1 & - & 9x_2 & + & 2x_3 & & & \leq & -87
\end{array}$$

$$x_1 \;,\; x_2 \;,\; x_3 \;,\; x_4 \;\geq\; 0$$

a. Find all extreme points of P. *[HINT: Only 12 of the 210 bases are feasible. Using one of your algorithms from Exercise 2.10.46 would help here.]*

b. For each extreme point $\mathbf{x}$ of P write an objective function that is maximized at $\mathbf{x}$.

8.5.5 Let $P \subseteq \mathbb{R}^2$ be a polyhedron. Prove that P has no extreme points if and only if it contains a line. *[HINT: Recall Workout 2.1.2.]*

8.5.6 Let $X = \{\mathbf{x}_1, \ldots, \mathbf{x}_m\} \subset \mathbb{R}^2$ be a set of points in convex position, and let $P = \mathrm{vhull}(X)$. Let $\Pi = \{P_1, \ldots, P_k\}$ be a partition of P into X-triangles. Find k in terms of m.

8.5.7 Let $X = \{\mathbf{x}_1, \ldots, \mathbf{x}_5\} \subset \mathbb{R}^2$ be a set of 5 points in convex position, and let $P = \mathrm{vhull}(X)$. Compute the number of partitions of P into X-triangles.

8.5.8 Let $X = \{\mathbf{x}_1, \ldots, \mathbf{x}_6\} \subset \mathbb{R}^2$ be a set of 6 points in convex position, and let $P = \mathrm{vhull}(X)$. Compute the number of partitions of P into X-triangles.

8.5.9 Let $X = \{\mathbf{x}_1, \ldots, \mathbf{x}_7\} \subset \mathbb{R}^2$ be a set of 7 points in convex position, and let $P = \mathrm{vhull}(X)$. Compute the number of partitions of P into X-triangles.

8.5.10 Let P be the polyhedron defined as the region of solutions of the system S below.

$$\begin{array}{rcrcrcr}
-18x_1 & - & 8x_2 & + & 11x_3 & \leq & 9 \\
4x_1 & - & 20x_2 & + & 3x_3 & \leq & 47 \\
33x_1 & - & 26x_2 & - & 10x_3 & \leq & 214 \\
-19x_1 & + & 54x_2 & - & 4x_3 & \leq & 150
\end{array}$$

Let u be a nonzero solution to the system T below.

$$\begin{array}{rcrcrcr}
-18x_1 & - & 8x_2 & + & 11x_3 & \leq & 0 \\
4x_1 & - & 20x_2 & + & 3x_3 & \leq & 0 \\
33x_1 & - & 26x_2 & - & 10x_3 & \leq & 0 \\
-19x_1 & + & 54x_2 & - & 4x_3 & \leq & 0
\end{array}$$

Use u to find a ray in P.

8.5. Exercises

8.5.11 Let $P \subseteq \mathbb{R}^3$ be a polyhedron. Prove that P has no extreme points if and only if it contains a line. [HINT: See Exercises 8.5.5 and 8.5.10.]

8.5.12 Prove that the product of DS matrices is DS.

8.5.13 Write the following matrix as a convex combination of permutation matrices.
$$\frac{1}{12}\begin{pmatrix} 5 & 3 & 4 \\ 4 & 1 & 7 \\ 3 & 8 & 1 \end{pmatrix}$$

8.5.14 Find all solutions to Exercise 8.5.13. [HINT: Use WebSim.]

8.5.15 Show that every cone is convex.

8.5.16 Let C be a nontrivial (contains a nonzero point) cone and define $C_1 = \{\mathbf{x} \in C \mid |\mathbf{x}| = 1\}$. Prove that $C = \cup_{\mathbf{x} \in C_1} \overrightarrow{\mathbf{0x}}$.

8.5.17 Let C be the polycone defined by $\mathbf{Ax} \le 0$, where
$$\mathbf{A} = \begin{pmatrix} -14 & -11 & 18 \\ -13 & 16 & -19 \\ 15 & -12 & -17 \\ 5 & 4 & -21 \end{pmatrix}.$$

Find the extreme rays of C.

8.5.18 Find all the extreme rays in the polyhedra defined by the system of constraints in

    a. Exercise 1.5.1(c).

    b. Problem 2.5.1.

    c. Problem 2.7.1.

    d. Problem 2.7.3.

    e. Exercise 4.5.14(a).

    f. Problem 6.4.8.

8.5.19 Define a ray $R = \overrightarrow{\mathbf{uv}}$ to be an **extreme ray** of a polyhedron P if $\mathbf{u}$ is an extreme point of P and no line segment, with both its endpoints in $P - R$, intersects R. The vector $\mathbf{w} = \mathbf{v} - \mathbf{u}$ is called an **extreme direction** of P. Find all extreme points and extreme directions (up to scalar multiples) of the polyhedron defined by the following system.

extreme ray of polyhedron

extreme direction

$$\begin{array}{rcrcr} -2x_1 & + & x_2 & \le & 11 \\ x_1 & + & 6x_2 & \ge & 60 \\ x_1 & - & 4x_2 & \le & 30 \\ -4x_1 & + & 5x_2 & \le & 105 \\ 3x_1 & + & 2x_2 & \ge & 36 \\ x_1 & , & x_2 & \ge & 0 \end{array}$$

8.5.20 *Find all extreme points and extreme directions (see Exercise 8.5.19) of the polyhedron defined by the following system.*

$$\begin{aligned} -14x_1 &- 11x_2 + 18x_3 \leq 92 \\ 13x_1 &- 16x_2 + 19x_2 \geq 127 \\ 15x_1 &- 12x_2 - 17x_3 \leq 115 \\ 5x_1 &+ 4x_2 - 21x_3 \leq 108 \end{aligned}$$

$$x_1, \quad x_2, \quad x_3 \geq 0$$

8.5.21 *Let Δ^n be the set of stochastic n-vectors, and denote by ∇ some partition of it into simplices, such as in the diagram below for $n = 3$. The vertices of Δ^n are denoted by $X = \{\mathbf{x}_1, \ldots, \mathbf{x}_n\}$, and the vertices of ∇, which include the $\mathbf{x}_i$s, are denoted by V.*

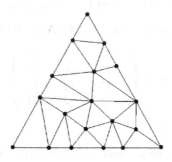

Sperner labeling

*A **Sperner labeling** of ∇ is a function $\psi : V \to \{1, \ldots, n\} = N$ so that, for every $S \subseteq N$, $\psi(\mathbf{v}) \in S$ whenever $\mathbf{v} \in \text{vhull}(X_S)$, where $X_S = \{\mathbf{x}^i \mid i \in S\}$. In particular, $\psi(\mathbf{x}^i) = i$. The figure below shows an example of a Sperner labeling on Δ^3.*

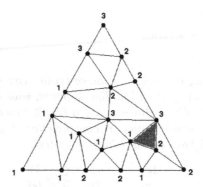

Sperner's Lemma

A simplex $\Delta \in \nabla$ is full if its vertices are labeled distinctly, using all the numbers in N, such as the shaded triangle, above. **Sperner's Lemma** *states that every Sperner labeling of Δ^n contains a full simplex. Prove this for $n = 2$.*

8.5.22 *Prove Sperner's Lemma for $n = 3$. [HINT: Use Exercise 8.5.21 and meditate on the following picture.]*

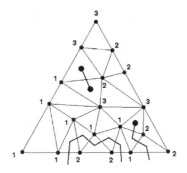

Challenges

8.5.23 *Let P be a polytope in $\mathbb{R}^n$ with k extreme points (and that is full dimensional). At most how many facets does P have? [HINT: See Exercises 8.5.2 and 8.5.3.]*

8.5.24 *Let $\mathcal{F}$ be an $(n+1)$-intersecting family of convex regions in $\mathbb{R}^n$. In the proof of Helly's Theorem 8.1.5, how many facets does each P_i have? [HINT: See Workout 8.3.1 and Exercise 8.5.23.]*

8.5.25 *Consider the following LOP.*

$$
\begin{aligned}
\text{Max.} \quad z = \quad & 64x_1 + 66x_2 + 73x_3 \\
\text{s.t.} \quad & 35x_1 + 38x_2 + 11x_3 \leq 565 \\
& 12x_1 + 36x_2 + 39x_3 \leq 520 \\
& 37x_1 + 13x_2 + 34x_3 \leq 600 \\
\& \quad & x_1, \ x_2, \ x_3 \geq 0
\end{aligned}
$$

  a. *Find a conic combination of the constraints (converted to equalities) that is parallel to the objective function.*

  b. *Find one that also contains $\mathbf{x}^*$.*

  c. *What does the set of all conic combinations of constraints (converted to equalities) that pass through $\mathbf{x}^*$ look like?*

  d. *Notice that x^* is not integral. For each value of b_4 below, add $64x_1 + 66x_2 + 73x_3 \leq b_4$ to the constraints and re-solve.*

  (i) $b_4 = 1271$
  (ii) $b_4 = 1275$
  (iii) $b_4 = 1280$

  e. *What is the optimum if $\mathbf{x}$ must be integral?*

8.5.26 Write the following matrix as a convex combination of permutation matrices.
$$\frac{1}{10}\begin{pmatrix} 4 & 1 & 3 & 2 \\ 4 & 2 & 2 & 2 \\ 2 & 4 & 2 & 2 \\ 0 & 3 & 3 & 4 \end{pmatrix}$$

8.5.27 Consider the matrix A, below.
$$\begin{pmatrix} 2 & -3 & 0 & 1 \\ -1 & 4 & -1 & -1 \\ 0 & -2 & 2 & -1 \end{pmatrix}$$

Let $\mathbf{v}_i$ be the ith column of $\mathbf{A}$. Find the maximum value x_0 such that some $\mathbf{v} \in \text{vhull}\{\mathbf{v}_1, \ldots, \mathbf{v}_4\}$ satisfies $\mathbf{v} \geq x_0 \mathbf{J}_3$.

8.5.28 Let $X = \{\mathbf{x}^1, \ldots, \mathbf{x}^m\} \subset \mathbb{R}^3$ be a set of points in convex position, and let $P = \text{vhull}(X)$. Let $\Pi = \{P_1, \ldots, P_k\}$ be a partition of P into X-tetrahedron. Prove that k is not necessarily unique. [HINT: Consider the 3-dimensional cube.]

8.5.29 Let P be the polyhedron defined as the region of solutions to the system $\sum_{j=1}^n a_{i,j} x_j \leq b_i$ ($1 \leq i \leq m$). Suppose P contains a line. Show that there is a nonzero solution to the system $\sum_{j=1}^n a_{i,j} x_j = 0$ ($1 \leq i \leq m$). Prove the converse as well.

8.5.30 Let P be a polyhedron in $\mathbb{R}^n$. Use Exercise 8.5.29 to prove that P has no extreme point if and only if it contains a line. [HINT: See also Exercise 1.5.16.]

8.5.31 Let S be the feasible region of a LOP and suppose that S has an extreme point. Use Exercise 8.5.30 to prove that if the LOP is optimal then one of the extreme points of S is optimal. [HINT: See Exercise 2.1.4.]

8.5.32 Let R be a ray of a polycone C. Prove that R is an extreme ray of C if and only if the region $C - R$ is convex.

8.5.33 State and prove a conic analog of Carathéodory's Theorem 3.4.1. (This is a repeat of Exercise 3.5.36.)

8.5.34 Consider the system of Exercise 8.5.19. Find m, k, and vectors $\mathbf{v}_1, \ldots, \mathbf{v}_m$ and $\mathbf{w}_1, \ldots, \mathbf{w}_k$ so that $\mathbf{A}\mathbf{u} \leq \mathbf{b}$ if and only if $\mathbf{u} = \sum_{i=1}^m r_i \mathbf{v}_i + \sum_{j=1}^k s_j \mathbf{w}_j$ with every $r_1, \ldots, r_m, s_1, \ldots, s_k \geq 0$ and $\sum_{j=1}^k s_j = 1$.

8.5.35 Consider the system of Exercise 8.5.20. Find m, k, and vectors $\mathbf{v}_1, \ldots, \mathbf{v}_m$ and $\mathbf{w}_1, \ldots, \mathbf{w}_k$ so that $\mathbf{A}\mathbf{u} \leq \mathbf{b}$ if and only if $\mathbf{u} = \sum_{i=1}^m r_i \mathbf{v}_i + \sum_{j=1}^k s_j \mathbf{w}_j$ with every $r_1, \ldots, r_m, s_1, \ldots, s_k \geq 0$ and $\sum_{j=1}^k s_j = 1$.

8.5.36 *Mimic the proof of the Minkowski–Weyl Theorem 8.4.5 to prove that, for every system $\mathbf{Ax} \leq \mathbf{b}$ there exist vectors $\mathbf{v}_1, \ldots, \mathbf{v}_m$ and $\mathbf{w}_1, \ldots, \mathbf{w}_k$ so that $\mathbf{Au} \leq \mathbf{b}$ if and only if $\mathbf{u} = \sum_{i=1}^{m} r_i \mathbf{v}_i + \sum_{j=1}^{k} s_j \mathbf{w}_j$ with every $r_1, \ldots, r_m, s_1, \ldots, s_k \geq 0$ and $\sum_{j=1}^{k} s_j = 1$. [HINT: See Exercises 8.5.34 and 8.5.35.]*

Finite Basis Theorem

8.5.37 *Prove Sperner's Lemma for all n.*

Projects

8.5.38 *Present the **Erdős–Ko–Rado** and **Hilton–Milnor** theorems.*

Erdős–Ko–Rado Theorem

8.5.39 *Present the relationship between the **Catalan numbers** and the number of ways to partition $\text{vhull}(X)$ into X-triangles.*

Hilton–Milnor Theorem

8.5.40 *Present the characterization of **f-vector** of convex polyhedra in $\mathbb{R}^n$.*

Catalan number

8.5.41 *Present the **Hirsch conjecture**.*

8.5.42 *Present the complexity of finding the partition of a convex polytope in $\mathbb{R}^n$ into the minimum number of simplices.*

f-vector

Hirsch conjecture

8.5.43 *Present the **Hyperplane Separation Theorem**.*

Hyperplane Separation Theorem

Chapter 9

Game Theory

9.1 Matrix Games

Which (if any) of the players in the following game holds the advantage?

Game 9.1.1 *Each of two players, Ken and Barbie, holds a large cache of dimes and quarters. At the same time, each shows the other one of their coins. Ken wins both coins if they match; if they differ Barbie wins them. This is repeated until one of them wins $100.* ⋄

Two things are useful when analyzing Game 9.1.1. First, it is handy to write a chart describing all possible plays and their outcomes. Consider a matrix whose columns are indexed by Ken's choices, whose rows are indexed by Barbie's choices, and whose entries are the payoffs from Barbie to Ken or the given plays by each player. We call this the **payoff matrix** for the game (with respect to Ken).[1] Note that if $\mathbf{A}$ is the payoff matrix with respect to Ken, then $-\mathbf{A}^\mathsf{T}$ is the payoff matrix with respect to Barbie.

payoff matrix

Workout 9.1.2 *Write the payoff matrix $\mathbf{A}$ for Game 9.1.1, with respect to Ken.*

Second, it is important to ignore games of few rounds, where anything might happen, and think instead about long games, where consistent strategy yields long-term averages via Bernoulli's Weak Law of Large Numbers, below. This average we call the **expectation**, or **expected value** of a function f on a random variable X, defined by $\mathbf{Ex}[X] = \sum_X f(X)\mathbf{Pr}[X]$, where $\mathbf{Pr}[X]$ is the probability of the event X.

expectation/ expected value

Theorem 9.1.3 *Let X_1, X_2, ... be independent, identically distributed random variables with expectation $\mathbf{Ex}[X_i] = \mu$, and define $S_n = \frac{1}{n}\sum_{i=1}^n X_i$. Then $\mathbf{Ex}[S_n] = \mu$ and, for all $\epsilon > 0$, $\mathbf{Pr}[|S_n - \mu| < \epsilon] \to 1$ as $n \to \infty$.* ⋄

Weak Law of Large Numbers

---

[1] Note that our presentation differs from most presentations that transpose (or, equivalently, negate) the payoff matrix in order to state winnings with respect to the row player. But we have made this choice in order to remain consistent with notations such as $\mathbf{y}^\mathsf{T}\mathbf{Ax}$.

G. H. Hurlbert, *Linear Optimization*, Undergraduate Texts in Mathematics,
DOI: 10.1007/978-0-387-79148-7_9, © Springer Science+Business Media LLC 2010

Workout 9.1.4 *Suppose that they played 100 rounds and that, in those rounds, Ken played 30 dimes and 70 quarters, while Barbie played 55 dimes and 45 quarters.*

  a. *What is the most that Ken could have won in those rounds?*

  b. *What is the least that Ken could have won in those rounds?*

  c. *What is the average, over all possible orders of play, that Ken could have won in those rounds?*

Workout 9.1.5

  a. *Suppose that Ken plays his coins randomly according to certain percentages* $\mathbf{x} = (x_1, x_2)$. *Find the percentage of dimes he must play in order to break even in the long run, regardless of what Barbie plays.*

  b. *Suppose that Barbie plays her coins randomly according to certain percentages* $\mathbf{y} = (y_1, y_2)$. *Find the percentage of dimes she must play in order to break even in the long run, regardless of what Ken plays.*

We've seen that Ken has a long-term strategy $\mathbf{x}^*$ that guarantees (on average) he won't lose money; that is, $\mathbf{y}^T \mathbf{A} \mathbf{x}^* \geq 0$ for all stochastic $\mathbf{y}$. We've also seen that Barbie has a long-term strategy $\mathbf{y}^*$ that guarantees (on average) she won't lose money; that is, $\mathbf{y}^{*T} \mathbf{A} \mathbf{x} \leq 0$ for all stochastic $\mathbf{x}$. Therefore if both players play their strategies then we should expect no payoff in the long run; indeed $\mathbf{y}^{*T} \mathbf{A} \mathbf{x}^* = 0$. Of course, there may be some payoff one way or the other in an actual playing of the game, but the Weak Law of Large Numbers 9.1.3 says that the expected payoff per round $\mathbf{Ex}[V(k)/k]$ will tend to an average of 0 when the number k of rounds tends to infinity, supposing both players are playing optimally. This limiting value **game value** $V(\mathbf{A}) = \lim_{k \to \infty} \mathbf{Ex}[V(k)/k]$ is called the **value** of the game (having payoff matrix $\mathbf{A}$). Of course, $V(\mathbf{A})$ is not always zero for every $\mathbf{A}$ — consider a game in which Barbie pays Ken in every possible situation — but games **fair/** with value zero are called **fair**. Some games are obviously fair because they **symmetric** are **symmetric**: the game doesn't change if Ken and Barbie reverse roles. **game** In such a game the players have the same set of strategies and, whatever the payoff for the play (K,B), the payoff for the play (B,K) is the opposite. **anti-** That is, the payoff matrix is **anti-symmetric**: $\mathbf{A}^T = -\mathbf{A}$. For example, **symmetric** Game 9.1.6, below, is symmetric and hence fair. (You'll be asked to verify **matrix** this in Workout 9.3.1.)

Game 9.1.6 *Let the following be the payoff matrix for a game.*

$$\begin{pmatrix} 0 & 3 & -5 & -2 \\ -3 & 0 & 4 & 0 \\ 5 & -4 & 0 & 6 \\ 2 & 0 & -6 & 0 \end{pmatrix}$$

It is interesting to ponder the practical considerations of playing a fair game. For example, Game 9.1.1 is supposed to continue until someone wins $100. Does that mean the game will never end? See Exercises 9.5.35, 9.5.36 and 9.5.37.

9.2 Minimax Theorem

Notice in Game 9.1.1 that $\mathbf{A}\mathbf{x}^* = (0,0)^\mathsf{T}$, so that $\mathbf{y}^\mathsf{T}\mathbf{A}\mathbf{x}^* = 0$ for all stochastic $\mathbf{y}$. This means that Ken can reveal his randomized strategy to Barbie and there is no way for her to use the knowledge to her advantage. Likewise, $\mathbf{y}^{*\mathsf{T}}\mathbf{A} = (0,0)$, and so $\mathbf{y}^{*\mathsf{T}}\mathbf{A}\mathbf{x} = 0$ for all stochastic $\mathbf{x}$, so she can also safely reveal her strategy to him. Here is a game that is slightly more general.

Game 9.2.1 *Let the following be the payoff matrix, with respect to Ken, for a game against Barbie.*

$$\begin{pmatrix} -1 & 1 & 3 & -1 & 0 \\ -1 & 0 & 1 & 0 & -1 \\ 0 & -2 & -1 & -2 & 3 \\ 2 & 3 & 0 & 2 & -5 \end{pmatrix}$$

In the case of Game 9.2.1, suppose that Barbie knows that Ken's strategy is to play his columns in the proportion $\mathbf{x}^* = (12, 0, 16, 0, 5)^\mathsf{T}/33$. Because $\mathbf{A}\mathbf{x}^* = (36, -1, -1, -1)^\mathsf{T}/33$, Barbie would be unwise to ever risk a big pay-out by playing her first row. However, she could play her remaining rows without caution since they share a common expected value. For example she could play solely row 3. Because any stochastic combination of these values is at least $-1/33$, Ken's strategy guarantees himself at least that much of a win per round, on average.

Similarly, if Ken knows that Barbie will play her rows in the proportion $\mathbf{y}^* = (0, 13, 14, 6)^\mathsf{T}/33$, then since $\mathbf{y}^{*\mathsf{T}}\mathbf{A} = (-1, -10, -1, -16, -1)/33$, Ken would be wise to avoid his second and fourth columns, playing only his columns 1, 3 and 5 in any stochastic manner. For example he could play row 5 only. Barbie's strategy guarantees her losses to be at most $-1/33$ per round, on average. Hence the value of Game 9.2.1 is exactly $-1/33$.

Let's move on to discuss how one might find such strategies $\mathbf{x}^*$ and $\mathbf{y}^*$. We say that a strategy $\mathbf{x}$ (or $\mathbf{y}$) is **pure** if one of its entries is a 1 and the rest are 0; it is **mixed** otherwise. The key observation is the following lemma.

pure/mixed strategy

Lemma 9.2.2 *Given any strategy $\mathbf{x}$ there exists a pure strategy $\mathbf{y}_\mathbf{x}$ that minimizes the value of $\mathbf{y}^\mathsf{T}\mathbf{A}\mathbf{x}$ over all stochastic $\mathbf{y}$.*

Proof. As we have seen above, the notion is to pick the smallest value out of $\mathbf{b} = \mathbf{A}\mathbf{x}$. Formally, let the vector $\mathbf{u}_i$ be the pure strategy with its 1 in position i. Then $\mathbf{u}_i^\mathsf{T}\mathbf{A}\mathbf{x} = b_i$. Now define $b_\mathbf{x} = \min_i b_i$.

Workout 9.2.3 *Prove that every stochastic vector* $\mathbf{y}$ *satisfies* $\mathbf{y}^T\mathbf{A}\mathbf{x} \geq b_\mathbf{x}$.

Thus the pure response is given by $\mathbf{y}_\mathbf{x} = \mathbf{u}_{i^*}$, where $i^* = \operatorname{argmin}\{b_i\}_{i=1}^m$.
◇

The partner to Lemma 9.2.2 the following lemma, which we state without proof.

Lemma 9.2.4 *Given any strategy* $\mathbf{y}$ *there exists a pure strategy* $\mathbf{x}_\mathbf{y}$ *that maximizes the value of* $\mathbf{y}^T\mathbf{A}\mathbf{x}$ *over all stochastic* $\mathbf{x}$. ◇

In light of Lemma 9.2.4, Ken will consider all of his possible strategies $\mathbf{x}$ and find each resulting $b_\mathbf{x}$. His best option, then, is to choose that $\mathbf{x}$ for which $b_\mathbf{x}$ is maximum. If we delve into the proof of Lemma 9.2.3 again, we see that there is another way to define $b_\mathbf{x}$. For a finite set S recall that its minimum $\min(S)$ equals its greatest lower bound $\operatorname{glb}(S)$. In the case of $S = \{b_i\}_{i=1}^m$ we have

$$b_\mathbf{x} = \min(S) = \operatorname{glb}(S) = \max\{x_0 \mid x_0 \leq b_i, i=1,\ldots,m\}. \quad (9.1)$$

We can rephrase even this. Say that $\mathbf{A}$ is $m \times n$. Then Equation (9.1) becomes

$$b_\mathbf{x} = \max\{x_0 \mid x_0\mathbf{J}_m \leq \mathbf{A}\mathbf{x}\}. \quad (9.2)$$

Now, since Ken seeks $\max b_\mathbf{x}$ over all stochastic $\mathbf{x}$, we derive from Equation (9.2) his **GLOP** (game LOP)

GLOP
$$\max\{x_0 \mid \mathbf{J}_n\mathbf{x} = 1,\ x_0\mathbf{J}_m - \mathbf{A}\mathbf{x} \leq 0,\ \mathbf{x} \geq 0\}. \quad (9.3)$$

For example, applying this approach to Game 9.1.1 yields the following GLOP for Ken.

Problem 9.2.5

$$\begin{aligned}
\text{Max.} \quad z &= x_0 \\
\text{s.t.} \quad & x_1 + x_2 = 1 \\
& x_0 - 10x_1 + 25x_2 \leq 0 \\
& x_0 + 10x_1 - 25x_2 \leq 0 \\
\& \quad & x_1,\ x_2 \geq 0
\end{aligned}$$

Principle of Indifference

This approach is based on what is often called the **Principle of Indifference**, meaning that, as noted in Workout 9.1.5, Ken has found a strategy $\mathbf{x}$ and value x_0 so that Barbie is indifferent regarding her response. That is, it doesn't matter how she plays, the result will be a payoff of x_0 to Ken. (Of course, she pulls of the same feat, making Ken indifferent as well.) The one caveat, as described in the paragraph discussing Game 9.2.1, is that coordinates of $\mathbf{A}\mathbf{x}$ with value higher than x_0 are never played in Barbie's optimal strategy, so she is only indifferent among her remaining options. In essence, this is really just a rephrasing of the Complementary Slackness Theorem 4.2.6.

9.2. Minimax Theorem

Workout 9.2.6

a. *Write Ken's GLOP for Game 9.2.1.*

b. *Use part a to find the value of Game 9.2.1 and find Ken's optimal strategy.*

Workout 9.2.7 *Use Lemma 9.2.4 to show that Barbie's GLOP is*

$$\min\{y_0 \mid \mathbf{J}_m \mathbf{y} = 1, \; y_0 \mathbf{J}_n - \mathbf{A}^\mathsf{T} \mathbf{y} \geq \mathbf{0}, \; \mathbf{y} \geq \mathbf{0}\} \; . \tag{9.4}$$

Workout 9.2.8

a. *Write Barbie's GLOP for Game 9.2.1.*

b. *Use part a to find the value of Game 9.2.1 and find Barbie's's optimal strategy.*

Workout 9.2.9 *Let $\mathbf{A} = (a_{i,j})$ be an $m \times n$ payoff matrix, with respect to Ken, for a game against Barbie.*

a. *Write Ken's GLOP for this game using summation notation.*

b. *Write Barbie's GLOP for this game using summation notation.*

c. *Prove that these GLOPs are dual to each other.*

d. *Use the General Fundamental Theorem 6.2.9 to prove that both GLOPs are optimal.*

Now we can etch in stone the realization that each player can give away their strategy without repurcussion.

Theorem 9.2.10 *For every $m \times n$ matrix $\mathbf{A}$ there are stochastic $\mathbf{x}^*$ and $\mathbf{y}^*$ such that $\min_\mathbf{y} \mathbf{y}^\mathsf{T} \mathbf{A} \mathbf{x}^* = \max_\mathbf{x} \mathbf{y}^{*\mathsf{T}} \mathbf{A} \mathbf{x}$, where min and max are taken over all stochastic m- and n-vectors, respectively.* Minimax Theorem

Proof. We claim that x^* and y^* are the optimal strategies for Ken and Barbie, respectively, in the game defined by payoff matrix $\mathbf{A}$. By Equation (9.3) we have $\mathbf{A}\mathbf{x}^* \geq z^* \mathbf{J}_m$ (with equality in at least one coordinate by the maximization of z^*), so that $\min_\mathbf{y} \mathbf{y}^\mathsf{T} \mathbf{A} \mathbf{x}^* = z^*$, the optimal value of Ken's GLOP. Similarly, Equation (9.4) implies $\max_\mathbf{x} \mathbf{y}^{*\mathsf{T}} \mathbf{A} \mathbf{x} = w^*$, the optimal value of Barbie's GLOP. By Workout 9.2.9c and the General Duality Theorem 6.4.1, $z^* = w^*$. ◇

Another formulation of this result is that $\max_\mathbf{x} \min_\mathbf{y} \mathbf{y}^\mathsf{T} \mathbf{A} \mathbf{x} = \min_\mathbf{y} \max_\mathbf{x} \mathbf{y}^\mathsf{T} \mathbf{A} \mathbf{x}$, which can also be proven in similar fashion with the ideas presented here; see Exercise 9.5.43.

Let's reconsider Game 9.2.1 and it optimal strategies. In hindsight, it makes perfect sense that $x_4^* = 0$. Compare column 4 to column 2 — every entry that differs is worse for Ken in column 4. Thus, if Ken plays an optimal strategy that includes playing column 4, the strategy derived from it by replacing every such play by column 2 can only improve his payoffs, and thus is also optimal. We say that the jth column $\mathbf{A}_{.,j}$ is **dominated** by the kth column $\mathbf{A}_{.,k}$ if $a_{i,j} \leq a_{i,k}$ for every $1 \leq i \leq m$.

dominated column

Workout 9.2.11 *Let column k dominate column j in payoff matrix $\mathbf{A}$, and let $\mathbf{A}'$ be the matrix $\mathbf{A}$ with its jth column removed. Compare the duals of their respective GLOPs to conclude that the value of the game has been preserved.*

dominated row

Likewise, we say that the ith row $\mathbf{A}_{i,.}$ is **dominated** by the kth row $\mathbf{A}_{k,.}$ if $a_{i,j} \geq a_{k,j}$ for every $1 \leq j \leq n$. Here the inequality is reversed because it is Barbie who is trying to avoid paying Ken more than is necessary.

Workout 9.2.12 *Let row k dominate row i in payoff matrix $\mathbf{A}$, and let $\mathbf{A}'$ be the matrix $\mathbf{A}$ with its ith row removed. Compare their respective GLOPs to conclude that the value of the game has been preserved.*

Thus the payoff matrix $\mathbf{A}$ for Game 9.2.1 can be reduced to the equivalent payoff matrix $\mathbf{A}'$, below, by removing the dominated column 4.

$$\begin{pmatrix} -1 & 1 & 3 & 0 \\ -1 & 0 & 1 & -1 \\ 0 & -2 & -1 & 3 \\ 2 & 3 & 0 & -5 \end{pmatrix}$$

Moreover, $\mathbf{A}'$ can further be reduced to $\mathbf{A}''$, below, by removing the dominated row 1.

$$\begin{pmatrix} -1 & 0 & 1 & -1 \\ 0 & -2 & -1 & 3 \\ 2 & 3 & 0 & -5 \end{pmatrix}$$

Interestingly, row 2 did not dominate row 1 until column 4 was deleted. While Workout 9.2.12 ensures that this operation preserves the value, it is worth putting this near paradox in context informally. In this case, Barbie realizes that an optimal Ken strategy will avoid column 4, and so the initial nondomination of row 1 by row 2 is somewhat of a decoy or mirage. One can imagine players alternating deleting dominated rows and columns several more times, if necessary.

9.3 Bitte Praxis

Workout 9.3.1 *Use LO to verify that Game 9.1.6 has value 0.*

Game 9.3.2 *Let the following payoff matrix define a game.*

$$\begin{pmatrix} 1 & -1 & 4 & 0 \\ 4 & -5 & -3 & 2 \\ -2 & 1 & 1 & -2 \\ 6 & -3 & 0 & 5 \end{pmatrix}$$

9.4. Saddles

Workout 9.3.3 *Reduce Game 9.3.2 to an equivalent 2×2 game by deleting dominated rows and columns. Then find the value of the game and both optimal strategies.*

$$\begin{pmatrix} 1 & -1 & 4 & 0 \\ 4 & -5 & -3 & 2 \\ -2 & 1 & 1 & -2 \\ 6 & -3 & 0 & 5 \end{pmatrix}$$

Game 9.3.4 *Define a game by the payoff matrix $\mathbf{A}$ below.*

$$\begin{pmatrix} 1 & -1 & 0 \\ 2 & 0 & -2 \\ 0 & 3 & -1 \\ -3 & -2 & 4 \end{pmatrix}$$

Workout 9.3.5 *Consider Game 9.3.4.*

a. *Find the value of the game and the optimal strategies.*

b. *Compute Row Barbie's best response to Column Ken's strategy $\mathbf{x} = (15, 4, 7)^\mathsf{T}/26$.*

c. *Compute Ken's best response to Barbie's strategy $\mathbf{x} = (5, 12, 19, 1)^\mathsf{T}/37$.*

d. *Suppose that Ken and Barbie do not know how to compute their optimal mixed strategies using LO, as in part a, but that they do know how to compute their best responses to the other's given strategy, as in parts b and c. What happens if they each adopt the following dynamic strategy (called the **Best Response Algorithm**): play their best response to the strategy given by their opponents plays so far? That is, if Ken has played 15 column 1s, 4 column 2s, and 7 column 3s (in any order) in his first 26 rounds, then Barbie will play row 4 on her 27th round. [HINT: Write MAPLE code to play a thousand or million rounds, with an arbitrary first round.]*

Best Response Algorithm

9.4 Saddles

Game 9.4.1 *Define a game by the payoff matrix $\mathbf{A}$ below.*

$$\begin{pmatrix} 5 & -4 & 0 & -7 & 4 \\ -6 & -2 & -2 & -3 & -1 \\ -2 & 1 & 4 & -7 & 3 \\ 3 & 6 & -5 & 5 & 0 \end{pmatrix}$$

Workout 9.4.2 *Find the value of Game 9.4.1, along with the optimal strategies.*

○ **Game 9.4.3** *Panch and Jody play the following game on a matrix* **A**. *Panch chooses a column and Jody chooses a row, simultaneously, determining a payoff entry in* **A**. *Then, with either player going first, players take turns changing their decision in order to improve their payoff. They play until one of them whacks the other.*

Workout 9.4.4 *Play Game 9.4.3 with the matrices from the games below, and describe the behavior under the assumption that no one gets whacked for a long time.*

a. *Game 9.3.4.*

b. *Game 9.4.1.*

○ **Game 9.4.5** *Boris (column player) and Natasha (row player) play the following game on a matrix* **A**. *One player chooses a column (or row) and then, after seeing the choice, the other player chooses a row (or column), determining a payoff entry in* **A**. *They play just once.*

Workout 9.4.6 *Play Game 9.4.5 with the matrices from the games below, and describe the results when Boris goes first versus when Natasha goes first.*

a. *Game 9.3.4.*

b. *Game 9.4.1.*

saddle point   Given a matrix **A**, the entry $a_{i,j}$ is called a **saddle point** if $a_{i,j} \geq a_{i,k}$ for all $k \neq j$ and if $a_{i,j} \leq a_{k,j}$ for all $k \neq i$.[2] In other words, it is the best choice in row i for the column player, and it is the best choice in column j for the row player. Put still another way, it is the column player's best response to the row player's pure strategy i, and the row player's best response to the column player's pure strategy j. The difference in behavior between the various Games played on the matrix of Game 9.3.4 and those played on the matrix of Game 9.4.1, for example, is that the latter matrix contains a saddle point, namely -1.

Workout 9.4.7 *Write an algorithm for finding all saddle points of a matrix* **A**, *or determining that none exist.*

In some sense, Game 9.4.3 is the dynamic version of Game 9.4.5, in which the players learn what their optimal strategy is by playing best responses. One can find the optimum for Game 9.4.5 on a matrix with a saddle dynamically, by playing Game 9.4.3, or statically, using Workout 9.4.7 ahead of time. This is the same dynamic learning that Workout 9.3.5(d)

---

[2]It is instructive to view this definition in the context of the following **MAPLE** commands.

```
with(plots):   plot3d(cos(y)-cos(x),x=-Pi/2..Pi/2,y=-Pi/2..Pi/2);
```

9.4. Saddles

outlines in the context of mixed strategies. One can either play dynamically to find (actually approximate — it is a theorem, whose proof is beyond the scope of this book, that the best-response strategies of Workout 9.3.5(d) converge to optimal strategies x^* and y^*) the optimum mixed strategy, or statically, using LO ahead of time.

We can generalize the notion of a saddle as follows. A saddle can be thought of as a pair (i, j) of pure strategies which neither player has any incentive to change; that is, i is a best response to j and j is a best response to i. Similarly, a pair $(\mathbf{x}°, \mathbf{y}°)$ of mixed strategies is called an **equilibrium** for the matrix $\mathbf{A}$ if $\mathbf{y}°$ is a best response to $\mathbf{x}°$ and $\mathbf{x}°$ is a best response to $\mathbf{y}°$. Of course, this idea is not new to us: for the traditional matrix game we have $(\mathbf{x}°, \mathbf{y}°) = (\mathbf{x}^*, \mathbf{y}^*)$. As we have seen, one can reveal their equilibrium strategy without compromising their results. Of course, not every matrix has a saddle, although the Minimax Theorem 9.2.10 says that every matrix has an equilibrium. The fact that equilibrium seems to mean optimal is about to change.

equilibrium

In some games, what one player wins is not always equal to what the other loses. One famous example is called the **Prisoner's Dilemma**, Game 9.4.8, below. First, let's describe the general format. Let $\mathbf{A} = (a_{i,j})$ be the matrix of column player winnings, and $\mathbf{B} = (b_{i,j})$ be the (same size) matrix of row player winnings. Denote by $\mathbf{M} = (\mathbf{A}, \mathbf{B})$ the game played under these conditions; that is, if Ken plays column c and Barbie plays row r, then Ken wins $a_{r,c}$ (from some source, not necessarily Barbie) and Barbie wins $b_{r,c}$ (from some source, not necessarily Ken). Up to this point, we have only played games for which $\mathbf{B} = -\mathbf{A}$. Such games are called **zero-sum** because $\mathbf{A} + \mathbf{B} = 0$. From now on we will study **general-sum** games, for which $\mathbf{A} + \mathbf{B}$ could be anything. The object $\mathbf{M}$ is called a **bimatrix**, and can be written either as an ordered pair $\mathbf{M} = (\mathbf{A}, \mathbf{B})$ of matrices or as a matrix $\mathbf{M} = ((a_{i,j}, b_{i,j}))$ of ordered pairs. You say potāto, I say potāto, ...

Prisoner's Dilemma

zero/general-sum game

bimatrix

Game 9.4.8 *Bunny and Clive are arrested on suspicion of cheating on their LO final exam. The dean presents them with the bimatrix $\mathbf{M}$, below,*

○

	F	U
F	(70, 70)	(90, 0)
U	(0, 90)	(20, 20)

and lays out the rules of the game. The dean will interrogate each of them individually (and secretly) and each has the choice of being faithful (F) or unfaithful (U) to the other, by saying nothing or turning the other in. Their choices determine the score each will receive on their final exam.

For example, if (Clive,Bunny) choose (column,row)=(F,U), then they score $(0, 90)$, which would suit Bunny well but cause Clive to get expelled (and break up with her).

Game 9.4.8 brings up two issues that were not present in zero-sum games. First, we find an equilibrium that is not optimal — in fact the notion of optimality no longer makes sense in this context because the players are playing on separate payoff matrices. Both players find that being unfaithful dominates being faithful (90 and 20 are better than 70 and 0, respectively), but by both being unfaithful they each flunk their final and lose their scholarships, a much worse outcome for each than passing the course with a C by being faithful to each other. Second, a **cooperative game** is one that allows the players to try to improve their individual payoffs by forming coalitions. That can get exciting and complicated when there are many people trying to survive on an island, for example. With just two prisoners they could decide to cooperate in order to claim passing scores of 70 each. But we do not allow it: our games are **noncooperative**. Zero-sum games are by nature noncooperative — what's good for one is bad for the other in equal measure — general-sum games introduce this new possibility. But by declaring noncooperativeness we ensure the selfishness of the players, which results in the outcome of $(20, 20)$ at the saddle $(\mathbf{x}^\circ, \mathbf{y}^\circ) = (U, U)$.

(non) cooperative game

Two important questions remain: does every bimatrix have an equilibrium and, if so, how can we find it? The first can be answered by generalizing the Best Response Algorithm of Workout 9.3.5(d), which we illustrate on the bimatrix of Game 9.4.9, with the current strategies $\mathbf{x} = (2, 5, 3)^\mathsf{T}/10$ and $\mathbf{y} = (2, 1, 3, 4)^\mathsf{T}/10$ for Ken and Barbie. (We shall see in Workout 9.4.10 that this example raises a third issue not present in zero-sum games, namely, that different equilibria can yield different payoffs.)

Game 9.4.9 *Let the following be a payoff bimatrix for a general-sum game.*

$$\mathbf{M} = \begin{pmatrix} (1,2) & (-1,3) & (0,-1) \\ (2,-3) & (0,1) & (-2,-2) \\ (0,-1) & (3,-2) & (-1,0) \\ (-3,1) & (-2,0) & (4,1) \end{pmatrix}$$

Since Ken's current payoff is $\mathbf{y}^\mathsf{T}\mathbf{A}\mathbf{x} = 12/100$, he could improve upon it by playing more often the columns corresponding to entries of $\mathbf{y}^\mathsf{T}\mathbf{A}$ that beat $12/100$. He notices that the third entry of $\mathbf{y}^\mathsf{T}\mathbf{A} = (-8, -1, 11)/10$ is $98/100$ larger than his current payoff, so he defines $\mathbf{x}^\Delta = (0, 0, 98)^\mathsf{T}/100$. Then he creates his new strategy $\widehat{\mathbf{x}}$ from the weighted average

$$\widehat{\mathbf{x}} = \frac{\mathbf{x} + \mathbf{x}^\Delta}{1 + \mathbf{J}_3^\mathsf{T}\mathbf{x}^\Delta} = \frac{(20, 50, 30)^\mathsf{T} + (0, 0, 98)^\mathsf{T}}{100 + 98}$$
$$= (20, 50, 128)^\mathsf{T}/198 \,.$$

In the same manner, Barbie wants to beat her current payoff of $\mathbf{y}^\mathsf{T}\mathbf{B}\mathbf{x} = 9/100$, so she calculates $\mathbf{B}\mathbf{x} = (16, -7, -12, 5)^\mathsf{T}/10$, $\mathbf{y}^\Delta = (151, 0, 0, 41)/100$, and

$$\widehat{\mathbf{y}} = \frac{\mathbf{y} + \mathbf{y}^\Delta}{1 + \mathbf{J}_4^\mathsf{T}\mathbf{y}^\Delta} = \frac{(20, 10, 30, 40)^\mathsf{T} + (151, 0, 0, 41)^\mathsf{T}}{100 + 192}$$
$$= (171, 10, 30, 81)^\mathsf{T}/292 \,.$$

Note that the new payoffs $\hat{\mathbf{z}} = (\hat{\mathbf{y}}^\mathsf{T} \mathbf{A} \hat{\mathbf{x}}, \hat{\mathbf{y}}^\mathsf{T} \mathbf{B} \hat{\mathbf{x}})$ are significant improvements for each player over the old payoffs.

Denote by Δ^n the set of all stochastic n-vectors. Then define the **improvement function** $f : \Delta^n \times \Delta^m \to \Delta^n \times \Delta^m$ by $f(\mathbf{x}, \mathbf{y}) = (\hat{\mathbf{x}}, \hat{\mathbf{y}})$. While the sequence $(\mathbf{x}, \mathbf{y})$, $f(\mathbf{x}, \mathbf{y})$, $f^2(\mathbf{x}, \mathbf{y})$, ... may not converge to an equilibrium $(\mathbf{x}^\circ, \mathbf{y}^\circ)$ of $\mathbf{M}$, we can still use f to characterize equilibria as fixed points of f (see Exercise 9.5.27).

improvement function

Workout 9.4.10 *Let* $\mathbf{x} = (2, 0, 1)^\mathsf{T}/3$ *and* $\mathbf{y} = (7, 0, 0, 1)^\mathsf{T}/8$.

  a. Verify that $(\mathbf{x}, \mathbf{y})$ *is an equilibrium of the bimatrix* $\mathbf{M}$, *above.*

  b. Verify that $f(\mathbf{x}, \mathbf{y}) = (\mathbf{x}, \mathbf{y})$.

  c. Repeat parts a and b for the two saddles of $\mathbf{M}$.

  d. Evaluate the payoffs of the three equilibria above.

This characterization is used to show that every bimatrix has at least one equilibrium — see Exercise 9.5.27. The key ingredient is the Brouwer Fixed Point Theorem (found in Exercise 9.5.26). It is interesting to note that the 1994 Nobel Prize in Economics was awarded to John Nash for his development of equilibrium theory in n-person games and for proving that equilibria exist.

The second question is a little harder in that no efficient algorithms are known for finding equilibria. Exercise 9.5.21 outlines one approach, a generalization of the method of Exercise 4.5.12, based on the Principle of Indifference, the bimatrix version of Complementary Slackness.

9.5 Exercises

Practice

9.5.1 *Here we modify Game 9.1.1 somewhat. Suppose that Ken carries pennies and quarters, while Barbie carries nickels and dimes. Let's say that Ken wins them if they are both persons' most valuable or both persons' least valuable coins, and Barbie wins them otherwise.*

  a. Write the payoff matrix with respect to Ken.

  b. Write Ken's GLOP.

  c. Use the Simplex Algorithm to find his optimal strategy, as well as the value of the game.

  d. Use LO to find Barbie's optimal strategy.

9.5.2 *Compare the values and optimal strategies of the following three matrix games.*

a. $\begin{pmatrix} 2 & 0 & -1 \\ -1 & 3 & 1 \\ -2 & -4 & 5 \end{pmatrix}$

b. $\begin{pmatrix} 1 & -1 & -2 \\ -2 & 2 & 0 \\ -3 & -5 & 4 \end{pmatrix}$

c. $\begin{pmatrix} 6 & 0 & -3 \\ -3 & 9 & 3 \\ -6 & -12 & 15 \end{pmatrix}$

9.5.3 *Solve the problems below in the context of matrix games.*

a. Exercise 3.5.28.

b. Exercise 3.5.29.

c. Exercise 6.5.7.

d. Exercise 8.5.27.

9.5.4 *Repeat the following exercise N times.*

a. For some $2 \leq m \leq n \leq 5$, generate an arbitrary $m \times n$ payoff matrix $\mathbf{A}$, with respect to Ken, for a game against Barbie.

b. Write Ken's GLOP and find the game's value and Ken's optimal strategy.

c. Write Barbie's GLOP and find the game's value and Barbie's optimal strategy.

9.5.5 *Consider Game 9.1.6.*

a. Compute Ken's optimal response to Barbie's strategies:
   (i) $\mathbf{y}' = (1,1,1,1)^\mathsf{T}/4$.
   (ii) $\mathbf{y}' = (1,0,3,2)^\mathsf{T}/6$.

b. Compute Barbie's optimal response to Ken's strategies:
   (i) $\mathbf{x}' = (1,1,0,1)^\mathsf{T}/3$.
   (ii) $\mathbf{x}' = (4,2,1,3)^\mathsf{T}/10$.

9.5.6 *Consider Game 9.2.1.*

a. Compute Ken's optimal response to Barbie's strategies:
   (i) $\mathbf{y}' = (2,1,0,4)^\mathsf{T}/7$.
   (ii) $\mathbf{y}' = (3,7,5,1)^\mathsf{T}/16$.

b. Compute Barbie's optimal response to Ken's strategies:
   (i) $\mathbf{x}' = (1,1,1,1,1)^\mathsf{T}/5$.
   (ii) $\mathbf{x}' = (1,0,1,0,1)^\mathsf{T}/3$.

9.5. Exercises

9.5.7 Use row and column dominations to reduce the following matrix as much as possible.
$$\begin{pmatrix} 41 & -88 & -13 & -17 & 0 \\ 24 & -65 & -51 & -44 & -56 \\ -32 & 0 & 27 & 59 & 21 \\ 30 & -79 & -24 & -32 & -73 \end{pmatrix}$$

9.5.8 The childhood game of Rock-Paper-Scissors is a symmetric game, and hence fair. (The game has two players showing each other their hand simultaneously. A fist represents rock, a flat hand palm downward represents paper, and two fingers extended represents scissors. Rock is covered by paper, paper is cut by scissors, and scissors is crushed by rock. For our purposes, replace "is verbed by" with "pays \$1 to".) Use LO to deduce the same result.

9.5.9 John and Björn are avid fantasy tennis players. They compiled the following chart of head-to-head results that Roger Federer, Rafael Nadal, Novak Djokovic, Andy Roddick, Andy Murray, Nikolay Davydenko, James Blake, and David Nalbandian have played against each other in their careers,[3] with entries signifying the number of matches that the column player has beaten the row player. For example, Blake has beaten Nadal 3 times, and Nadal has beaten Blake twice.

	Fed	Nad	Djo	Rod	Mur	Dav	Bla	Nal
Fed	0	12	2	1	2	0	1	3
Nad	6	0	4	2	0	1	3	2
Djo	5	10	0	1	2	1	0	1
Rod	10	3	1	0	4	1	2	1
Mur	1	5	4	2	0	3	1	0
Dav	11	3	0	2	3	0	4	5
Bla	7	2	1	1	0	0	0	0
Nal	8	0	2	0	0	1	1	0

John and Björn each choose a player and do nothing if they have chosen the same player. Otherwise, the winner is the one who has chosen the player that has beaten the other person's (the loser's) player more often. In the case that they tie (the players have beaten each other equally often), nothing happens; otherwise the loser pays the winner according to one of the rules below. They repeat these fantasy games indefinitely. For each rule, find the value of the game to Björn, as well as his optimal strategy for achieving that value.

a. The loser pays the winner \$1.

b. The loser pays the winner the difference between the number of matches each has won against each other. For example, whoever chooses Murray would pay whoever chooses Djokovic \$2 = 4 − 2.

---
[3] Before the 2008 U.S. Open.

c. The loser pays the winner the difference between the percentage of matches each has won against each other. For example, whoever chooses Nalbandian would pay whoever chooses Federer $5/11 = (8 - 3)/(8 + 3)$.

9.5.10 Repeat Exercise 9.5.9(a) with the modified rule that, in the case that John and Björn each chose the same player, then John pays Björn \$1.

9.5.11 Repeat Exercise 9.5.9 with other sports, e.g., top ten women's tennis players, Major League Baseball teams with last year's results, etc.

○ **9.5.12** Consider the following version of Battleship. Lieutenant McHale hides a mine under one of the points of 2×3 grid (shown below). Captain Binghamton, not able to see the mine, places a ship along one of the 7 horizontal or vertical lines joining two adjacent points. McHale wins \$1 from Binghamton if the ship lands on the mine (in which case McHale has the right to make appropriate sound effects), and pays Binghamton \$1 otherwise. Find the game value and optimal strategies for both players.

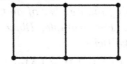

9.5.13 Consider playing Battleship (see Exercise 9.5.12) on a 4×5 grid. Find the game value and optimal strategies for both players.

9.5.14 Consider playing Battleship (see Exercise 9.5.12) on a 3×3 grid. Find the game value and optimal strategies for both players.

9.5.15 Consider playing Battleship (see Exercise 9.5.12) on a 5×5 grid. Find the game value and optimal strategies for both players.

9.5.16 Use Workout 9.4.7 to find all saddle points of the following matrix.

$$\begin{pmatrix} 4 & 2 & -3 & -1 \\ -1 & 2 & 0 & 1 \\ 1 & 3 & -2 & 0 \\ -3 & 2 & 1 & 2 \\ 0 & 3 & 4 & 2 \end{pmatrix}$$

9.5.17 Suppose that $a_{i,j}$ and $a_{r,c}$ are both saddles in the matrix **A**.

  a. Prove that $a_{i,j} = a_{r,j} = a_{r,c} = a_{i,c}$.

  b. Prove that $a_{r,j}$ and $a_{i,c}$ are also saddles.

9.5.18 Construct a 5×6 matrix with exactly three saddles.

9.5. Exercises

9.5.19 Let $\mathbf{M} = (\mathbf{A}, \mathbf{B})$ be the game bimatrix below.

$$\begin{pmatrix} (2,2) & (1,3) \\ (3,4) & (5,1) \end{pmatrix}$$

Use the Principle of Indifference to find the unique equilibrium $(\mathbf{x}^\circ, \mathbf{y}^\circ)$ for $\mathbf{M}$. [HINT: Since $\mathbf{M}$ has no saddle, what must the entries of $\mathbf{Bx}$ (and of $\mathbf{y}^T\mathbf{A}$) satisfy?]

9.5.20 Prove that every 2×2 payoff bimatrix $\mathbf{M} = (\mathbf{A}, \mathbf{B})$ has an equilibrium.

9.5.21 Generalize the method of Exercise 4.5.12 to verify that Game 9.4.9 has exactly the three equilibria mentioned in Workout 9.4.10.

9.5.22 Use the method of Exercise 9.5.21 to find all equilibria of the following bimatrix $\mathbf{M} = (\mathbf{A}, \mathbf{B})$.

$$\begin{pmatrix} (4,1) & (0,2) & (3,3) \\ (2,2) & (3,0) & (1,4) \\ (0,5) & (1,3) & (2,1) \end{pmatrix}$$

9.5.23 Consider the general-sum version of Exercise 9.5.9, in which each person who chooses player X while his opponent chooses player Y wins (from Frank) the number of dollars equal to the number of times that X beats Y. For example, if John chooses Roddick and Björn chooses Davydenko then Frank pays John \$2 and Björn \$1. Find all equilibria strategy and their corresponding game values.

9.5.24 Repeat Exercise 9.5.23 with your scenarios from Exercise 9.5.11.

9.5.25 Consider the general-sum game of Exercise 9.5.9. Find John's and Björn's strategies that maximize the total amount that Frank pays them.

9.5.26 The **Brouwer Fixed Point Theorem** states that every continuous function on Δ^n has a fixed point. Prove the theorem for $n = 2$. [HINT: Use the Intermediate Value Theorem on a related function.]

> Brouwer Fixed Point Theorem

9.5.27 Let $\mathbf{M} = (\mathbf{A}, \mathbf{B})$ be a payoff bimatrix for a game.

   a. Prove that if $(\mathbf{x}^\circ, \mathbf{y}^\circ)$ is an equilibrium for $\mathbf{M}$ then it is a fixed point of the improvement function f.

   b. Prove that if $(\mathbf{x}', \mathbf{y}')$ is a fixed point of the improvement function f then it is an equilibrium for $\mathbf{M}$.

   c. Prove that the improvement function f is continuous on $\Delta^n \times \Delta^m$.

   d. Use the Brouwer Fixed Point Theorem (see Exercise 9.5.26) to prove that $\mathbf{M}$ has an equilibrium.

Challenges

9.5.28 Let $\mathbf{A}$ be an $m \times n$ payoff matrix. Prove that, for any $r \in \mathbb{R}$, the game on $\mathbf{A} + r\mathbf{J}$ has the same optimal strategies as the game on $\mathbf{A}$, with value $z^* + r$, where z^* is the value on $\mathbf{A}$.

9.5.29 Let $\mathbf{A}$ be an $m \times n$ payoff matrix and $r \in \mathbb{R}^+$. Prove that the game on $r\mathbf{A}$ has the same optimal strategies as the game on $\mathbf{A}$, with value rz^*, where z^* is the value on $\mathbf{A}$.

9.5.30 Analyze Game 9.1.1 with arbitrary values a and b instead of 10 and 25 cents. What is the value of the game, and what are the optimal strategies?

9.5.31 Let $\mathbf{x} = (2, 3)^\mathsf{T}/5$ and $\mathbf{y} = (1, 2, 4)^\mathsf{T}/7$. Find a nonconstant 3×2 matrix $\mathbf{A}$, having no all-zero row or column, so that $\mathbf{x}$ and $\mathbf{y}$ are Ken's and Barbie's optimal strategies for the game played on $\mathbf{A}$, and so that

   a. $\mathbf{A} \geq \mathbf{0}$ and $\mathbf{A}$ is integral.

   b. $\mathbf{A}$ is integral and the resulting game is fair.

9.5.32 Prove that Game 9.4.3 terminates on an entry $a_{i,j}$ of $\mathbf{A}$ if and only if $a_{i,j}$ is a saddle of $\mathbf{A}$.

○ **9.5.33** Monty presents you with three doors, saying that behind one of them is hidden $1,000, and behind the other two are his brother Darryl and his other brother Darryl, each of whom is of no value to either of you. You choose one of the doors, let's call it A, in hopes of winning the money. Then Monty either opens door A and you win what is behind it, or offers you the chance to switch to one of the other two doors, B or C. In the latter case, he also opens one of those two, say B, to reveal one of the Darryls. Then you make your final choice between door A and door C, winning whatever is behind it. (Note that this differs slightly from a popularized version in which Monty always shows a Darryl door.)

   a. Model this game by a single payoff matrix. [HINT: The pure strategies of each player must take into account the actions of the other.]

   b. Find each player's optimal strategies and the value of the game to you.

9.5.34 Use Sperner's Lemma (see Exercise 8.5.21) to prove Brouwer's Fixed Point Theorem (see Exercise 9.5.26).

9.5.35 Let $X_n \in \{-1, +1\}$ be chosen uniformly at random for each $n \geq 1$ and denote $S_n = \sum_{i=1}^n X_i$.

   a. Show that $\mathbf{Ex}[|S_{2k+1}|] = \mathbf{Ex}[|S_{2k+2}|] = (2k+1)\binom{2k}{k}/4^k$.

b. Use part a to prove that $\mathbf{Ex}[|S_n|] \sim \sqrt{2n/\pi}$. *[HINT: Use Stirling's formula.]*

9.5.36 *Given Barbie's optimal strategy in Game 9.1.1, suppose that Ken's response is to always play his dime.*

  a. Use Exercise 9.5.35 to find the expected absolute value of the payoff after $1,000$ plays.

  b. Find the same computation for his quarter instead of his dime.

9.5.37 *Consider Ken's and Barbie's optimal strategies in Game 9.1.1.*

  a. Use Exercise 9.5.36 to find the expected absolute value of the payoff after $1,000$ plays.

  b. When does the expected absolute value of the payoff reach $\$100$?

9.5.38 *Consider playing Battleship (see Exercises 9.5.12–9.5.15) on a $s \times t$ grid. Find the game value and optimal strategies for both players. [HINT: Pay attention to whether st is even or odd.]*

9.5.39 *Suppose Ken and Barbie simultaneously choose integers k and b from 1 to 100. If $k = b$ then there is no payoff. If $k < b - 1$ or $k = b + 1$ then Barbie pays Ken $\$1$. Otherwise Ken pays Barbie $\$1$. What is the value of this game and what are their optimal strategies?*

9.5.40 *Suppose Ken and Barbie simultaneously choose integers k and b from 1 to 100. If $k = b$ then there is no payoff. If $k < b - 1$ or $k = b + 1$ then Barbie pays Ken $\$|b - k|$. Otherwise Ken pays Barbie $\$|k - b|$. What is the value of this game and what are their optimal strategies?*

9.5.41 *Suppose Ken and Barbie simultaneously choose positive integers k and b, with a payoff of $k - b$ from Barbie to Ken.*

  a. Prove that this is a fair game.

  b. Prove that Ken's strategy $\mathbf{x}$, with $x_k = 1/2k$ for $k = 2^t$ and $x_k = 0$ otherwise, provides him infinite expected payoff.

  c. How do you resolve this paradox?

9.5.42 *A **tournament** T is a set of points V, called vertices, along with a set of arrows A, called arcs, between each pair of vertices (only one arc per pair, either pointing one way or the other, but not both) — an example with 5 vertices is below, shown with a set of stochastic weights given to its vertices.*

tournament

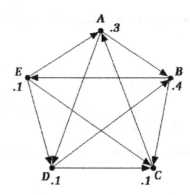

Given weight function $W : V \to \mathbb{R}$, define the function $D_W : V \to \mathbb{R}$ by $D_W(v) = W(v) + \sum_{u \to v} W(u)$. For example, $D_W(C) = .7$, above. Prove that, for every tournament T, there exists a stochastic weight function W such that every $v \in V$ has $D_W(v) \geq 1/2$. [HINT: See Exercise 9.5.10.]

9.5.43 Prove the following alternative version of the Minimax Theorem 9.2.10: $\max_x \min_y y^T A x = \min_y \max_x y^T A x$, where min and max are taken over all stochastic vectors of the appropriate length.

Projects

9.5.44 Present the reduction of LO to matrix games.

Nash's Theorem **9.5.45** Present **Nash's Theorem** on the existence of equilibria for n-person games.

Shapley value **9.5.46** Present the **Shapley value**.

Chapter 10

Network Environment

10.1 Shipping

Consider the following Transshipment Problem.

Problem 10.1.1 *Eumerica makes bottled air at three plants in Vienna, Athens, and Moscow, and ships crates of their products to distributors in Venice, Frankfurt, and Paris. Each day the Athens plant produces 23 thousand crates, while Vienna can produce up to 15 thousand, and Moscow can produce up to 20 thousand. In addition, Venice must receive 17 thousand and Paris must receive 25 thousand crates, while Frankfurt can receive up to 27 thousand. The company pays Arope Trucking to transport their products at the following per-crate Eurodollar costs.*

100	Vienna to Frankfurt		200	Venice to Paris
120	Frankfurt to Athens		210	Paris to Venice
130	Athens to Frankfurt		260	Frankfurt to Venice
140	Frankfurt to Paris		280	Venice to Moscow
150	Paris to Athens		290	Moscow to Frankfurt
170	Athens to Vienna			

Eumerica would like to tell Arope which shipments to make between cities so as to minimize cost.

It is helpful in this case to interpret the given information visually. In Figure 10.1, we've drawn the cities about where they are in the United States — each city in the diagram is called a **node**. Every direct trucking route from one city to another is drawn as an **arc** in the diagram. Each node is labeled by its **demand** in the problem (we will think of a **supply** as a negative demand), and each arc is labeled by the per item **cost** of shipping along its route. The entire structure of nodes, arcs, demands and costs is what we call a **network**. (Without the demand and cost labels, it is just a **directed graph** (or **digraph**), and further, without direction on

node
arc
demand/
supply
cost
network
directed
graph
digraph

G. H. Hurlbert, *Linear Optimization*, Undergraduate Texts in Mathematics,
DOI: 10.1007/978-0-387-79148-7_10, © Springer Science+Business Media LLC 2010

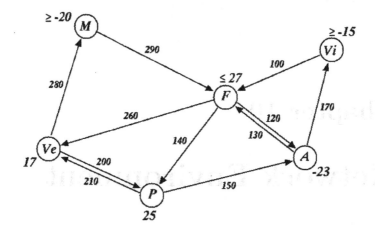

Figure 10.1: Network for Problem 10.1.1

the arcs, the structure is called a graph; its undirected arcs are called edges and its nodes are typically called vertices.)

Our immediate aim here is to visualize the networks that correspond to the tableaux during each step of the Simplex Algorithm. Then we will attempt to interpret the tableau pivot decision-making process in the network environment in the hopes of being able to ignore tableaux entirely. As in Chapter 5, where we implemented the Simplex Algorithm in the Matrix Environment, the result here will not be a new algorithm, but instead merely the same Simplex Algorithm in the Network Environment. Note that each node A corresponds to a constraint, while each arc AB (from the tail/head **tail** A to the **head** B) corresponds to the variable y_{AB} whose value equals the number of (thousands of) items shipped on the arc. We arrive at the following general LOP.

Problem 10.1.2

Min. $w = 130y_{AF} + 170y_{AVi} + 120y_{FA} + 140y_{FP} + 260y_{FVe} + 290y_{MF}$
$\qquad\qquad + 150y_{PA} + 210y_{PVe} + 280y_{VeM} + 200y_{VeP} + 100y_{ViF}$

s.t. $\quad -y_{AF} - y_{AVi} + y_{FA} + y_{PA} = -23$
$\qquad y_{AF} - y_{FA} - y_{FP} - y_{FVe} + y_{MF} + y_{ViF} \leq 27$
$\qquad -y_{MF} + y_{VeM} \geq -20$
$\qquad y_{FP} - y_{PA} - y_{PVe} + y_{VeP} = 25$
$\qquad y_{FVe} + y_{PVe} - y_{VeM} - y_{VeP} = 17$
$\qquad y_{AVi} - y_{ViF} \geq -15$

& $\quad y_{\text{all}} \geq 0$

We purposely ordered the constraints (rows) and variables (columns) alphabetically so as to make least subscript decisions easily identifiable. We also avoid standard form here by adding or subtracting appropriate slack

10.1. Shipping

variables. The resulting initial tableau is as follows. (Note that the infinity (slack) arcs are also listed alphabetically, ignoring the infinity symbol, and that these are listed after all original (problem) arcs — keep this in mind for later.)

Tableau 10.1.3

$$\left[\begin{array}{cccccccccc|cccc|c} -1 & -1 & 1 & 0 & 0 & 0 & 1 & 0 & 0 & 0 & 0 & 0 & 0 & 0 & -23 \\ 1 & 0 & -1 & -1 & -1 & 1 & 0 & 0 & 0 & 0 & 1 & 0 & 0 & 0 & 27 \\ 0 & 0 & 0 & 0 & 0 & -1 & 0 & 0 & 1 & 0 & 0 & -1 & 0 & 0 & -20 \\ 0 & 0 & 0 & 1 & 0 & 0 & -1 & -1 & 0 & 1 & 0 & 0 & 0 & 0 & 25 \\ 0 & 0 & 0 & 0 & 1 & 0 & 0 & 1 & -1 & -1 & 0 & 0 & 0 & 0 & 17 \\ 0 & 1 & 0 & 0 & 0 & 0 & 0 & 0 & 0 & -1 & 0 & 0 & -1 & 0 & -15 \\ \hline 130 & 170 & 120 & 140 & 260 & 290 & 150 & 210 & 280 & 200 & 100 & 0 & 0 & 0 & 1 & 0 \end{array}\right]$$

There are several things worth noting about Tableau 10.1.3. First, every nonslack column has exactly one -1 and 1. Of course, this makes sense because every column represents an arc *from* one node *to* another. Second, this is the kind of sparse example alluded to in Section 5.2 — the density of nonzero entries is only $2/n$, where n is the number of nodes. Third, we can think of the slack columns as arcs as well by imagining an invisible **infinity node**, whose corresponding tableau row equals the negative of the sum of all other node rows. The resulting tableau is below.

infinity node

Tableau 10.1.4

$$\left[\begin{array}{cccccccccc|cccc|c} -1 & -1 & 1 & 0 & 0 & 0 & 1 & 0 & 0 & 0 & 0 & 0 & 0 & 0 & -23 \\ 1 & 0 & -1 & -1 & -1 & 1 & 0 & 0 & 0 & 0 & 1 & 0 & 0 & 0 & 27 \\ 0 & 0 & 0 & 0 & 0 & -1 & 0 & 0 & 1 & 0 & 0 & -1 & 0 & 0 & -20 \\ 0 & 0 & 0 & 1 & 0 & 0 & -1 & -1 & 0 & 1 & 0 & 0 & 0 & 0 & 25 \\ 0 & 0 & 0 & 0 & 1 & 0 & 0 & 1 & -1 & -1 & 0 & 0 & 0 & 0 & 17 \\ 0 & 1 & 0 & 0 & 0 & 0 & 0 & 0 & 0 & -1 & 0 & 0 & -1 & 0 & -15 \\ 0 & 0 & 0 & 0 & 0 & 0 & 0 & 0 & 0 & 0 & -1 & 1 & 1 & 0 & -11 \\ \hline 130 & 170 & 120 & 140 & 260 & 290 & 150 & 210 & 280 & 200 & 100 & 0 & 0 & 0 & 1 & 0 \end{array}\right]$$

The corresponding minimization problem, having only equalities, nonnegative variables, and demand sum 0, is in **standard network form**.

standard network form

Problem 10.1.5

Min. $w = 130y_{AF} + 170y_{AVi} + 120y_{FA} + 140y_{FP} + 260y_{FVe} + 290y_{MF}$
$\phantom{\text{Min. } w =} + 150y_{PA} + 210y_{PVe} + 280y_{VeM} + 200y_{VeP} + 100y_{ViF}$

s.t.
$-y_{AF} - y_{AVi} + y_{FA} + y_{PA} = -23$
$y_{AF} - y_{FA} - y_{FP} - y_{FVe} + y_{MF} + y_{ViF} + y_{\infty F} = 27$
$-y_{MF} + y_{VeM} - y_{M\infty} = -20$
$y_{FP} - y_{PA} - y_{PVe} + y_{VeP} = 25$
$y_{FVe} + y_{PVe} - y_{VeM} - y_{VeP} = 17$
$y_{AVi} - y_{ViF} - y_{Vi\infty} = -15$
$-y_{\infty F} + y_{M\infty} + y_{Vi\infty} = -11$

& $y_{\text{all}} \geq 0$

We say that Tableau 10.1.3 is the initial **reduced tableau** for Problem 10.1.5, while Tableau 10.1.4 is its initial **augmented tableau**. There may be several reduced tableaux for a given network problem — the point is

reduced/ augmented tableau/ network

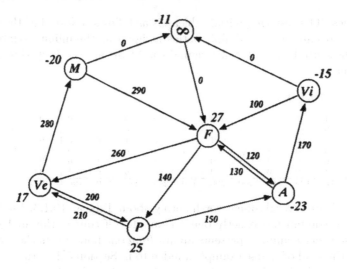

Figure 10.2: Augmented network for Problem 10.1.1

that we must remove a (any) redundant row from a tableau since it needs to have full row rank (see Chapter 5). The initial **augmented network** is Figure 10.2. Here, our concern is to include the redundancy, as we shall soon see.

Starting from the reduced Tableau 10.1.3, we need to fill the basis using Phase 0 (see Chapter 6). However, for networks, we will modify the Phase 0 rules by ignoring any current partial basis (some of whose coefficients are negative, by the way) and instead pivoting in least subscript variables greedily from scratch, without any other regard.

Workout 10.1.6 *Starting with Tableau 10.1.3, use modified Phase 0 to find the first basis.*

    a. At each stage, starting with no arcs, draw each basic arc pivoted in.

    b. In the third step, arc FA cannot be chosen for the basis. Explain why not.

       (i) algebraically.

       (ii) visually.

    c. In the sixth step, arc PA cannot be chosen for the basis. Explain why not

       (i) algebraically.

       (ii) visually.

One thing we will learn that the Network Environment has in common with the Matrix Environment is that they share their dependence on initial information rather than current information. In both cases one can specify any basis possible and work from there.

Workout 10.1.7 *Let* $\beta = \{AVi, FP, MF, PA, VeM, M\infty\}$ *in Problem 10.1.2.*

 a. *Draw the network corresponding to* β.

 b. *Use Figure 10.1 to label the arcs of your result in part a by the shipments needed to satisfy all demands. [HINT: Start with finding the shipment across AVi, VeM or $M\infty$ (why?).]*

 c. *Verify your results from part b by finding the tableau associated with basis* β.

Workout 10.1.8 *Let* $\beta = \{AVi, FP, FVe, M\infty, Vi\infty, \infty F\}$ *in Problem 10.1.2.*

 a. *Draw the network corresponding to* β.

 b. *Use Figure 10.1 to label the nodes of your result in part a so that the node labels differ by the costs of the basic arcs. [HINT: Start anywhere with any value you like, and work outwards.]*

 c. *Find the tableau associated with basis β and compare it to your results from part b as follows. For each arc not in the basis, subtract your tail label from your head label. Then subtract the result from the arc cost. What do you notice?*

10.2 Trees

One may have noticed by now that each network basis has a very recognizable structure. In this section we will describe this structure in detail.

A **path** (resp. **chain**) in a graph (resp. network) is a sequence of distinct vertices $A_0, A_1, \ldots, A_k$ for which each pair $\{A_{i-1}, A_i\}$ ($1 \leq i \leq k$) is an edge (resp. arc). We say that the path (resp. chain) is *from* A_0 *to* A_k, and that it has **length** k. Furthermore, the chain is **oriented** if each $A_{i-1}A_i$ (as opposed to A_iA_{i-1}) is an arc. In Figure 10.2, for example, $MVeP$ is a chain of length 2 from M to P, and $AVi\infty F$ is an oriented chain of length 3 from A to F. If $A_k = A_0$ then we say instead that the sequence is a **cycle** (resp. **circuit**) in the graph (resp. network). The circuit is **oriented** when the sequence is oriented.

We say that a graph is **connected** if, for every pair of vertices A and B there is a path of edges that joins them. We call a network connected when its underlying graph is connected, and **strongly connected** when every ordered pair of vertices is joined by an oriented path. A (strongly) connected **component** of a graph or network is a maximal set of (strongly) connected vertices. Different components necessarily share no vertices, and there is no edge or arc that joins vertices of different components. The following Workouts 10.2.1–10.2.6 are stated for graphs for simplicity, but have obvious counterparts for networks.

Workout 10.2.1 *Prove that if a graph has exactly k components then the removal of one of its edges results in at most $k+1$ components.*

bridge

(spanning) subgraph/ subnetwork

acyclic graph/ network

(spanning) tree

An edge or arc in a graph or network whose removal increases the number of components by one is called a **bridge**. A graph $\mathcal{H}$ is a **subgraph** of the graph $\mathcal{G}$ if every edge (resp. vertex) of $\mathcal{H}$ is an edge (resp. vertex) of $\mathcal{G}$. A **subnetwork** is defined analogously. We say that such a subgraph (resp. subnetwork) **spans** $\mathcal{G}$ if every vertex of $\mathcal{G}$ is a vertex of $\mathcal{H}$. A subgraph or subnetwork is **acyclic** if it contains no cycles, and is called a **tree** if it is acyclic and connected. You may have observed in Workout 10.1.6 that each step produced a collection of trees, and that the final step resulted in a spanning tree. Indeed, we soon will prove that every basis corresponds to a spanning tree and vice versa (the Network Basis Theorem 10.2.7). First, we note an immediate consequence of the definition of a tree.

Workout 10.2.2 *Prove that every pair of vertices of a tree is joined by a unique path.*

The following two workouts are fundamental to the workings of the network Simplex Algorithm.

Workout 10.2.3 *Prove that every edge of a tree is a bridge.*

Workout 10.2.4 *Let T be a spanning tree of a connected graph $\mathcal{G}$. Prove that the addition of any edge of G not in T creates a cycle.*

leaf

pendant edge/arc

Next we discover the size of a tree, which will be useful in making the connection between trees and bases. A **leaf** is a vertex (resp. node) incident with exactly one edge (resp. arc), which is called **pendant**.

Workout 10.2.5 *Prove that every tree with at least two vertices has at least two leaves.*

Workout 10.2.6 *Let T be a spanning tree of a connected graph $\mathcal{G}$. Use induction and Workout 10.2.5 to prove that the number of edges of T is one less than the number of vertices of $\mathcal{G}$.*

We are now ready to prove the Network Basis Theorem.

Network Basis Theorem

Theorem 10.2.7 *Let T be a subnetwork of a connected network $\mathcal{N}$ in standard form. Then T is a basis for $\mathcal{N}$ if and only if it is a spanning tree of $\mathcal{N}$.*

If the network $\mathcal{N}$ is not connected, then we can run Simplex on each component independently. In that case a basis will consist of one tree per component.

Proof. Because of standard form, we know that the reduced tableau has full row rank (equal to $n-1$ when there are n nodes). Indeed, suppose that

10.2. Trees

some linear combination of the rows equals zero: $\sum_{i=1}^{n-1} \alpha_i \mathbf{A}_i^\top = \mathbf{0}$, where $\mathbf{A}^\top$ is the network tableau. Let node v_n correspond to the deleted row $\mathbf{A}_n^\top$, and let $\mathcal{T}$ be any spanning tree of $\mathcal{N}$. Then if $v_i v_n$ or $v_n v_i$ is an arc, we must have $\alpha_i = 0$, since the column corresponding to the arc has only the single nonzero entry in row i. Having $\alpha_i = 0$ is equivalent to deleting row $\mathbf{A}_i^\top$. Now $\mathcal{T} - v_n$ has several components, and this argument can be repeated on each one of them recursively, eventually deleting every row; i.e., every coefficient $\alpha_1 = \cdots = \alpha_{n-1} = 0$. Hence the rows of the reduced tableau are independent. Thus every basis has size $n - 1$.

Workout 10.2.8 *Consider the circuit $AF, AVi, MF, M\infty, Vi\infty$ from Figure 10.2. Find a linear dependence among their respective columns in Tableau 10.1.3. [HINT: What would it take to orient the circuit?]*

Workout 10.2.9 *Prove that a network basis contains no circuit.*

Now that we know that a network basis has no circuits, and that it contains $n - 1$ arcs, we refer the reader to Exercise 10.5.10, which proves that the basis is a spanning tree. For the converse argument, the arcs of a spanning tree form an independent set of size $n - 1$, as noted above. ◊

Phase I. It is time, finally, to turn our attention to Phase I.

Finding Shipments. Given any spanning tree, one can discover the amounts shipped along each of its arcs by iteratively finding the shipment necessary along a pendant arc in order to satisfy the demand on its leaf, and then updating the demand on its other node accordingly. For example, in the spanning tree from Workout 10.1.7, we might begin with pendant arc AVi. Because the parameter (missing) arcs ship zero, the original (standard form) node Vi constraint $y_{AVi} - y_{ViF} - y_{Vi\infty} = -15$ from Problem 10.1.5 reduces to $y_{AVi} = -15$. Consequently, the original node A constraint $-y_{AF} - y_{AVi} + y_{FA} + y_{PA} = -23$, becomes $-y_{AF} + y_{FA} + y_{PA} = -38$. Repeating this analysis allows one to compute the values of $y_{PA} = -38$, $y_{FP} = -13$, $y_{MF} = 14$, $y_{VeM} = -17$ and $y_{M\infty} = -11$, in that order (other orders are possible as well, for example $M\infty$ AVi, PA, VeM, MF and FP).

Economic Interpretation. Another interpretation of this arithmetic is that Vi has 15 to give, and it gives it to A along a *wrong way street* — the wrong way negates the 15. Once that happens, the demand at A becomes $-23 + (-15) = -38$ — its supply was 23 and it just got 15 more.

Leaving and Entering Arcs. As Least Subscript prescribes, we remove the most negative arc, namely PA. As if in a civil war, we have blown up one bridge, separating the country into two components, according to Workout 10.2.1. In order to create the next spanning tree, the entering arc must rejoin the two components. In fact, because the tableau instructions require a negative pivot entry, in particular a sign opposite from the current basic entry, such an arc must be oppositely directed from PA. The only three such arcs in the network are AF, ViF, and $Vi\infty$ — note that these

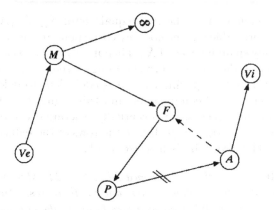

Figure 10.3: Phase I pivot from Workout 10.1.7

are also the only three choices in the tableau from Workout 10.1.7c. The resulting pivot yields the spanning tree in Figure 10.3.

Economic Interpretation. The two components each have cumulative demands. Together, $\{F, M, P, Ve, \infty\}$ have a demand of $27 - 20 + 25 + 17 - 11 = 38$, while $\{A, Vi\}$ has a supply of $23 + 15 = 38$. That is why the shipment along PA was negative, and why an arc from $\{A, Vi\}$ to $\{F, M, P, Ve, \infty\}$ is needed. If no such arc exists, then the partition is a simple certificate of infeasibility.

Phase II. Now we turn our attention to Phase II.

Finding Relative Prices. Given a feasible tree T we can choose any node u and label it with any number x_u. Then we can iterate the process of choosing any unlabeled node u that shares an arc e with a labeled node v. If $e = uv$ then set $x_u = x_v - b_{uv}$ to be the label on u; otherwise set $x_u = x_v + b_{uv}$. For example, in the spanning tree from Workout 10.1.8, we could start with label 0 at node ∞. Then each of the nodes F, M and Vi gets the label 0 also, since all three (slack) arcs ∞F, $M\infty$ and $Vi\infty$ cost nothing. Finally, $x_A = x_{Vi} - b_{AVi} = -170$, and similarly $x_P = 140$ and $x_{Ve} = 280$.

Economic Interpretation. Suppose that uv is an arc of T. Then we should expect that the price of items sold at v would be b_{uv} more than their corresponding price at u. That is, the cost of shipping should be reflected in the price. Since we don't actually know what the prices are at any location, we can only compute prices relative to each other.

Entering Arc. Given a feasible tree T suppose we have found labels $x'_1, \ldots, x'_n$ for the nodes $v_1, \ldots, v_n$, as in Workout 10.1.8b, and let $\mathbf{x}' = (x'_1, \ldots, x'_n)^\mathsf{T}$. We write the network LOP as Min. $w = \mathbf{b}^\mathsf{T}\mathbf{y}$ s.t. $\mathbf{A}^\mathsf{T}\mathbf{y} = \mathbf{c}$, and let $\mathbf{y}'$ denote the current solution on T. Define $\mathbf{b}' = \mathbf{b} - \mathbf{A}\mathbf{x}'$, and note that this yields $b'_{uv} = b_{uv} + x'_u - x'_v$ for every arc uv. In particular, if $uv \in T$ then $b'_{uv} = 0$ by the definition of $\mathbf{x}'$. On the other hand, if $uv \notin T$

10.2. Trees

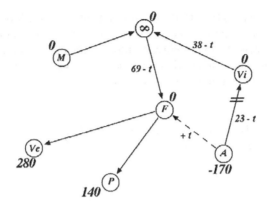

Figure 10.4: Phase II pivot from Workout 10.1.8

then $y'_{uv} = 0$. Together, these facts yield $\mathbf{b'^T y'} = \mathbf{0}$. Also, whenever $\mathbf{A^T y} = \mathbf{c}$ we have $\mathbf{b^T y} = \mathbf{b'^T y} + \mathbf{x'^T A^T y} = \mathbf{b'^T y} + \mathbf{x'^T c}$, implying that $\mathbf{b^T y'} = \mathbf{x'^T c}$. This means that $\mathbf{b^T y} = \mathbf{b^T y'} + \mathbf{b'^T y}$, which is a disguise for $w = w(\mathbf{y'}) + \mathbf{b'^T y}$; i.e., the current objective row. Hence, the tableau method of finding a negative entry in $\mathbf{b'^T y}$ corresponds to finding an arc $uv \notin \mathcal{T}$ so that $b_{uv} + x'_u - x'_v < 0$; in other words $x'_u + b_{uv} < x'_v$. The first such arc from the spanning tree of Workout 10.1.8 is $AF : -170 + 130 < 0$. If no such arc exists then $\mathbf{b'^T y} > \mathbf{0}$, so that $\mathbf{b^T y} \geq \mathbf{b^T y'}$ — the minimum has been found. In fact, the verification that no such arc exists is a certificate that the variables $\mathbf{x'}$ are dual-feasible, which proves optimality because of the equality $\mathbf{b^T y'} = \mathbf{x'^T c}$ above.

Economic Interpretation. Finding such an arc means that one can reduce the price at v by shipping along uv instead of the route determined by $\mathcal{T}$ — the price at u plus the cost along uv is less than the current price at v. If your competitor finds the arc then their items will sell for less than yours at v unless you include the arc in your shipping route.

Leaving Arc. By Workout 10.2.4 there exists now a unique circuit $\mathcal{C}$, as shown for our example in Figure 10.4. Consider an orientation $\vec{\mathcal{C}}$ of $\mathcal{C}$ that agrees with the direction on the entering arc uv (AF). As the value of y_{uv} changes, the values along other arcs of $\mathcal{C}$ must also change in order that the demands at the nodes on $\mathcal{C}$ remain satisfied. In particular, as y_{uv} increases by t, so does every arc whose direction agrees with $\vec{\mathcal{C}}$, while every arc whose direction disagrees with $\vec{\mathcal{C}}$ decreases by t. The arc whose shipment first hits zero leaves the $\mathcal{T}$. If no such arc exists, then we have the nifty certificate of unboundedness given by letting $t \to \infty$, sending the total cost to $-\infty$ (the "number", not the node). Such an oriented circuit is called a **negative cycle**.

Economic Interpretation. Then as t more is shipped along uv, t more is shipped on every other arc whose direction agrees with $\vec{\mathcal{C}}$, and t less is shipped on every arc whose direction disagrees with $\vec{\mathcal{C}}$.

It is worth reiterating that the original information labels the original network by node demands and arc costs. Thereafter, the labels on current spanning trees arise from the current information of arc shipments and node prices. It is the node demands that determine the arc shipments, and the arc costs that determine the node prices.

Workout 10.2.10 *Prove that the spanning tree resulting from the pivot operation in Figure 10.4 is optimal, and find the optimum for Problem 10.1.1.*

10.3 Nilai!

Workout 10.3.1 *Consider the network below.*

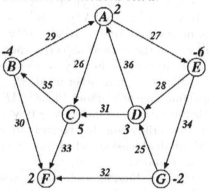

a. *Draw the initial spanning tree.*

b. *Starting from the spanning tree below, perform one iteration of the Network Simplex Algorithm.*

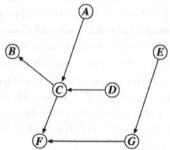

c. *Starting from the spanning tree below, perform one iteration of the Network Simplex Algorithm.*

10.4. Integrality

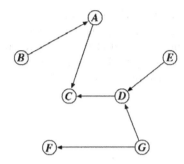

Workout 10.3.2 *Starting with the spanning tree from Workout 10.1.6, use Network Simplex to solve Problem 10.1.1.*

Workout 10.3.3 *Starting with the spanning tree from Workout 10.1.7, use Network Simplex to solve Problem 10.1.1.*

10.4 Integrality

Surely, you have noticed by now that every basic network solution encountered so far has been integral — this can be no coincidence. Either this is a property of all networks, or two evil professors somewhere are chuckling. It turns out that the former is true (not that the latter isn't ...).

Theorem 10.4.1 *Given a network LOP with integer demands, every one of its basic solutions is integral.* Integrality Theorem

For those who studied Chapter 5, especially Section 5.4, this theorem is an application of Theorem 5.4.7 to a special case of Exercise 5.5.21. Here we will give a more direct proof that network matrices are TU (recall: totally unimodular means every square submatrix has determinant 0, 1 or −1), and then give an even more straightforward proof of the Integrality Theorem 10.4.1 that avoids determinants altogether. Both of these arguments rely on the Network Basis Theorem 10.2.7 and the shipment labeling algorithm alluded to in the Phase I description above. Now we make this algorithm explicit.

Lemma 10.4.2 *Let T be a tree on n vertices. It is possible to order the edges of T $e_1, \ldots, e_{n-1}$ so that every e_k is a pendant edge of the subtree $T - \{e_1, \ldots, e_{k-1}\}$. Moreover, defining v_k to be the leaf of e_k, one can stipulate for any vertex u of T that $u \notin \{v_1, \ldots, v_{n-1}\}$.*

Workout 10.4.3 *Prove Lemma 10.4.2.*

First Proof. Let $\mathbf{A}^\mathsf{T}$ be the reduced network matrix, and let $\mathbf{B}$ be a square submatrix of it. If $\det(\mathbf{B}) \neq 0$ then the columns of $\mathbf{B}$ correspond to an acyclic set of arcs, which is a collection of disjoint trees. A simple counting argument similar to Exercise 10.5.10 shows that it is just one tree (on k vertices). Now reorder the columns by the order $e_1, \ldots, e_{k-1}$, and the rows by the order $v_1, \ldots, v_{k-1}$, as guaranteed by Lemma 10.4.2. Now it is easy to see that the reordered $\mathbf{B}$ is lower triangular, so its determinant is equal to the product along its diagonal. Since its entries are always 0, 1 or -1, then so is its determinant. Hence $\mathbf{A}^\mathsf{T}$ is TU. ◇

Workout 10.4.4 *Illustrate the first proof by reordering the rows and columns of the submatrix corresponding to the arcs of the spanning tree in Workout 10.1.7.*

Second Proof. Let T be a spanning tree of a network. Let $e_1, \ldots, e_{n-1}$ and $v_1, \ldots, v_{n-1}$ be as guaranteed by Lemma 10.4.2, with v_n the deleted row. Since the demand at v_1 is integral, the shipment across e_1 is integral. Thus the demand at the other end of e_1 is updated to remain integral. Iterate this process (or use induction), and the result follows. ◇

10.5 Exercises

Practice

10.5.1 *The Network Simplex Algorithm halts for each of the following networks at the spanning tree shown. In each case use the tree to prove that the output (infeasible, unbounded, optimal) is correct.*

a.

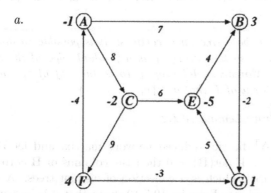

10.5. Exercises

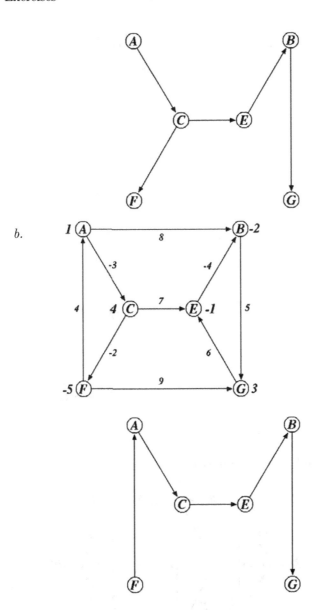

c.

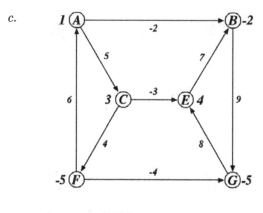

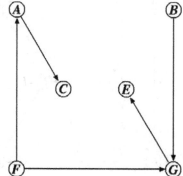

10.5.2 *Consider the network below.*

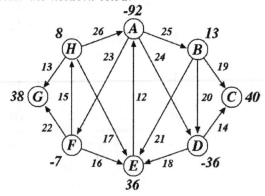

a. *Draw the initial spanning tree.*

b. *Starting from the spanning tree below, perform one iteration of the Network Simplex Algorithm.*

10.5. Exercises

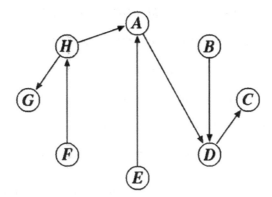

c. Starting from the spanning tree below, perform one iteration of the Network Simplex Algorithm.

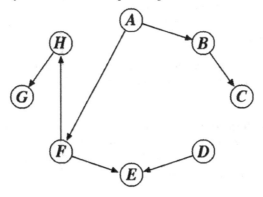

d. Find the optimal solution.

10.5.3 Consider the network below.

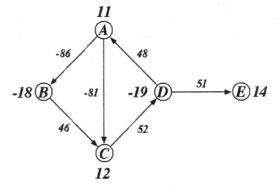

a. Write its reduced tableau.

b. Without using the Network Simplex Algorithm, prove that, regardless of the arc costs, there is only one basic feasible solution (which is therefore optimal).

10.5.4 *Use a network to solve the LOP Exercise 1.5.22.*

10.5.5 *Use a network to solve the LOP Exercise 1.5.38.*

10.5.6 *Use a network to solve the LOP Exercise 1.5.39.*

10.5.7 *Use a network to solve the LOP Exercise 5.3.5.*

10.5.8 *Repeat the following 3-step exercise N times.*

   a. *Draw an arbitrary network on 6–8 nodes (with demand sum zero).*

   b. *Use the Network Simplex Algorithm to solve the network.*

   c. *Prove that the output of the algorithm is correct.*

10.5.9 *Let T be a connected subgraph of a graph $\mathcal{G}$ having n vertices. Use Workout 10.2.6 to prove that if T has n vertices and $n-1$ edges then T is a spanning tree.*

10.5.10 *Let T be an acyclic subgraph of a graph $\mathcal{G}$ having n vertices. Use induction and Workout 10.2.6 to prove that if T has $n-1$ edges then T is a spanning tree.*

degree **10.5.11** *The **degree** $\deg(v)$ of a vertex v in a graph $\mathcal{G}$ equals the number of edges of $\mathcal{G}$ incident with v. Use Workouts 10.2.3 and 10.2.5 to prove that the number of leaves of a tree T is at least its maximum degree. [HINT: Remove a vertex of maximum degree.]*

Handshaking Lemma **10.5.12** *Prove that $\sum_{v \in V(\mathcal{G})} \deg(v) = 2|E(\mathcal{G})|$, where $V(\mathcal{G})$ (resp. $E(\mathcal{G})$) is the set of vertices (resp. edges) of $\mathcal{G}$.*

in/out-degree **10.5.13** *The **indegree** $\deg^+(v)$ (resp. **outdegree** $\deg^-(v)$) of a node v in a digraph $\mathcal{G}$ equals the number of arcs of $\mathcal{G}$ for which v is the head (resp. tail). Prove that every digraph $\mathcal{G}$ satisfies $\sum_{v \in \mathcal{G}} \deg^+(v) = |E(\mathcal{G})| = \sum_{v \in \mathcal{G}} \deg^-(v)$, where $E(\mathcal{G})$ is the set of all arcs of $\mathcal{G}$.*

10.5.14 *Let T be a tree. Prove that every pair of maximum length paths of T intersect.*

10.5.15 *Show that one can start at any node of Figure 10.1, traverse every arc exactly once, and return to the starting node. Such a traversing is called* Euler tour *an **Euler tour**.*

10.5.16 *For your tour in Exercise 10.5.15, any for every node except the starting node, color in red the last arc left each node. Such a structure is* rooted spanning tree *called a **rooted** spanning tree. How many rooted spanning trees are there in Figure 10.1?*

10.5.17 *Use networks to solve Exercise 1.5.39.*

Challenges

10.5.18 *Use networks to solve the dual to the LOP in Exercise 5.5.1.*

10.5.19 *Let $\mathbf{C}$, $\mathbf{r}$ and $\mathbf{s}$ be the following matrices.*

$$\mathbf{C} = \begin{pmatrix} 2 & 4 & -1 & 3 & 4 \\ 3 & 5 & 2 & 0 & 2 \\ 1 & 1 & 0 & 4 & 3 \end{pmatrix}, \quad \mathbf{r} = \begin{pmatrix} 13 \\ 16 \\ 9 \end{pmatrix}, \quad \mathbf{s} = \begin{pmatrix} 7 \\ 8 \\ 6 \\ 7 \\ 10 \end{pmatrix}$$

Use the Network Simplex Algorithm to solve the following LOP.

$$\text{Max.} \quad z = \sum_{i=1}^{3} \sum_{j=1}^{5} c_{i,j} x_{i,j}$$

$$\text{s.t.} \quad \sum_{j=1}^{5} x_{i,j} = r_i \quad (1 \leq i \leq 3)$$

$$\sum_{i=1}^{3} x_{i,j} = s_j \quad (1 \leq j \leq 5)$$

$$\& \quad x_{i,j} \geq 0 \quad (\forall i, j)$$

[HINT: Draw arcs from each of three nodes to each of 5 other nodes.]

10.5.20 *Use the Network Simplex Algorithm to solve Problem 1.2.4.*

10.5.21 *Write an algorithm (code or pseudocode) that takes a spanning tree and node demands (in standard network form) as input and gives all the arc shipments as output. [HINT: A recursive algorithm may be the simplest.]*

10.5.22 *Write an algorithm (code or pseudocode) that takes a spanning tree $\mathcal{T}$ and an arc e of $\mathcal{T}$ as input and computes the two components of $\mathcal{T} - e$ as output.*

10.5.23 *Write an algorithm (code or pseudocode) that takes a spanning tree and arc costs as input and gives all the relative node prices as output. [HINT: A recursive algorithm may be the simplest.]*

10.5.24 *Write an algorithm (code or pseudocode) that takes a spanning tree $\mathcal{T}$ and an arc e not in $\mathcal{T}$ as input and computes the circuit in $\mathcal{T} + e$ as output.*

10.5.25 *Using the algorithms from Exercises 10.5.21–10.5.24 as procedures, write an algorithm (code or pseudocode) that takes a network and one of its spanning trees as input and performs a Network Simplex pivot to give the subsequent spanning tree (or message of infeasible, unbounded or optimal) as output.*

source
sink

10.5.26 A node u in a network $\mathcal{N}$ is a **source** (resp. **sink**) if $b(u) < 0$ (resp. $b(u) > 0$), where $b(u) = b_{\mathcal{N}}(u)$ denotes the demand at u. For given source u and sink v in a network $\mathcal{N}$ define the network $\mathcal{N}' = \mathcal{N}_{u,v}$ to be identical to $\mathcal{N}$ except that $b'(u) = b(u) + 1$ and $b'(v) = b(v) - 1$, where $b' = b_{\mathcal{N}'}$. Construct a network $\mathcal{N}$ with positive arc costs and with optimal costs $w^*(\mathcal{N}) < w^*(\mathcal{N}')$.

10.5.27 Let W be the set of vertices of the tree T that have degree at least 3, and let $d(v) = \deg(v) - 2$. Use Exercise 10.5.12 and Workout 10.2.6 to prove that if the number of vertices of T is at least 2 then the number of leaves of T equals $2 + \sum_{v \in W} d(v)$.

10.5.28 Relate the number of Euler tours in a digraph G to the number of rooted spanning trees of G. [HINT: See Exercise 10.5.16.]

10.5.29 Let T be a tree. Prove that the intersection of three maximum length paths of T is nonempty. [HINT: See Exercise 10.5.14.]

10.5.30 Let T be a tree. Prove that the intersection of all maximum length paths of T is a nonempty path. [HINT: See Exercise 10.5.29.]

even

10.5.31 A graph is **even** if every node v has even degree

    a. Prove that a graph with an Euler tour (see Exercise 10.5.15) is connected and even.

Euler's Theorem

    b. Use induction to prove **Euler's Theorem**, that every connected, even graph has an Euler tour.

balanced digraph

10.5.32 A digraph is **balanced** if every node v has $\deg^+(v) = \deg^-(v)$.

    a. Prove that a digraph with an Euler tour (see Exercise 10.5.15) is strongly connected and balanced.

    b. Use induction to prove that every strongly connected, balanced digraph has an Euler tour.

nearly balanced

10.5.33 A digraph is **nearly balanced** if every node v has $|\deg^+(v) - \deg^-(v)| \leq 1$. Use Exercise 10.5.31 to prove that the arcs of any digraph can be redirected, if necessary, so that the resulting digraph is nearly balanced.

10.5.34 Let $\mathbf{A}^\mathsf{T} = (a_{i,j})$ be a $\{0, 1, -1\}$ matrix. Use Exercise 10.5.33 to prove that $\mathbf{A}^\mathsf{T}$ is TU (see Section 5.4) if and only if, for every set J of columns there is some subset $J_1 \subseteq J$ (with $J_2 = J - J_1$) so that every row i satisfies

$$\left| \sum_{j \in J_1} a_{i,j} - \sum_{j \in J_2} a_{i,j} \right| \leq 1.$$

Projects

10.5.35 *Present the Depth-First and Breadth-First Search Algorithms and their relation to computing the components of a graph.*

10.5.36 *Present the* **Network Flow Problem** *and the* **Ford–Fulkerson Algorithm**.

10.5.37 *Present* **Cayley's Theorem** *and* **Prufer's code**.

10.5.38 *Present the* **Matrix Tree Theorem**.

Chapter 11

Combinatorics

11.1 Matchings

In this chapter we investigate an array of combinatorial applications of networks, beginning with the problem of finding large matchings in bipartite graphs, mentioned in Section 1.2 (see that section for relevant graph definitions).

Problem 11.1.1 *CarbonDating.com keeps a database of their clients and their love interests (for unrealism, we assume symmetry: if A loves B then B loves A). Annette loves David, John and Warren, Kathy loves Bill, John and Regis, Monica loves Bill, David and Warren, Teresa loves Bill, John and Regis, and Victoria loves David, Regis and Warren. A marriage of a woman and a man is* good *if the couple love each other. The company would like to find as many pairwise disjoint, good marriages as possible.*

Workout 11.1.2 *Solve Problem 11.1.1.*

As in Chapter 1 we can visualize the setup of Problem 11.1.1 as the bipartite graph in Figure 11.1. In this form it is not difficult to write down many sets of 5 marriages. In fact, it is easy to imagine a related network by replacing each woman-man edge by an arc from woman to man, with a supply of 1 at each woman node and a demand of 1 at each man node. Recall from Exercise 10.5.11 that the degree of a vertex is the number of its incident edges. A graph is k-**regular** if every vertex has degree k, and is **regular** if it is k-regular for some k. Thus the graph in Figure 11.1 is 3-regular. Consequently, its corresponding network is feasible, regardless of the arc costs: send $1/3$ on every arc.

(k-)regular graph

Workout 11.1.3 *Find an integral solution to the network corresponding to graph in Figure 11.1.*

These ideas lead us to König's Theorem 11.1.4, below. Recall from Section 1.2 that a matching is a set of pairwise disjoint edges. A matching is **perfect** if every vertex is in some edge of the matching.

perfect matching

G. H. Hurlbert, *Linear Optimization*, Undergraduate Texts in Mathematics,
DOI: 10.1007/978-0-387-79148-7_11, © Springer Science+Business Media LLC 2010

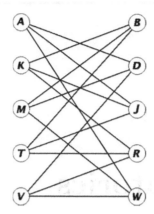

Figure 11.1: Bipartite graph for Problem 11.1.1

König's Theorem

Theorem 11.1.4 *Every regular bipartite graph has a perfect matching.*

Proof. Let $\mathcal{B}$ be a bipartite graph with bipartition $U = \{u_1, \ldots, u_r\}$, $V = \{v_1, \ldots, v_s\}$, and let k be its degree of regularity.

Workout 11.1.5 *Prove that $r = s$.*

Define the network $\mathcal{N}(\mathcal{B})$ to have nodes equal to the vertices of $\mathcal{B}$, with an arc $u_i v_j$ if and only if $\mathcal{B}$ has the edge $\{u_i, v_j\}$. Let all nodes u_i (resp. v_j) have supply (resp. demand) 1, and let the arc costs be arbitrary.

Workout 11.1.6 *Prove that $\mathcal{N}(\mathcal{B})$ has a feasible solution.*

Workout 11.1.7 *Use the Integrality Theorem 10.4.1 to finish the proof.*

◊

Corollary 11.1.8 *The edges of every regular bipartite graph can be partitioned into perfect matchings.*

Workout 11.1.9 *Prove Corollary 11.1.8.*

Notice that we can also formulate the Problem 11.1.1 in terms of matrices. Define the $\{0,1\}$-matrix $\mathbf{A} = \mathbf{A}(\mathcal{B}) = (a_{i,j})$ to have a row (resp. column) for every left (resp. right) vertex of $\mathcal{B}$, with an entry of $a_{i,j} = 1$ for every edge $\{u_i, v_j\}$ in $\mathcal{B}$. Then two things stand out. First, if every entry is divided by 3 then the resulting matrix $\mathbf{D} = \mathbf{A}/3$ is doubly stochastic. Second, Corollary 11.1.8 yields a partition of $\mathbf{A}$ into permutation matrices, such as

$$\begin{pmatrix} 0 & 1 & 1 & 0 & 1 \\ 1 & 0 & 1 & 1 & 0 \\ 1 & 1 & 0 & 0 & 1 \\ 1 & 0 & 1 & 1 & 0 \\ 0 & 1 & 0 & 1 & 1 \end{pmatrix} = \begin{pmatrix} 0 & 0 & 0 & 0 & 1 \\ 0 & 0 & 0 & 1 & 0 \\ 1 & 0 & 0 & 0 & 0 \\ 0 & 0 & 1 & 0 & 0 \\ 0 & 1 & 0 & 0 & 0 \end{pmatrix}$$

11.2. Covers

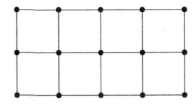

Figure 11.2: Street map for Gridburg

$$+\begin{pmatrix} 0 & 1 & 0 & 0 & 0 \\ 0 & 0 & 1 & 0 & 0 \\ 0 & 0 & 0 & 0 & 1 \\ 1 & 0 & 0 & 0 & 0 \\ 0 & 0 & 0 & 1 & 0 \end{pmatrix} + \begin{pmatrix} 0 & 0 & 1 & 0 & 0 \\ 1 & 0 & 0 & 0 & 0 \\ 0 & 1 & 0 & 0 & 0 \\ 0 & 0 & 0 & 1 & 0 \\ 0 & 0 & 0 & 0 & 1 \end{pmatrix}.$$

Coincidentally, this is precisely what is claimed by the Birkhoff–von Neumann Theorem 8.2.7. We revisit this theorem from a new perspective in Exercises 11.5.10 and 11.5.11.

11.2 Covers

Problem 11.2.1 *The Police Chief of Gridburg decides to place 15 policemen at the 15 street corners of his town, as shown in the map of Figure 11.2. A policeman has the ability to see the activity of people on the streets leading from his street corner, but only as far as one block. The Mayor fires the Chief for spending over budget, and hires Joseph Blough to make sure that every street can be seen by some policeman, using the fewest possible policemen.*

Workout 11.2.2 *Become the new Police Chief.*

Problem 11.2.3 *The Commerce Secretary of Gridburg decides to place 3 hot dog vendors on the 3 East-West streets of Gridburg (see the map of Figure 11.2). The Hotdogger's Union requires that no two vendors can be on street blocks that share an intersection. The Mayor fires the Secretary for not generating enough commerce in town, and hires Anna Benannaugh to place the maximum number of vendors, subject to union rules.*

Workout 11.2.4 *Become the new Commerce Secretary.*

Recall from Section 1.2 that a cover $\mathcal{C}$ in a graph $\mathcal{G}$ is a set of vertices such that every edge of $\mathcal{G}$ has at least one of its endpoints in $\mathcal{C}$. For example, the set of all vertices is a cover, and the job of the Police Chief of Gridburg in Problem 11.2.1 is to place policemen at the vertices of a cover in the Figure 11.2 graph. It is not difficult to find large covers in graphs; the challenge is to find small ones. Similarly, it is the job of the Commerce Secretary of Gridburg in Problem 11.2.3 to place vendors on the edges of a matching in the Figure 11.2 graph. It is likewise not difficult to find small matchings (such as the empty matching); the challenge is in finding large ones.

Workout 11.2.5 *Prove that every graph has $|\mathcal{M}| \leq |\mathcal{C}|$ for every matching $\mathcal{M}$ and cover $\mathcal{C}$.*

maximum matching

minimum cover

A matching $\mathcal{M}$ is **maximum** if no other matching contains more edges than $\mathcal{M}$. The new Commerce Secretary in Problem 11.2.3 must find a maximum matching in the Figure 11.2 graph. A cover $\mathcal{C}$ is **minimum** if no other cover contains fewer vertices than $\mathcal{C}$. The new Police Chief in Problem 11.2.1 must find a minimum cover in the Figure 11.2 graph. As usual, we use the notations $\mathcal{M}^*$ and $\mathcal{C}^*$ to denote any maximum matching and minimum cover, respectively.

Workout 11.2.6 *For each of the following two graphs, find a maximum matching $\mathcal{M}^*$ and a minimum cover $\mathcal{C}^*$ possible.*

a.

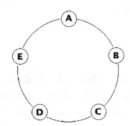

b.

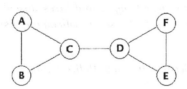

Workout 11.2.7 *Find a sequence of graphs $\{\mathcal{G}_n\}$ for which $|\mathcal{C}^*(\mathcal{G}_n)| - |\mathcal{M}^*(\mathcal{G}_n)| \to \infty$ as $n \to \infty$.*

Workout 11.2.6 illustrates that odd cycles in a graph may cause $|\mathcal{M}^*| < |\mathcal{C}^*|$. Exercises 11.5.3 and 11.5.20 show that a graph is bipartite if and only if it contains no odd cycles. Thus, Workouts 11.2.2 and 11.2.4 illustrate the following theorem.

König–Egerváry Theorem

Theorem 11.2.8 *Every bipartite graph has $|\mathcal{M}^*| = |\mathcal{C}^*|$.*

Proof. In light of Workout 11.2.5, we need only prove that $|\mathcal{C}^*| \leq |\mathcal{M}^*|$. We will do this by associating a network to a given bipartite graph. From the optimal solution to the network we will find both $\mathcal{M}^*$ and $\mathcal{C}^*$, and then show that the inequality holds. Several illustrations along the way should help prove the various steps.

11.2. Covers

Let $\mathcal{B}$ be a bipartite graph with left vertices $L_1, \ldots L_s$ and right vertices $R_1, \ldots, R_t$. Without loss of generality we assume that $\mathcal{B}$ is connected. We define the network $\mathcal{N} = \mathcal{N}(\mathcal{B})$ to have supply nodes $L_1, \ldots, L_s$ (each of supply 1) and demand nodes $R_1, \ldots, R_t$ (each of demand 1), with two extra "infinity" nodes L_∞ (having demand s) and R_∞ (having supply t). There is a 0-cost arc $L_iR_j \in \mathcal{N}$ for every edge $\{L_i, R_j\} \in \mathcal{B}$, and there are 0-cost arcs L_iL_∞ and $R_\infty R_j$ in $\mathcal{N}$ for every $1 \le i \le s$ and $1 \le j \le t$. Finally, there is a special "infinity" arc $R_\infty L_\infty \in \mathcal{N}$, having cost -1.

Workout 11.2.9 *Draw the network $\mathcal{N}_0$ that corresponds to the bipartite graph $\mathcal{B}_0$, below.*

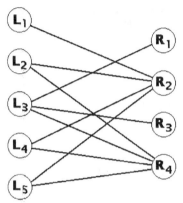

By shipping 1 along each of the arcs L_iL_∞ and $R_\infty R_j$, we see that $\mathcal{N}$ is feasible. Also, there are no negative cycles (since there are no oriented circuits), and so $\mathcal{N}$ is bounded. Thus, by the General Fundamental Theorem 6.2.9 and the Integrality Theorem 10.4.1, $\mathcal{N}$ has an optimal integral solution S^*.

Workout 11.2.10 *Find an optimal integral solution S_0^* to the network $\mathcal{N}_0$.*

Workout 11.2.11

    a. *For each feasible integral solution S to $\mathcal{N}$ find a corresponding matching $\mathcal{M}$ in $\mathcal{B}$.*

    b. *For each matching $\mathcal{M}$ in $\mathcal{B}$ find a corresponding feasible integral solution S to $\mathcal{N}$.*

    c. *Prove that the size of the matching $\mathcal{M}$ in $\mathcal{B}$ corresponding to a feasible integral solution S to $\mathcal{N}$ equals the negative of the cost of S.*

Workout 11.2.12 *Find a minimum cover $\mathcal{C}_0^*$ in $\mathcal{B}_0$. Compare $\mathcal{C}_0^*$ with your relative prices from S_0^*.*

Given the optimal integral solution S^*, define the relative prices of the nodes by starting with the price of 0 at L_∞. By the Network Basis Theorem

10.2.7, the basis for S^* is a spanning tree T^* of $\mathcal{N}$, and by Exercise 10.2.2 every node v is connected to node R_∞ by a unique path $\mathcal{P}_v$ in T^*. Then the relative price at v equals 0 if $\mathcal{P}_v$ contains L_∞ and equals 1 otherwise. Now define $\mathcal{C}^*$ to be the left vertices of $\mathcal{B}$ whose corresponding nodes have price 1 in T^* and the right vertices of $\mathcal{B}$ whose corresponding nodes have price 0 in T^*.

Workout 11.2.13 *Prove that $\mathcal{C}^*$ is a cover in $\mathcal{B}$.*

Workout 11.2.14 *Finish the proof by finding an injection from $\mathcal{C}^*$ into some maximum matching of $\mathcal{B}$.*

◇

Note that this theorem does not say that being bipartite is a requirement for having $|\mathcal{M}^*| = |\mathcal{C}^*|$.

Workout 11.2.15 *Find the smallest nonbipartite graph having $|\mathcal{M}^*| = |\mathcal{C}^*|$.*

11.3 もっと練習しましょう

Workout 11.3.1 *Use the network methods of Section 11.1 to partition the edges of the bipartite graph below into perfect matchings.*

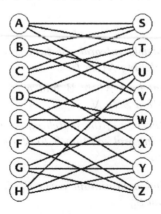

Workout 11.3.2 *Consider the following bipartite graph $\mathcal{B}$.*

11.4. Systems of Distinct Representatives

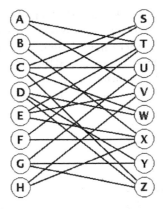

a. Find by hand a matching $\mathcal{M}$ in $\mathcal{B}$ that is as large as possible. Argue why it is of maximum size.

b. Find by hand a set $\mathcal{C}$ of vertices of $\mathcal{B}$, as small as possible, so that every edge of $\mathcal{B}$ has at least one endpoint in $\mathcal{C}$. Argue why it is of minimum size.

11.4 Systems of Distinct Representatives

Problem 11.4.1 *Consider the following six committees of students. Committee 1: Clifford, Kara; Committee 2: Ben, Donyell, Jake, Rebecca; Committee 3: Clifford, Kara, Sue; Committee 4: Ben, Jake, Kara, Nykesha, Rebecca; Committee 5: Clifford, Sue; Committee 6: Kara, Sue. Each committee must choose a representative to send to the school Senate, but the Senate requires that no person represent more than one committee. Is this possible?*

An ordered family $\mathcal{F} = (F_1, \ldots, F_k)$ of sets has a **system of distinct representatives** (SDR) $E = (e_1, \ldots, e_k)$ if $e_i \in F_i$ for every $1 \leq i \leq k$ and $e_i \neq e_j$ whenever $i \neq j$. Thus Problem 11.4.1 is asking for an SDR, where F_i is the set of people on Committee i.

system of distinct representatives

Workout 11.4.2 *Let $\mathcal{F} = (A, B, C, D, E, F, G, H)$, where $A = \{S, T, V\}$, $B = \{S.U, V\}$, $C = \{S, T, X\}$, $D = \{T, W, Y\}$, $E = \{U, W, Z\}$, $F = \{V, X, Z\}$, $G = \{W, Y, Z\}$ and $H = \{U, X, Y\}$. Find an SDR for $\mathcal{F}$.*

It is simple to see that the committees in Problem 11.4.1 have no SDR: the union of Committees 1, 3, 5 and 6 contain only 3 people (Clifford, Kara and Sue). Such a collection $\mathcal{K} = \{F_1, F_3, F_5, F_6\}$ is called an SDR blocker. More formally, a collection $\mathcal{K} = \{F_i \mid i \in I\}$ is called an **SDR blocker** for the family $\mathcal{F} = \{F_1, \ldots, F_k\}$ if $|\cup_{i \in I} F_i| < |I|$.

SDR blocker

Workout 11.4.3 *Use network methods to find the largest set of distinct representatives possible in Problem 11.4.1.*

Notice that the SDR blocker for Problem 11.4.1 consists of those left vertices of its bipartite graph that are missing from its minimum cover C^*. This suggests that, while there may be many different ways to prove that a given set system might not have an SDR, there will always be the nice way: find an SDR blocker (construct one from a minimum cover). This is the set system version of the Theorem of the Alternative 7.2.5.

Hall's Theorem

Theorem 11.4.4 *A family of sets has no SDR if and only if it contains an SDR blocker.*

Proof. Let $\mathcal{F} = \{F_1, \ldots, F_s\}$ be a family of sets. Without loss of generality we may relabel the elements so that $\cup_{i=1}^{s} F_i = \{R_1, \ldots, R_t\}$. Now define the bipartite graph $\mathcal{B} = \mathcal{B}(\mathcal{F})$ as having left vertices $F_1, \ldots, F_s$ and right vertices $R_1, \ldots, R_t$, with an edge $\{F_i, R_j\}$ precisely when $F_i \ni R_j$. Let $\mathcal{M}^*$ and $\mathcal{C}^*$ be a maximum matching and minimum cover, respectively, in $\mathcal{B}$. By the König-Egerváry Theorem 11.2.8 and the hypothesis, we have $|\mathcal{C}^*| = |\mathcal{M}^*| < s$. Finally, define $\mathcal{K} = \{F_i \mid F_i \notin \mathcal{C}^*\} = \mathcal{F} - \mathcal{C}^*$.

Workout 11.4.5 *Prove that $\mathcal{K}$ is an SDR blocker for $\mathcal{F}$.*

◇

11.5 Exercises

Practice

11.5.1 *Use the Network Environment to solve Problem 1.2.4.*

11.5.2 *How many sets of 5 good marriages are there in Problem 11.1.1?*

11.5.3 *Prove that a bipartite graph has no odd cycles.*

maximal matching

11.5.4 *A matching $\mathcal{M}$ is **maximal** if no larger matching contains $\mathcal{M}$. Find a maximal matching in the graph from Workout 11.2.6b that is not maximum.*

11.5.5 *What do Problems 11.2.2 and 11.2.4 have to do with the game of Battleship? Use the König-Egerváry Theorem 11.2.8 to prove your result from Exercise 9.5.38.*

11.5.6 *Let $\mathcal{F} = (A, B, C, D, E, F, G, H)$, where $A = \{T, U\}$, $B = \{S, V, W\}$, $C = \{S, T\}$, $D = \{U, V, W, Y\}$, $E = \{X, Y\}$, $F = \{U, X\}$, $G = \{Z\}$ and $H = \{X, Y, Z\}$. Find an SDR or prove that none exists.*

11.5.7 *Let $\mathcal{F} = (A, B, C, D, E, F, G, H)$, where $A = \{T, V\}$, $B = \{T, X\}$, $C = \{S, U, W, Z\}$, $D = \{S, W, Y, Z\}$, $E = \{T, V, X\}$, $F = \{T, X\}$, $G = \{U, Y, Z\}$ and $H = \{V, X\}$. Find an SDR or prove that none exists.*

11.5. Exercises

11.5.8 Let $\mathcal{X}$ (resp. $\mathcal{Y}$) be the collection of all subsets of $\{1,\ldots,n\}$ of size 2 (resp. 3), where $n \geq 5$. For $X \in \mathcal{X}$ define $F_X = \{Y \in \mathcal{Y} \mid X \subset Y\}$, and let $\mathcal{F} = \{F_X \mid X \in \mathcal{X}\}$. Use Hall's Theorem 11.4.4 to prove that $\mathcal{F}$ has an SDR.

11.5.9 Prove that every 2×2 doubly stochastic matrix can be written uniquely as a convex combination of permutation matrices.

11.5.10 Let $\mathbf{A} = (a_{i,j})$ be a nonnegative $n \times n$ matrix whose rows and columns sum to a fixed constant c. Define the matrix $\mathbf{D} = \mathbf{A}/c$ (then $\mathbf{D} = (d_{i,j})$ is doubly stochastic — see Chapter 8). Now define the bipartite network $\mathcal{N} = \mathcal{N}(\mathbf{D})$ as having (left) supply nodes $u_1, \ldots, u_n$, each of supply 1 (representing the rows of $\mathbf{D}$) and (right) demand nodes $v_1, \ldots, v_n$, each of demand 1 (representing the columns of $\mathbf{D}$), with an arc $u_i v_j$ (of arbitrary cost) if and only if $d_{i,j} > 0$.

    a. Prove that $\mathcal{N}$ is feasible.

    b. Use the Integrality Theorem 10.4.1 and part a to prove that there is a permutation matrix $\mathbf{P} = (p_{i,j})$ so that $a_{i,j} > 0$ whenever $p_{i,j} = 1$. [HINT: See Workout 11.1.7.]

11.5.11 Let $\mathbf{D}$ be a doubly stochastic matrix and consider the following algorithm.

1. Let $\mathbf{A}_1 = \mathbf{D}$.
2. Let $i = 1$.
3. While $v A_i \neq \mathbf{0}$ do

    a. Let $\mathbf{P}_i$ be a permutation matrix guaranteed by Lemma 11.5.10.

    b. Let t_i be the minimum of the entries of $\mathbf{A}_i$ that correspond to the 1s in $\mathbf{P}_i$.

    c. Define $\mathbf{A}_{i+1} = \mathbf{A}_i - t_i \mathbf{P}_i$.

    d. Increment i by 1.

    a. Prove for each i that the rows and columns of $\mathbf{A}_i$ sum to a fixed constant.

    b. Prove that the algorithm in Exercise 11.5.11 halts by proving that the number of iterations of Step 3 is finite.

    c. Prove that $\mathbf{D}$ is a convex combination of the $\mathbf{P}_i$s.

    d. Use parts (a,b,c) to prove the Birkhoff–von Neumann Theorem 8.2.7.

11.5.12 Find a 3×3 doubly stochastic matrix for which the number of iterations of Step 3 in the algorithm in Exercise 11.5.11 varies depending on the permutation matrices found in Step 3a.

Workout 11.5.1 *Illustrate the algorithm in Exercise 11.5.11 on the following matrix.*

$$\mathbf{D} = \begin{pmatrix} .5 & .1 & .2 & .2 \\ .3 & .2 & .2 & .3 \\ .1 & .4 & .3 & .2 \\ .1 & .3 & .3 & .3 \end{pmatrix}$$

11.5.13 *Provide an example of a 4×4 doubly stochastic matrix $\mathbf{D}$ that could require 10 iterations of Step 3 in the algorithm in Exercise 11.5.11.*

11.5.14 *Prove that the number of iterations of Step 3 in the algorithm in Exercise 11.5.11 is at most n^2.*

11.5.15 *Prove that the number of iterations of Step 3 in the algorithm in Exercise 11.5.11 is at most $n^2 - n + 1$.*

11.5.16 *Prove that if a 3×3 doubly stochastic matrix $\mathbf{D}$ can be written as a convex combination of the permutation matrices $\mathbf{P}_1, \ldots, \mathbf{P}_k$, then the coefficients $t_1, \ldots, t_k$ are unique.*

Challenges

11.5.17 *Let $\mathbf{A} = (a_{i,j})$ be an $n \times n$ matrix. Recall (from any linear algebra text) the definition of the determinant $\det(\mathbf{A}) = \sum_{\sigma \in \mathcal{S}_n} \mathrm{sign}(\sigma) \prod_{i=1}^n a_{i,\sigma_i}$, where $\mathcal{S}_n$ is the set of all permutations of $\{1, \ldots, n\}$ and $\mathrm{sign}(\sigma)$ is the sign of the permutation σ. Define the* **permanent** $\mathrm{per}(\mathbf{A}) = \sum_{\sigma \in \mathcal{S}_n} \prod_{i=1}^n a_{i,\sigma_i}$. *Now consider the $\{0,1\}$-matrix $\mathbf{A}$ associated with Problem 11.1.1.*

 a. *Compute $\mathrm{per}(\mathbf{A})$.*

 b. *Relate your answer to part a to your answer to Exercise 11.5.2.*

11.5.18 *Find the number of perfect matchings in the bipartite graph of Workout 11.3.1.*

11.5.19 *Prove that a tree is bipartite.*

11.5.20 *Use Exercise 11.5.19 to prove that a graph having no odd cycles is bipartite.*

11.5.21 *One way of generating a random k-regular bipartite graph is to take the union of k random perfect matchings. The problem, however, is that some of the matchings may share edges, so that their union is not regular. Let $k = 2$ and suppose there are n vertices in each part.*

 a. *Compute the expected number of edges two random matchings share. [HINT: Restate in terms of permutations.]*

11.5. Exercises

b. Compute the probability that two random perfect matchings share no edges. [HINT: This requires the Principle of Inclusion-Exclusion.]

11.5.22 A cover $\mathcal{C}$ is **minimal** if no smaller cover contains $\mathcal{C}$. Find a sequence of graphs $\{\mathcal{G}_n\}$ for which there exist minimal covers $\{\mathcal{C}_n\}$ with $|\mathcal{C}_n|/|\mathcal{C}^*(\mathcal{G}_n)| \to \infty$.

minimal cover

11.5.23 Use Carathéorody's Theorem 3.4.1 and the Birkhoff–von Neumann Theorem 8.2.7 to prove that every $n \times n$ doubly stochastic matrix is a convex combination of at most $n^2 - 2n + 2$ permutation matrices. [HINT: It may be most useful to think in terms of the proof of Theorem 8.2.7.]

11.5.24 Prove for all $n \geq 4$ that the algorithm in Exercise 11.5.11 never requires more than $n^2 - 2n + 2$ iterations of Step 3 on an $n \times n$ doubly stochastic matrix $\mathbf{D}$.

11.5.25 Prove for all $n \geq 4$ that there exists an $n \times n$ doubly stochastic matrices $\mathbf{D}$ that could require $n^2 - 2n + 2$ iterations of Step 3 in the algorithm in Exercise 11.5.11.

11.5.26 Prove that if the algorithm in Exercise 11.5.11 writes the doubly stochastic matrix $\mathbf{D}$ as the convex combination of the permutation matrices $\mathbf{P}_1, \ldots, \mathbf{P}_k$, then the coefficients $t_1, \ldots, t_k$ are unique, regardless of the order that the $\mathbf{P}_i$ may have arisen.

11.5.27 Let $\mathcal{M}$ be a matching in a graph $\mathcal{G}$. A path $\mathcal{P}$ in is called $\mathcal{M}$-**alternating** if $\mathcal{P} - \mathcal{M}$ is a also a matching in $\mathcal{G}$, and $\mathcal{P}$ is $\mathcal{M}$-**augmenting** if it is $\mathcal{M}$-alternating and no edge of $\mathcal{M}$ touches its endpoints. Prove that if $\mathcal{P}$ is an augmenting path for $\mathcal{M}$ then $\mathcal{M} \Delta \mathcal{P}$ is a larger matching than $\mathcal{M}$.

alternating/augmenting path

11.5.28 Let $\mathcal{M}$ be a maximal matching of $\mathcal{G}$ that is not maximum. Prove that there is an $\mathcal{M}$-augmenting path. [HINT: First prove there is an $\mathcal{M}$-alternating path.]

11.5.29 Use Exercises 11.5.27 and 11.5.28 to design an algorithm for finding a maximum matching in a graph $\mathcal{G}$.

11.5.30 Use Exercise 11.5.27 to prove the König–Egerváry Theorem 11.2.8.

11.5.31 Use Hall's Theorem 11.4.4 to prove the König–Egerváry Theorem 11.2.8. [HINT: Consider an SDR blocker $\mathcal{K} = \{F_i \mid i \in I\}$ with largest difference $\delta(\mathcal{K}) = |I| - |\cup_{i \in I} F_i|$.]

11.5.32 *Consider the following matrix* $M = (m_{i,j})$.

$$M = \begin{pmatrix} 6 & 3 & 7 & 1 & 2 & 9 \\ 5 & 0 & 1 & 8 & 6 & 4 \\ 2 & 7 & 5 & 3 & 8 & 1 \\ 1 & 4 & 8 & 0 & 5 & 3 \\ 7 & 2 & 5 & 9 & 0 & 6 \\ 4 & 9 & 3 & 6 & 3 & 7 \end{pmatrix}$$

Find a permutation matrix whose corresponding entries in M have the greatest sum.

11.5.33 *Let $\mathcal{X}$ (resp. $\mathcal{Y}$) be the collection of all subsets of $\{1, \ldots, n\}$ of size r (resp. s), where $r < s \leq n - r$. For $X \in \mathcal{X}$ define $F_X = \{Y \in \mathcal{Y} \mid X \subset Y\}$, and let $\mathcal{F} = \{F_X \mid X \in \mathcal{X}\}$. Use Hall's Theorem 11.4.4 to prove that $\mathcal{F}$ has an SDR. [HINT: See Exercise 11.5.8.]*

Projects

11.5.34 *Present the* **Stable Marriage Theorem**.

11.5.35 *Present* **Dilworth's Theorem**.

11.5.36 *Present the* **Hungarian Method** *for the* **Assignment problem**.

11.5.37 *Present the* **fractional chromatic number** *of a graph and its relation to LO.*

Chapter 12

Economics

12.1 Shadow Prices

In this section we begin to perform **sensitivity** (or **post-optimality**) **analysis**, which amounts to figuring out how the optimum solution might change under small changes in the LOP. For example, an investment company might need to solve roughly the same problem every day, with only the stock prices changing. Or maybe an airline might consider adding a new route because the current schedule suggests that the resulting income might outweigh its costs. Also, a production company might realize the optimal solution isn't valid because they forgot to include a necessary constraint — could the new problem be solved without starting again from scratch? It may help first to figure out what the dual variables really mean.

sensitivity (post-optimality) analysis

Consider the plight of folk legend Öreg MacDonald.

Problem 12.1.1 *Öreg MacDonald owns* 1,000 *acres of land and is contemplating conserving, farming, and/or developing it. His annual considerations are as follows. It will only cost him* $1 *per acre in registration fees to own conservation land, and he will reap* $30 *per acre in tax savings. Farming will cost him* $50 *per acre for seeds, from which he can earn* $190 *per acre by selling vegetables. He can earn* $290 *per acre by renting developed land, which costs* $85 *per acre in permits. Öreg has only* $40,000 *to use, and is also bound by having only* 75 *descendants, each of whom can work at most* 2,000 *hours. How should he apportion his acreage in order to maximize profits, if conservation, farming, and development uses* 12, 240, *and* 180 *hours per acre, respectively?*

The LOP he needs to solve is the following.

Problem 12.1.2

$$\text{Max. } z = 29x_1 + 140x_2 + 205x_3$$

$$\text{s.t. } \begin{array}{rcrcrcr} x_1 & + & x_2 & + & x_3 & \leq & 1{,}000 \\ x_1 & + & 50x_2 & + & 85x_3 & \leq & 40{,}000 \\ 12x_1 & + & 240x_2 & + & 180x_3 & \leq & 150{,}000 \end{array}$$

$$\& \quad x_1, \quad x_2, \quad x_3 \geq 0$$

Here, x_1, x_2, and x_3 are the amounts of land (in acres) that Öreg will use for conservation, farming, and development, respectively. A few Simplex pivots will reveal the optimal solution $z^* = 1251000000/10920$ (about $\$114{,}560.44$) at $\mathbf{x}^* = (3750000, 5040000, 2130000)/10920$ (roughly $(343.41, 461.54, 195.05)$). The optimal tableau also displays the optimal dual solution at $\mathbf{y}^* = (286800, 21480, 700)/10920 \approx (26.26, 1.97, .06)$. But, other than providing a certificate of optimality as primal constraint multipliers, what do these dual variables mean?

Take a closer look at the third constraint. The 12 represents the annual number of hours per acre required to work conservation land. The other coefficients represent similar hours-to-acres ratios, meaning that the dimensions of the constraint look something like

$$\left(\frac{\text{hours}}{\text{cons acres}}\right)(\text{cons acres}) + \left(\frac{\text{hours}}{\text{farm acres}}\right)(\text{farm acres})$$

$$+ \left(\frac{\text{hours}}{\text{dev acres}}\right)(\text{dev acres}) \leq \text{hours},$$

activity/ since x_j is the number of acres put to the jth **activity** (or **product**).
product   We can analyze the second dual constraint $y_1 + 50y_2 + 240y_3 \geq 140$ similarly. The coefficients, in order, represent the ratio of acres, dollars, and hours to farm acres, with the right-hand result of profit per farm acre. Thus we have constraint

$$\left(\frac{\text{acres}}{\text{farm acres}}\right)y_1 + \left(\frac{\text{dollars}}{\text{farm acres}}\right)y_2 + \left(\frac{\text{hours}}{\text{farm acres}}\right)y_3 \geq \left(\frac{\text{profit}}{\text{farm acres}}\right),$$

giving the impression in this case that y_i must be the profit per i^{th} re-
resource **source**. Indeed, this interpretation works for the other two constraints as well.

Just as the primal objective function has dimension

$$\left(\frac{\text{profit}}{\text{cons acres}}\right)(\text{cons acres}) + \left(\frac{\text{profit}}{\text{farm acres}}\right)(\text{farm acres})$$

$$+ \left(\frac{\text{profit}}{\text{dev acres}}\right)(\text{dev acres}) = \text{profit},$$

12.1. Shadow Prices

the dual objective function has dimension

$$(\text{acres})\left(\frac{\text{profit}}{\text{acre}}\right) + (\text{dollars})\left(\frac{\text{profit}}{\text{dollar}}\right) + (\text{hours})\left(\frac{\text{profit}}{\text{hour}}\right) = \text{profit}.$$

Hence we may view the primal as maximizing the profit of activities, subject to resource constraints, and the dual as minimizing the profit of resources, subject to activity constraints. Each constraint coefficient $a_{i,j}$ measures the amount of resource i used up by one unit of activity j, and each dual variable y_i measures the worth of one unit of resource i. Let us define the **shadow price** (or **marginal value**) of resource i to be the amount that the optimum objective value increases per unit increase in resource i, just to see how this compares with our concept of worth.

shadow price (marginal value)

Workout 12.1.3 *Re-solve Problem 12.1.2 with each of the following revised resource values.*

a. Öreg has 999, 1,001, and 1,002 acres.

b. Öreg has 39,999, 40,001, and 40,002 dollars.

c. Öreg has 149,999, 150,001, and 150,002 hours.

d. Öreg has (simultaneously) 999 acres, 40,001 dollars, and 150,002 hours.

Öreg could decide to pay extra workers in order to increase the number of hours available, but according to these calculations he would have to pay them less than 6¢/hr to increase his profits. He could also take out a loan to increase his available cash, which seems smart from the above results — an 18% loan still leaves him an extra 79¢ profit for every extra dollar borrowed. It looks like his most valuable resource, as every real estate agent will tell you, is land, although it is doubtful he can buy extra acres at less than 26 bucks a pop. But can he simply max out his credit card and make as much money as he wants?

Well, not so fast, take a look at what happens when he borrows $30,000. At this point he doesn't farm any land. Once he passes the $70,000 mark, the {conserve, farm, develop} basis is no longer feasible, extra pivots are necessary (the best idea is to perform a Dual Simplex pivot — see Section 12.4), and the optimal value stays the same. In fact, one observes that $y_2^* = 0$, so more money has no extra worth. Thus we should keep in mind that the nice formula developed in Workout 12.1.3 only applies locally, for "small" changes in resources (if you think of thirty grand as small!).

When all is said and done, it is financially prudent for Öreg MacDonald to take the $30,000 loan, develop the farm, and make an extra $53,611 ($59,011 minus interest). As Joni wrote, "They paved paradise and put up a parking lot." At least he still has almost 179 acres of conservation.

o

Shadow Approximation Theorem

Theorem 12.1.4 *Let P be the LOP*

$$\text{Max.} \sum_{j=1}^{n} c_j x_j \quad \text{s.t.} \sum_{j=1}^{n} a_{i,j} x_j \leq b_i \ (1 \leq i \leq m), \quad x_j \geq 0 \ (1 \leq j \leq n)$$

and, for $\boldsymbol{\delta} = (\delta_1, \ldots, \delta_m)^\mathsf{T}$ *let* $P_{\boldsymbol{\delta}}$ *be the LOP*

$$\text{Max.} \sum_{j=1}^{n} c_j x_j \quad \text{s.t.} \sum_{j=1}^{n} a_{i,j} x_j \leq b_i + \delta_i \ (1 \leq i \leq m), \quad x_j \geq 0 \ (1 \leq j \leq n).$$

If P has a nondegenerate basic optimal solution z^*, then there is some $\epsilon > 0$ so that, if every $|\delta_i| < \epsilon$, then $P_{\boldsymbol{\delta}}$ has optimum $z^{\boldsymbol{\delta}} = z^* + \sum_{i=1}^{m} y_i^* \delta_i$.

Proof. It is simplest to use the matrix formulation of solutions from Chapter 5. The optimal solution z^* can be written as $\mathbf{y}^{*\mathsf{T}} \mathbf{b}$, occurring at $\mathbf{x}^* = \mathbf{B}^{-1} \mathbf{b}$. Adding $\boldsymbol{\delta}$ to $\mathbf{b}$ ($\mathbf{b}^{\boldsymbol{\delta}} = \mathbf{b} + \boldsymbol{\delta}$) results in $z^{\boldsymbol{\delta}} = z^* + \mathbf{y}^{*\mathsf{T}} \boldsymbol{\delta}$ at $\mathbf{x}^{\boldsymbol{\delta}} = \mathbf{x}^* + \mathbf{B}^{-1} \boldsymbol{\delta}$. Since $\mathbf{x}^* > 0$ there is some $\epsilon > 0$ so that if each $|\delta_i| < \epsilon$ then $\mathbf{x}^* + \mathbf{B}^{-1} \boldsymbol{\delta} \geq 0$. Hence $\mathbf{x}^{\boldsymbol{\delta}}$ is feasible. Since the objective row is unchanged, $\mathbf{x}^{\boldsymbol{\delta}}$ is optimal, with optimum $z^{\boldsymbol{\delta}}$. ◇

To see an example of the boundedness of $\boldsymbol{\delta}$ in the Shadow Approximation Theorem 12.1.4, consider the following LOP, whose feasible region is shown in Figure 12.1.

Problem 12.1.5

$$
\begin{aligned}
\text{Max.} \quad z \ &= \ 28x_1 \ + \ 27x_2 \\
\text{s.t.} \quad \quad \quad & \quad \ \ 6x_1 \ + \ \ 5x_2 \ \leq \ 108 \\
& \quad \ \ 4x_1 \ + \ \ 7x_2 \ \leq \ 113 \\
& \ \ 10x_1 \ + \ \ \ x_2 \ \leq \ 110 \\
& \quad \ \ \ x_1 \ + \ 10x_2 \ \leq \ 130 \\
\& \quad \quad \quad \quad & \quad \ \ \ x_1 \ , \ \ x_2 \ \geq \ 0
\end{aligned}
$$

Its optimal solution occurs at $(191, 246)/22 \approx (8.7, 11.2)$, having basis $\{1, 2, 5, 6\}$. Adding δ_1 to 108 should increase the optimal value of $11{,}990/22 = 545$ by $4\delta_1$, since $88/22 = 4$ is the appropriate shadow price. But δ_1 cannot be too large for this to hold.

Note that a negative δ_1 moves the first constraint line toward the origin in a parallel fashion. This has the effect of sliding the optimal point up and to the left, until it reaches the $x_1 + 10x_2 \leq 130$ constraint boundary, when $\delta_1 = -19/3$. As δ_1 decreases further the optimal point slides along the line $x_1 + 10x_2 = 130$ instead, with basis $\{1, 2, 4, 5\}$, changing the rate at which the optimum value decreases (the new shadow price is $253/55$).

Similarly, for positive δ_1 the optimal point slides down and to the right until it reaches the $10x_1 + x_2 \leq 110$ constraint when $\delta_1 = 4$. Now a slightly different phenomenon occurs: as δ_1 increases further we find that

12.1. Shadow Prices

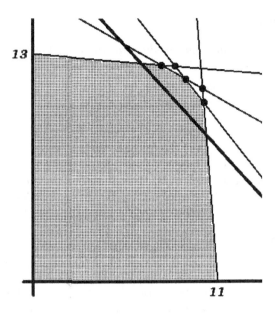

Figure 12.1: Feasible region for LOP 12.1.5

the optimal point stays fixed at the basis $\{1, 2, 3, 6\}$, incurring no more objective increase. This coincides with the new shadow price of 0, as the constraint is redundant now.

In both cases we witness the restricted range of δ in Theorem 12.1.4, as well as the necessity of the optimal solution being nondegenerate — in the degenerate cases, the rate of change in z^* depends on which direction you move.

Furthermore, one sees from the tableau that a change by δ_i in b_i adds $\delta_i(\mathbf{B}^{-1})_i$ (the column of the optimal tableau T^* corresponding to the ith slack variable x_{n+i}) to $\mathbf{b}^*$ (this is similar to our Phase II analysis). For example, a change in 108 incurs alterations in $\mathbf{x}^* = (191, 246 \mid 0, 0, 264, 209)^{\mathsf{T}}/22$ by the appropriate portion of $(7, -4 \mid 0, 0, -66, 33)^{\mathsf{T}}/22$. Thus if we expect $\mathbf{x}^\delta$ to remain feasible we must have (again, just like our Phase II ratio calculations)

$$
\begin{aligned}
191 + 7\delta_1 &\geq 0, \\
246 - 4\delta_1 &\geq 0, \\
264 - 66\delta_1 &\geq 0, \text{ and} \\
209 + 33\delta_1 &\geq 0,
\end{aligned}
$$

which implies that

$$-19/3 = \max\{-191/7, -209/33\} \leq \delta_1 \leq \min\{246/4, 264/66\} = 4,$$

the bounds we obtained geometrically, above.

Workout 12.1.6

a. Draw the appropriate diagram to verify the bounds on δ_1 claimed in the above paragraphs.

b. Provide similar analyses for changes to 113, 10, and 12, both geometrically and algebraically.

How does one take advantage of Theorem 12.1.4? In the case that $\boldsymbol{\delta}$ is small enough, we know that the optimal basis is still optimal and we use the theorem and go home happy. We will know that $\boldsymbol{\delta}$ is small enough if $\mathbf{x}^{\boldsymbol{\delta}}$ is feasible, and a quick check of $\mathbf{B}^{-1}\mathbf{b}^{\boldsymbol{\delta}}$ will tell us whether or not this is so. For example, if $\delta_1 = -3$ above then $\mathbf{x}^{\boldsymbol{\delta}} = \mathbf{B}^{-1}\mathbf{b}^{\boldsymbol{\delta}} = (258, 50, 170, 6)^\mathsf{T}/22 \geq 0$ and $z^{\boldsymbol{\delta}} = \mathbf{c}^\mathsf{T}\mathbf{x}^{\boldsymbol{\delta}} = 11726/22 \ (= z^* + \mathbf{y}^{*\mathsf{T}}\boldsymbol{\delta})$, and these are the optimal values for the new problem.

But if $\mathbf{x}^{\boldsymbol{\delta}}$ is infeasible — for example, if $\delta_1 = 6$ above then $\mathbf{x}^{\boldsymbol{\delta}} = (222, -13, 233, 42)^\mathsf{T}/22$ — then what shall we do? One option is to take the problem home along with your laundry on the weekend and hope your folks do that for you also. Another option is to read about the Dual Simplex Algorithm in Section 12.4. It turns out that we won't have to start the new problem over from scratch. Instead, it is likely in general that the old optimal point is fairly close to the new optimal point, and so a few well chosen pivots from the old optimal point might get us there quickly. (This is often referred to as a hot start.) But if we use the Primal Simplex Algorithm we know and love, we'll have to pass through Phase I and Phase II, possibly taking a long time (although the current problem luckily only requires one more pivot). Dual Simplex has a better idea.

Workout 12.1.7 *Illustrate the above discussion using the extended tableau below.*

Tableau 12.1.8

$$\begin{bmatrix} 6 & 5 & 1 & 0 & 0 & 0 & 108 & 1 & 105 & 114 \\ 4 & 7 & 0 & 1 & 0 & 0 & 113 & 0 & 113 & 113 \\ 10 & 1 & 0 & 0 & 1 & 0 & 110 & 0 & 110 & 110 \\ 1 & 10 & 0 & 0 & 0 & 1 & 130 & 0 & 130 & 130 \\ \hline -28 & -27 & 0 & 0 & 0 & 1 & 0 & 0 & 0 & 0 \end{bmatrix}$$

Workout 12.1.9 *Interpret the dual variables for Problem 1.1.3 appropriately. Use them to calculate the marginal values) of the USRDAs.*

12.2 Reduced Costs

So far we have considered changes in the **b**-column (resources). Now we consider changes in the objective function coefficients (profits or costs per activity). Be careful here: when we work with a minimization problem we need to remember that the objective function coefficients are held in **b**.

12.2. Reduced Costs

Problem 12.2.1 *Biff has infestations of crickets, ants, and moths in his house. He estimates that there are 50 ounces of crickets, 20 ounces of ants, and 15 ounces of moths, and realizes that they must be removed before his girlfriend Muffy arrives in an hour. He could buy cockroaches, a pound of which would cost 48¢ and eat 5 ounces of ants and 3 ounces of moths in an hour. He could also purchase a pound of black widow spiders that would eat 6 ounces of crickets, 3 ounces of ants, and 2 ounces of moths per hour for 73¢. A pound of scorpions would eat 8 ounces of crickets and 4 ounces of ants per hour and would cost Biff 56¢. His final choice is to spend 93¢ for a pound of rough green snakes that would eat 11 ounces of crickets and 1 ounce of moths per hour.*

Workout 12.2.2 *Write Biff's LOP and find his optimal solution. Make note of the optimal objective coefficients* $\mathbf{b}^*$.

Workout 12.2.3 *Resolve Problem 12.2.1 with 47¢ cockroaches instead.*

Note the decrease of one unit in b_1^* resulting from the cost reduction of 1¢ in cockroaches. Will another penny decrease force another pivot because b_1^* will be negative (and thus not be optimal)?

Workout 12.2.4 *Resolve Problem 12.2.1 with 46¢ cockroaches instead.*

It seems reasonable that an appropriate cost reduction in snakes might put them in the optimal basis as well.

Workout 12.2.5 *Resolve Problem 12.2.1 with 92¢ snakes instead.*

Now we define the **reduced cost** of optimal nonbasic activity i (decision variable y_i) in a minimization LOP P to be the amount that the objective coefficient b_i must be reduced in order that y_i is basic in some optimal solution of P. From the above workouts, the reduced costs seem to be related to $\mathbf{b}^*$.

reduced cost

Workout 12.2.6 *Resolve Problem 12.2.1 with $(93 - \epsilon)$¢ snakes instead. For what value of ϵ is the originally optimal basis $\beta^* = \{2, 3, 6\}$ degenerate? [HINT: Treat ϵ as a new variable (like y_0) by introducing an extra 0th column in WebSim.]*

Theorem 12.2.7 *For a given minimization LOP P, the reduced cost of variable y_i is equal to the value of the optimal objective coefficient b_i^*.*

Reduced Cost Theorem

(Note that the theorem even holds when $i \in \beta^*$, since in that case $\mathbf{b}_i^* = 0$ and the ith cost needs no reduction because i is already in the optimal basis.)

Proof. For optimal parameters, one could use the WebSim idea of Workout 12.2.6 and note that the ϵ-column is the negative of the w-column.

ϵ-column

Workout 12.2.8 *Carry out the proof using this WebSim idea.*

An alternative proof uses the matrix formulation from Chapter 5, where, for a maximization problem, $c_\beta^T B^{-1} \Pi - c_\pi^T$ calculates the nonbasic coefficients of c^* when $\beta = \beta^*$. For a minimization problem the calculation looks like $b_\beta^T B^{-1} \Pi - b_\pi^T$.

Workout 12.2.9 *Carry out the proof using this matrix formulation.*

◊

Note that reducing any or all costs of optimal parameters by less than their individual reduced costs maintains the optimality of y^*, and consequently incurs no changes in w^*. However, as we can see from the matrix calculation $b_\beta^T B^{-1} \Pi - b_\pi^T$, any change in the cost of an optimal basic variable affects every entry of b^*. For example, if spiders become too expensive, Biff might have to buy cockroaches or snakes instead, right? In fact, he'll buy both if spiders cost more that 74¢. How do we find that critical cutoff? This leads us to define the **increased cost** of optimal basic activity i (decision variable y_i) in a minimization LOP P to be the amount that the objective coefficient b_i must be increased in order that the current optimal basis β^* becomes suboptimal. Understand, however, that it is not necessary that i leaves β^*, as this cannot be controlled (maybe i can never be optimal basic), only that β^* changes (see Workout 12.3.3, for example).

increased cost

Workout 12.2.10 *Resolve Problem 12.2.1 with*

  a. *74¢ spiders instead.*

  b. *57¢ scorpions instead.*

In each case, compare the resulting objective row to the original b^*. Where in the tableau do you find the difference between the two? Use your result to find the increased costs of spiders and scorpions.

Workout 12.2.11 *State and prove the Increased Cost Theorem.*

Increased Cost Theorem

The moral of these discussions is that one should pay attention to the reduced and increased costs of an optimal solution — when they are near zero it means that a significant change may be around the corner: either a nonbasic activity is about to be cheap enough to include, or a basic activity is almost expensive enough to potentially exclude. Get ready to pivot!

12.3 Gyakoroljon egy Kicsit

Problem 12.3.1 *Aussie Foods Co. makes three different emu pet foods in 10-kg bags. The Premium bag is a mixture of 5 kgs of kiwi fruit, 2 kgs of wattle leaves, 2 kgs of boab seeds, and 1 kg of ground diamond weevil, and AFC makes a profit of 91¢ per bag sold. The Regular bag mixes 4, 4, 0, and 2 kgs of kiwi, wattle, boab, and diamond, respectively, making 84¢ for AFC. The corresponding numbers for the Bargain bag are 1, 2, 3, 4, and*

73¢, respectively. The weekly supply available to AFC is $1,000$ kgs of kiwi fruit, $1,200$ kgs of wattle leaves, $1,500$ kgs of boab seeds, and $1,400$ kgs of ground diamond weevil.

Workout 12.3.2 *Consider Problem 12.3.1*

    a. *What numbers of bags should AFC produce of each kind in order to maximize their weekly profit (assuming that all they make is sold[1])?*

    b. *What are the shadow prices of each commodity?*

    c. *The following per-kg market prices are found at obscurefoods.com: 15¢ (kiwi), 12¢ (wattle), 13¢ (boab), and 15¢ (diamond). What should AFC do?*

    d. *Use* MAPLE *to illustrate your answers.*

The same concepts regarding changes in the objective function can be applied to maximization problems. One can define the **increased profit** of activity j (decision variable x_j) in a maximization LOP P to be the amount that the objective coefficient c_j must be increased in order that x_j is basic in some optimal solution of P. Likewise, the **reduced profit** of activity j (decision variable x_j) in a maximization LOP P is the amount that the objective coefficient c_j must be reduced in order that the current basis β^* becomes suboptimal.

increased profit

reduced profit

Workout 12.3.3 *What are the increased and reduced profits of each type of emu food bag in Problem 12.3.1? Illustrate your results with* MAPLE *.*

12.4 Dual Simplex

The previous sections have identified situations in which a given problem must be re-solved after slight modifications. These include changes in the values of **A**, **b**, and **c**, as well as the introduction of new variables and constraints (see Exercises 12.5.15, 12.5.18, and 12.5.16). This section describes the Dual Simplex Algorithm in order to handle solving the modified problems more efficiently than solving them from scratch. Starting with a given point other than the origin, especially one that is potentially nearoptimal, is often referred to as a **hot start**. We'll see another opportunity for its use in Chapter 13, in particular, with the addition of new constraints (called cutting planes). First, let's continue with Problem 12.1.5, with the case of $\delta_1 = 5$. The optimal Tableau 12.4.1 for the original problem determines the starting Tableau 12.4.2 for the modified problem, since $5(-4, -66, 7, 33, 88)^\mathsf{T}$ gives the difference in the **b**-column.

hot start

---

[1] Wouldn't that be a nice assumption to make in the real world!

Tableau 12.4.1

$$\begin{bmatrix} 0 & 22 & -4 & 6 & 0 & 0 & 0 & 246 \\ 0 & 0 & -66 & 44 & 22 & 0 & 0 & 264 \\ 22 & 0 & 7 & -5 & 0 & 0 & 0 & 191 \\ 0 & 0 & 33 & -55 & 0 & 22 & 0 & 209 \\ \hline 0 & 0 & 88 & 22 & 0 & 0 & 22 & 11990 \end{bmatrix}$$

Tableau 12.4.2

$$\begin{bmatrix} 0 & 22 & -4 & 6 & 0 & 0 & 0 & 226 \\ 0 & 0 & -66 & 44 & 22 & 0 & 0 & -66 \\ 22 & 0 & 7 & -5 & 0 & 0 & 0 & 226 \\ 0 & 0 & 33 & -55 & 0 & 22 & 0 & 374 \\ \hline 0 & 0 & 88 & 22 & 0 & 0 & 22 & 12430 \end{bmatrix}$$

Workout 12.4.3 Verify that $\mathbf{B}^{-1}\mathbf{b}^\delta$ (with $\beta = \{1,2,5,6\}$) makes the appropriate calculation.

Now hearken back to the glorious learnings of Section 4.4, where we discovered the primal-dual pairings of bases. Since $\beta_P^* = \{1,2 \mid 5,6\}$ for Problem 12.1.5, we have $\beta_D^* = \{1,2\}$ for its dual. In general, for a set $S \subseteq [t]$ and integer k, denote by $S + k$ the set $\{s + k \mid s \in S\}$, and by $\overline{S}$ its complement $[t] - S = \{s \in [t] \mid s \notin S\}$. Then if $\beta_P^* = J \cup (I + n)$ for a primal problem P having n decision variables and m inequalities, its dual problem D has $\beta_D^* = \overline{I} \cup (\overline{J} + m)$. In fact, we can extend the primal-dual basis pairings to define, for any primal basis $\beta_P = J \cup (I + n)$, its **dual basis partner** $\beta_D = \overline{I} \cup (\overline{J} + m)$.

dual basis partner

Workout 12.4.4 Suppose the LOP P has 7 variables and 12 inequalities. Find the dual basis partner of $\beta_P = \{2,3,5,7 \mid 8,9,10,14,16,17,18,19\}$.

It is instructive to look at the dual Tableau 12.4.5, corresponding to the basis $\beta_D = \{1,2\}$ for the dual of the modified primal.

Tableau 12.4.5

$$\begin{bmatrix} 22 & 0 & 66 & -33 & -7 & 4 & 0 & 88 \\ 0 & 22 & -44 & 55 & 5 & -6 & 0 & 22 \\ \hline 0 & 0 & -66 & 374 & 226 & 226 & 22 & -12430 \end{bmatrix}$$

The similarities of values in Tableaux 12.4.2 and 12.4.5 are striking. It's a little tricky to write down the pattern correctly, but let's try. First we need some notation. For any $h \in [m+n]$ define the function $f : [m+n] \to [m+n]$ by $f(h) = m + h$ if $h \leq n$ and $f(h) = h - n$ if $h > n$. Notice that $h \in \beta_P$ if and only if $f(h) \notin \beta_D$, where β_D is the dual basis partner of β_P. (Make note also that $f^{-1}(k) = n + k$ if $k \leq m$ and $f^{-1}(k) = k - m$ if $k > m$.) Denote by T^P (resp. T^D) the primal (resp. dual) tableau with basis β_P (resp. β_D), and let $r(h)$ (resp. $s(k)$) be the row of tableau T^P (resp. T^D) in which x_h (resp. y_k) is basic. Then what we observe, respectfully, is the following theorem.

12.4. Dual Simplex

Theorem 12.4.6 *With the notation as described above, for all $h \in \beta_P$ and all $k \in \pi_P$ we have $T^P_{r(h),k} = -T^D_{s(f(k)),f(h)}$. Moreover, the basic value of x_h in T^P equals the objective coefficient of $y_{f(h)}$ in T^D, and the basic value of $y_{f(k)}$ in T^D equals the objective coefficient of x_k in T^P.*
Primal-Dual Correspondence Theorem

We defer the proof to Exercise 12.5.21. Note that the correspondence actually holds for all $h, k \in [n+m]$. However, it's only relevent in the range mentioned since all values other than the basic coefficients are zero elsewhere.

Workout 12.4.7 *Verify Theorem 12.4.6 on the following example for*

$$
\begin{array}{rrrrrrrrr}
\text{Max. } z & = & 85x_1 & + & 51x_2 & + & 76x_3 & & \\
\text{s.t.} & & x_1 & + & 4x_2 & + & 3x_3 & \leq & 13 \\
& & 3x_1 & + & 9x_2 & - & 8x_3 & \leq & 18 \\
& & -5x_1 & - & 7x_2 & & & \leq & -20 \\
& & 8x_1 & + & 2x_2 & + & x_3 & \leq & 28 \\
& & -4x_1 & & & + & 6x_3 & \leq & 15 \\
\& & & x_1 & , & x_2 & , & x_3 & \geq & 0
\end{array}
$$

a. $\beta = \{1,2,3,4,5\}$.
b. $\beta = \{1,2,4,5,8\}$.
c. $\beta = \{1,2,3,5,8\}$.

The second part of Theorem 12.4.6 implies that the basic solution $\mathbf{y}_{\beta^D}$ is dual feasible. That is because the objective coefficients in T^P are non-negative, which, in turn, is due to the fact that only the **b**-column of T^P differs from the optimal tableau T^*. Thus, the pivots on T^D are already in Phase II. For example, we would pivot next on the 7 in Tableau 12.4.5. But by Primal-Dual Correspondence again, that is equivalent to pivoting on the -7 in Tableau 12.4.2. Holy cow! This means that we don't need to actually pivot in the dual, we can merely reinterpret the dual Phase II rule in the primal tableau. Instead of pivoting on the positive entry in column k of T^D that has the smallest **b**-ratio, we can simply pivot on its dual partner in T^P — that is, on the negative entry in row $r(f^{-1}(h))$ having the greatest **c**-ratio. This is the Dual Simplex Algorithm.[2] c-ratio

After one Dual Simplex pivot on Tableau 12.4.2 we arrive at the optimal solution $z^* = 35706/66$ at $\mathbf{x}^* = (657, 690 \mid 66, 0, 0, 1023)^\top/66$ for the above modification of Problem 12.1.5 with $\delta_1 = 5$.

Workout 12.4.8 *Use the Dual Simplex Algorithm to solve the modification of Problem 12.1.5 with $\boldsymbol{\delta} = (16, 16, 0, 0)^\top$.*

---

[2] If we use the least subscript rule in the dual, we end up with an unusual interpretation of it in the primal, in which the slack variables have precedence over the decision variables. It will be easier to maintain the usual ordering in the primal, inferring the unusual one in the dual. This will cause no harm, since the least-subscript rule works on any given ordering, simply by thinking of such an ordering with relabeled subscripts.

12.5 Exercises

Practice

12.5.1 *Consider Problem 5.5.17. Find the reduced cost of each of its menu items.*

12.5.2 *Consider Problem 1.5.18.*

    a. *Describe the meaning of its dual variables.*

    b. *Find the shadow prices of each.*

    c. *Suppose the factory could pay for extra hours on one of the machines at $20/hour. Which machine would they rent and for how long?*

12.5.3 *Consider Problem 1.5.19.*

    a. *Describe the meaning of its dual variables.*

    b. *Find the shadow prices of each.*

    c. *Which should the company increase in order to decrease the amount of wood used, the number of hours required for finishing or the number of hours available for assembly? What should it be increased to?*

12.5.4 *Consider Problem 1.5.20.*

    a. *Describe the meaning of its dual variables.*

    b. *Find the shadow prices of each.*

12.5.5 *Consider Problem 1.5.21(a).*

    a. *Describe the meaning of its dual variables.*

    b. *Find the shadow prices of each resource.*

    c. *If Farmer Brown could buy extra acreage for planting, what is the maximum cost per acre she should pay? In that case, how much should she buy?*

    d. *If Farmer Brown could take out a loan to pay for extra capital and labor, what is the maximum interest rate she should pay? In that case, how much should she borrow?*

    e. *How much less profitable must arugula become (while broccoli profits remain fixed) in order that Farmer Brown begins converting some arugula acreage to broccoli?*

    f. *How much more profitable must broccoli become (while arugula profits remain fixed) in order that Farmer Brown begins converting some arugula acreage to broccoli?*

12.5. Exercises

g. Draw a graph that shows the region of arugula and broccoli profits per acre in which Farmer Brown makes no crop changes.

12.5.6 Consider Problem 12.1.5. What bounds on ϵ_1 and ϵ_2 are required in order that changes to c_1 and c_2 by those respective amounts ($c_j^\epsilon = c_j + \epsilon_j$) do not change the optimal basis?

12.5.7 Use the Dual Simplex Algorithm in Problem 1.5.18 from the hot start at the following bases (write the constraints in the order of the machines for the bases to make sense).

a. $\beta = \{1, 3, 4\}$.
b. $\beta = \{2, 4, 5\}$.

12.5.8 Use the Dual Simplex Algorithm in Problem 1.5.19 from the hot start at the following bases (write the finishing constraint first for the bases to make sense).

a. $\beta = \{3, 5\}$.
b. $\beta = \{4, 5\}$.

12.5.9 Use the Dual Simplex Algorithm in Problem 1.5.20 from the hot start at the following bases (write the constraint in the order of chads, construction paper, tissue paper, and ink for the bases to make sense).

a. $\beta = \{3, 4, 5, 6\}$.
b. $\beta = \{1, 3, 4, 6\}$.

12.5.10 Use the Dual Simplex Algorithm in Problem 1.5.21(b) from the hot start at the following bases (write the constraints in the order of acreage, total expenses, and capital expenses for the bases to make sense).

a. $\beta = \{2, 3, 5\}$.
b. $\beta = \{1, 2, 4\}$.

12.5.11 Use the Dual Simplex Algorithm in Problem 12.3.1 from the hot start at the following bases (order the constraints by kiwi, wattle, boab, and weevil for the bases to make sense).

a. $\beta = \{1, 4, 5, 6\}$.
b. $\beta = \{1, 2, 4, 7\}$.
c. $\beta = \{3, 5, 6, 7\}$.

12.5.12 Use the Dual Simplex Algorithm in Problem 12.1.1 (order the constraints by land, money, and hours) from the hot start required after modifying **b** by

a. $\delta_1 = -350$.

b. $\delta_2 = 31,000$.

12.5.13 Use the Dual Simplex Algorithm in Problem 12.1.5 from the hot start required after modifying **b** by

a. $\delta_3 = -20$.

b. $\delta_4 = -20$.

c. $\delta_1 = \delta_2 = 15$.

Challenges

12.5.14 Consider Problem 12.3.1. What bounds on ϵ are required in order that changes to **c** by ϵ do not change the optimal basis ; i.e., if **c** is replaced by $\mathbf{c}^\epsilon = \mathbf{c} + \epsilon$ then $\beta^*_\epsilon = \beta^*$? (See Exercise 12.5.6.)

12.5.15 Devise an algorithm or strategy for re-solving a LOP with modified **A**.

12.5.16 Devise an algorithm or strategy for re-solving a LOP with a new variable.

12.5.17 Give various reasons (in business or science or wherever) why new variables may need to be added to a LOP after it has been solved.

12.5.18 Devise an algorithm or strategy for re-solving a LOP with a new constraint.

12.5.19 Give various reasons (in business or science or wherever) why new constraints may need to be added to a LOP after it has been solved.

12.5.20 Write pseudocode for the Dual Simplex Algorithm.

12.5.21 Prove Theorem 12.4.6.

Projects

12.5.22 Present the economic application of the Hyperplane Separation Theorem.

Pareto efficiency

12.5.23 Present the economic notion of **Pareto efficiency**.

12.5.24 Present the Nobel Prize winning work of Kenneth Arrow.

Chapter 13

Integer Optimization

13.1 Cutting Planes

In this chapter we investigate integer linear optimization problems (ILOPs). We may find it useful to denote the set of all integers by $\mathbb{Z}$ and those that are nonnegative by $\mathbb{N}$. The **relaxation** of an ILOP P is the fractional LOP (**FLOP**)[1] Q that is identical to P but without the integer constraints. For a maximization ILOP, its FLOP provides an upper bound (**floptimal value**) on its optimum (**iloptimal value**) for sure. But we learned from Problem 1.2.5 that such an upper bound can be horrible.

Instead, we can find the optimal integer solution fairly quickly by rewriting the constraint in terms of x_2: $x_2 \leq (500 - x_1)/625$. But since x_2 must be an integer we can replace the right side by its floor $\lfloor (500 - x_1)/625 \rfloor = 0$. Thus we can add this new constraint to the problem without losing any of the original integer feasible points. In the new system we find $0 \leq x_2 \leq 0$, which means that every feasible solution has $x_2 = 0$, so we may as well delete x_2 from the ILOP entirely. Now we must maximize x_1 subject to $0 \leq x_1 \leq 500$. Pretty obvious answer there.

The ease with which we solved that is misleading — no fast algorithm for solving such problems is known.[2] However, the techniques we will learn here tend to work well in practice, but in the worst case could take exponentially many steps to complete. Of course, the same can be said of the Simplex Algorithm!

The first technique is related to the argument we gave above for generating a new constraint that holds for all integer feasible solutions but that is not explicitly stated. Such a constraint is called a **valid inequality** by algebraists, and a **cutting plane** by geometers, because of its two important

relaxation

FLOP

floptimal/ iloptimal value

valid inequality/ cutting plane

---

[1] There are other, lesser-known, related acronyms, in addition to LOP, BLOP, FLOP, GLOP, and ILOP, such as IHOP(it's not "International", you just can't order 2.5 pancakes off the menu) and JLOP (the Jennifer Lopez Problem).

[2] Fast means that it runs in time bounded by some polynomial of the input size (see Appendix C). If you discover a fast algorithm, let me know — I'd be glad to share the million dollars from the Clay Institute with you.

properties:

(1) every integer feasible solution satisfies the new constraint, and

(2) the (noninteger) optimum solution violates the new constraint.

It really does cut away the old optimum without harm to the integer system. The hope is that the new LOP has an integer optimum and, if not, we cut away over and over again until it does. This is the favorite method of Freddy, Jason, and Leatherface.

Here is a more interesting example.

Problem 13.1.1

$$\text{Max. } z = 10x_1 + 8x_2$$

$$\text{s.t. } \quad 11x_1 + 7x_2 \leq 38$$
$$7x_1 + 9x_2 \leq 35$$

$$\& \quad x_1, x_2 \in \mathbb{N}$$

Just as in Section 2.1, we can draw this in two dimensions, list the 12 feasible integer points, and pick the one having the greatest objective value (see Figure 13.1). But of course this isn't a reasonable method in general because the number of such points is usually too large — remember that we don't even have enough time to list all the extreme points, never mind all the interior integral points.

However, one might notice that the multipliers $\mathbf{y} = (3,1)^\mathsf{T}/10$ yield the valid inequality $4x_1 + 3x_2 \leq \lfloor 149/10 \rfloor = 14$, while the multipliers $\mathbf{y} = (1,2)^\mathsf{T}/25$ yield the valid inequality $x_1 + x_2 \leq \lfloor 108/25 \rfloor = 4$. Thus the optimum for Problem 13.1.1 is the same as for Problem 13.1.2, below.

Problem 13.1.2

$$\text{Max. } z = 10x_1 + 8x_2$$

$$\text{s.t. } \quad 11x_1 + 7x_2 \leq 38$$
$$7x_1 + 9x_2 \leq 35$$
$$4x_1 + 3x_2 \leq 14$$
$$x_1 + x_2 \leq 4$$

$$\& \quad x_1, x_2 \in \mathbb{N}$$

Now the multipliers $\mathbf{y} = (0,0,2,2)$ certify that $z^* \leq 36$; in fact, $z^* = 36$ at $\mathbf{x}^* = (2,2)$. These results come from solving the LOP below, the relaxation of Problem 13.1.2.

13.1. Cutting Planes

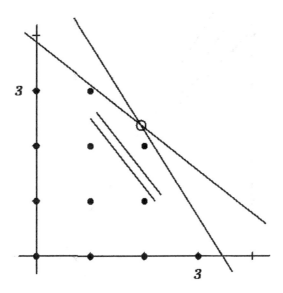

Figure 13.1: Feasible region for the relaxation of Problem 13.1.1

Problem 13.1.3

$$\text{Max.} \quad z = 10x_1 + 8x_2$$

$$\text{s.t.} \quad \begin{aligned} 11x_1 + 7x_2 &\leq 38 \\ 7x_1 + 9x_2 &\leq 35 \\ 4x_1 + 3x_2 &\leq 14 \\ x_1 + x_2 &\leq 4 \end{aligned}$$

$$\& \quad x_1, \; x_2 \geq 0$$

Figure 13.1 shows the optimum solution to the relaxation of Problem 13.1.1, at $\mathbf{x} = (97, 119)^T/50$, while Figure 13.2 shows the effect of the two cuts on it. I hope that you're curious enough to wonder how they were found. We follow the method of Gomory that lets the Simplex Algorithm find them for us. We begin with the optimal tableau for Problem 13.1.1.

Tableau 13.1.4

$$\left[\begin{array}{cc|ccc|c} 50 & 0 & 9 & -7 & 0 & 97 \\ 0 & 50 & -7 & 11 & 0 & 119 \\ \hline 0 & 0 & 34 & 18 & 50 & 1922 \end{array} \right]$$

The optimal solution occurs at $\mathbf{x} = (97, 119)^T/50$, which, for those of you who've studied number theory, is nonintegral. By taking a closer look at the final equation containing x_1, we will be able to derive a new valid

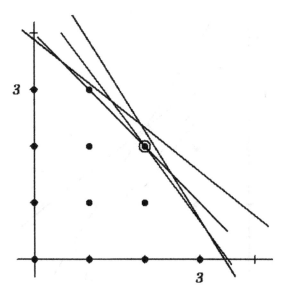

Figure 13.2: Optimal cuts for Problem 13.1.1

inequality. Indeed, we may rewrite $50x_1 + 9x_3 - 7x_4 = 97$ as

$$50(x_1 - x_4 - 1) = 47 - 9x_3 - 43x_4 \,. \tag{13.1}$$

Since $x_3, x_4 \geq 0$, the right-hand side of (13.1) is at most 47 and, in particular, strictly less than 50. Moreover, it is a multiple of 50, and so is at most 0. Hence we know that every feasible integer solution satisfies

$$-9x_3 - 43x_4 \leq -47 \,, \tag{13.2}$$

and so we add this valid constraint to our system. Note that we do not need to solve the revised problem from scratch; let's use a hot start instead. We simply introduce a new slack variable x_5 (with the same basic coefficient of 50) into (13.2) and slide the resulting equality into Tableau 13.1.4 to obtain the following.

Tableau 13.1.5

$$\begin{bmatrix} 50 & 0 & 9 & -7 & 0 & 0 & 97 \\ 0 & 50 & -7 & 11 & 0 & 0 & 119 \\ 0 & 0 & -9 & -43 & 50 & 0 & -47 \\ \hline 0 & 0 & 34 & 18 & 0 & 50 & 1922 \end{bmatrix}$$

decision/ original form
Two pivots return the tableau to its initial, standard form in Tableau 13.1.6, revealing the new cut in terms of the decision variables (see Figure 13.3): $8x_1 + 9x_2 \leq 36$. We call this the **decision form** (as opposed to the **original form**) of the valid inequality or cutting plane.

13.1. Cutting Planes

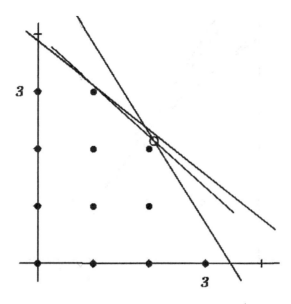

Figure 13.3: First cut for Problem 13.1.1

Tableau 13.1.6

$$\left[\begin{array}{rr|rrrr|r} 11 & 7 & 1 & 0 & 0 & 0 & 38 \\ 7 & 9 & 0 & 1 & 0 & 0 & 35 \\ 8 & 9 & 0 & 0 & 1 & 0 & 36 \\ \hline -10 & -8 & 0 & 0 & 0 & 1 & 0 \end{array}\right]$$

Of course, we don't want to pivot in this direction; we'd rather use the Dual Simplex Algorithm (see Section 12.4) on Tableau 13.1.5. To pivot in the negative row, we consider the c-ratios $-34/9$ and $-18/43$ and choose the one closest to zero. The pivot $4 \mapsto 5$ results in the temporarily optimal (optimal for the relaxed problem) tableau below.

Tableau 13.1.7

$$\left[\begin{array}{rr|rrrr|r} 43 & 0 & 9 & 0 & -7 & 0 & 90 \\ 0 & 43 & -8 & 0 & 11 & 0 & 92 \\ 0 & 0 & 9 & 43 & -50 & 0 & 47 \\ \hline 0 & 0 & 26 & 0 & 18 & 43 & 1636 \end{array}\right]$$

Since the relaxed optimal solution $\mathbf{x} = (90, 92 \mid 0, 47, 0))^\mathsf{T}/43$ is again nonintegral, we employ Gomory's trick again. We write $43x_2 - 8x_3 + 11x_5 = 92$ as $43(x_2 - x_3 - 2) = 6 - 35x_3 - 11x_5$, and derive the valid inequality

$$-35x_3 - 11x_5 \leq -6 \tag{13.3}$$

from the knowledge that $6 - 35x_3 - 11x_5$ is a multiple of 43 that is less than 43, and hence at most zero. As before, we slip this and the new slack

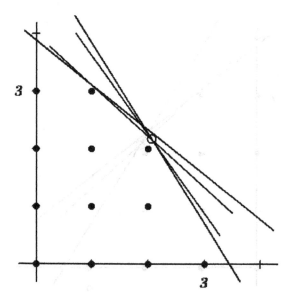

Figure 13.4: Second cut for Problem 13.1.1

into Tableau 13.1.7 (WebSim is really great for this — right click on the entry of the last constraint and last slack variable, then select Add Row and Column After) and follow Dual Simplex. Here we have Tableau 13.1.8, followed by Tableau 13.1.9.

Tableau 13.1.8

$$\begin{bmatrix} 43 & 0 & | & 9 & 0 & -7 & 0 & 0 & | & 90 \\ 0 & 43 & | & -8 & 0 & 11 & 0 & 0 & | & 92 \\ 0 & 0 & | & 9 & 43 & -50 & 0 & 0 & | & 47 \\ 0 & 0 & | & -35 & 0 & -11 & 43 & 0 & | & -6 \\ \hline 0 & 0 & | & 26 & 0 & 18 & 0 & 43 & | & 1636 \end{bmatrix}$$

Tableau 13.1.9

$$\begin{bmatrix} 35 & 0 & | & 0 & 0 & -8 & 9 & 0 & | & 72 \\ 0 & 35 & | & 0 & 0 & 11 & -8 & 0 & | & 76 \\ 0 & 0 & | & 0 & 35 & -43 & 9 & 0 & | & 37 \\ 0 & 0 & | & 35 & 0 & 11 & -43 & 0 & | & 6 \\ \hline 0 & 0 & | & 0 & 0 & 8 & 26 & 35 & | & 1328 \end{bmatrix}$$

Workout 13.1.10 *Confirm that the new cut is $11x_1 + 8x_2 \le 40$, as shown in Figure 13.4.*

It's difficult to tell whether anything was shaved off, but indeed there was. It may take forever if we can barely see the cuts, but be patient, some cuts are big and some are small.

Workout 13.1.11 *Use* MAPLE *to draw the current feasible region.*

Once again the relaxed optimum is not integral, so we need to find another cut. Before doing so, however, we should pause to notice a pattern in the tableaux that could save us the trouble of thinking. Consider an entry $T_{r,j}$, other than the basic coefficient d, of the new row r in a tableau T that corresponds to a newly determined cut derived from row i of T. By the manner in which we discovered the cut, $T_{i,j} + T_{r,j}$ equals the greatest multiple of d that is at most $T_{i,j}$. That is, $T_{r,j} = -(T_{i,j} \mod d)$. (Careful — this is different from $-T_{i,j} \mod d$: e.g., $-(2 \mod 7) = -2$, while $-2 \mod 7 = 5$.)

We should also take note of some of the decisions we've made so far. As we found in defining pivoting rules (least subscript, etc.), we have choices here that may matter none in theory but may matter some in practice. For example, at present, x_k is nonintegral for each $k \in \{1, 2, 3, 4\}$, so we are free to derive the valid equality $35x_7 - \sum_{j=1}^{6}(T_{i,j} \mod 35)x_j = -(b'_i \mod 35)$ from each of the current equalities $\sum_{j=1}^{6} T_{i,j} x_j = b'_i$, $i = 1, \ldots, 4$. The rule we have been following is to choose row $\operatorname{argmax}_i\{b'_i \mod d\}$; that is, the row i whose right-hand side needs the most subtraction. This is merely a heuristic that seems to behave well in practice. In this case, we cut from row 2.

Workout 13.1.12 *Finish solving Problem 13.1.1.*

  a. *Write each of the valid inequalities along the way, in original and decision form.*

  b. *Write the sequence of pivots $i \mapsto j$.*

  c. *Use* MAPLE *to draw the final feasible region with all the cutting planes.*

13.2 Branch-and-Bound

We return to Problem 13.1.1 (call it P) in order to consider a different method for finding its optimal integer solution. We first make note of the floptimal solution, which we denote $\widehat{\mathbf{x}}^* = (97, 119)/50$. The essence of the Branch-and-Bound technique is in splitting (branching) the problem (feasible region) into two subproblems (subregions), say those points with large x_1 value and those with small x_1 value. A natural split between large and small might be the value of $\widehat{x}_1^*$, but because of integrality this amounts to $x_1 \geq 2$ in one case and $x_2 \leq 1$ in the other. Thus we are led to solve the following two subproblems (PA and PB, respectively).

Problem 13.2.1 (PA)

$$\text{Max. } z = 10x_1 + 8x_2$$

$$\text{s.t. } \begin{aligned} 11x_1 + 7x_2 &\leq 38 \\ 7x_1 + 9x_2 &\leq 35 \\ x_1 &\geq 2 \end{aligned}$$

$$\& \quad x_1, x_2 \in \mathbb{N}$$

Problem 13.2.2 (PB)

$$\text{Max. } z = 10x_1 + 8x_2$$

$$\text{s.t. } \begin{aligned} 11x_1 + 7x_2 &\leq 38 \\ 7x_1 + 9x_2 &\leq 35 \\ x_1 &\leq 1 \end{aligned}$$

$$\& \quad x_1, x_2 \in \mathbb{N}$$

It will certainly be the case that the optimal value in P is the maximum of the optimal values in PA and PB — we haven't lost any integer-feasible solutions (see Figure 13.5). Let's now look at the relaxations of these ILOPs.

Workout 13.2.3 *For each of the Problems 13.2.1 and 13.2.2, add the appropriate constraints to the floptimal tableau of Problem 13.1.1 and use the Dual Simplex Algorithm to solve them.*

First, PA has $\widehat{\mathbf{x}}^* = (14, 16)^\mathsf{T}/7$ with $\widehat{z}^* = 268/7$. Nicely, $\widehat{x}_1^*$ is integral, although $\widehat{x}_2^*$ is not. However, we can split this time along $\widehat{x}_2^*$, which in essence means $x_2 \geq 3$ in Problem 13.2.4 (PAA) and $x_2 \leq 2$ in Problem 13.2.5 (PAB), below.

Problem 13.2.4 (PAA)

$$\text{Max. } z = 10x_1 + 8x_2$$

$$\text{s.t. } \begin{aligned} 11x_1 + 7x_2 &\leq 38 \\ 7x_1 + 9x_2 &\leq 35 \\ x_1 &\geq 2 \\ x_2 &\geq 3 \end{aligned}$$

$$\& \quad x_1, x_2 \in \mathbb{N}$$

13.2. Branch-and-Bound

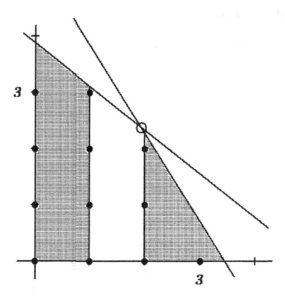

Figure 13.5: Splitting Problem 13.1.1 in two

Problem 13.2.5 *(PAB)*

$$
\begin{array}{rrrcl}
\text{Max.} \ z = & 10x_1 & + & 8x_2 & \\
\text{s.t.} & 11x_1 & + & 7x_2 & \leq 38 \\
 & 7x_1 & + & 9x_2 & \leq 35 \\
 & x_1 & & & \geq 2 \\
 & & & x_2 & \leq 2 \\
\& & x_1 & , & x_2 & \in \mathbb{N}
\end{array}
$$

Workout 13.2.6 *For each of the Problems 13.2.4 and 13.2.5, add the appropriate constraints to the floptimal tableau of Problem 13.2.1 and use the Dual Simplex Algorithm to solve them.*

Now we see that PAA is infeasible, which means that PA has the same integer optimum as PAB. Also, PAB has floptimum $\widehat{z}^* = 416/11$ at $\widehat{\mathbf{x}}^* = (24, 22)^\mathsf{T}/11$, which means we're going to have to branch on PAB as well. But let's not go too far too fast; we'll remember to do this later.

Second, we go back to Problem 13.2.2 to see that PB has $\widehat{\mathbf{x}}^* = (9, 28)^\mathsf{T}/9$ with $\widehat{z}^* = 314/9$. Repeating the same branching technique on x_2, we build the two subproblems PBA and PBB below, with $x_2 \geq 4$ and $x_2 \leq 3$, respectively.

Problem 13.2.7 *(PBA)*

$$\begin{aligned}
\text{Max. } z = {} & 10x_1 + 8x_2 \\
\text{s.t. } & 11x_1 + 7x_2 \le 38 \\
& 7x_1 + 9x_2 \le 35 \\
& x_1 \le 1 \\
& x_2 \ge 4 \\
\& \quad & x_1, \; x_2 \in \mathbb{N}
\end{aligned}$$

Problem 13.2.8 *(PBB)*

$$\begin{aligned}
\text{Max. } z = {} & 10x_1 + 8x_2 \\
\text{s.t. } & 11x_1 + 7x_2 \le 38 \\
& 7x_1 + 9x_2 \le 35 \\
& x_1 \le 1 \\
& x_2 \le 3 \\
\& \quad & x_1, \; x_2 \in \mathbb{N}
\end{aligned}$$

Workout 13.2.9 *For each of the Problems 13.2.7 and 13.2.8, add the appropriate constraints to the floptimal tableau of Problem 13.2.2 and use the Dual Simplex Algorithm to solve them.*

Now because PBA is infeasible, we know that the optimal solution to PB is the optimal solution to PBB. This "half" of the branching of Problem 13.1.1 is therefore complete, and we may return our thoughts to Problem PAB. Recall that PAB has floptimum $\widehat{z}^* = 416/11$ at $\widehat{\mathbf{x}}^* = (24, 22)^\mathsf{T}/11$, so it's time to branch on x_1 again. The new Problems $PABA$ and $PABB$ will include constraints $x_1 \ge 3$ and $x_1 \le 2$, respectively.

Problem 13.2.10 *(PABA)*

$$\begin{aligned}
\text{Max. } z = {} & 10x_1 + 8x_2 \\
\text{s.t. } & 11x_1 + 7x_2 \le 38 \\
& 7x_1 + 9x_2 \le 35 \\
& x_1 \ge 2 \\
& x_2 \le 2 \\
& x_1 \ge 3 \\
\& \quad & x_1, \; x_2 \in \mathbb{N}
\end{aligned}$$

Problem 13.2.11 *(PABB)*

$$\text{Max. } z = 10x_1 + 8x_2$$
$$\text{s.t. } \begin{array}{rcl} 11x_1 + 7x_2 & \leq & 38 \\ 7x_1 + 9x_2 & \leq & 35 \\ x_1 & \geq & 2 \\ x_2 & \leq & 2 \\ x_1 & \leq & 2 \end{array}$$
$$\& \quad x_1, x_2 \in \mathbb{N}$$

Workout 13.2.12 *For each of the Problems 13.2.10 and 13.2.11, add the appropriate constraints to the floptimal tableau of Problem 13.2.5 and use the Dual Simplex Algorithm to solve them.*

By the way, where's the *Bound* in *Branch-and-Bound*? Well, here it comes. Because $PABA$ has a fractional optimal solution, it looks like we need to branch on it. However, its floptimal value is only $250/7 < 36$, the iloptimal value of $PABB$, and its iloptimal value can only be smaller. Hence the optimal solution to PA equals that of PAB.

Furthermore, the iloptimal value in PA (36) exceeds that in PB (34), and so the optimal solution to P resides in PA; that is, P has optimal solution $z^* = 36$ at $\mathbf{x}^* = (2,2)^\mathsf{T}$.

Figure 13.6 shows the branching diagram (rooted tree — see Section 10.2 and Exercise 10.5.16) used to keep track of the ILOPs in the Branch-and-Bound Algorithm on Problem 13.1.1. Note the manner in which we traversed the tree, solving the ILOPs P, PA, PB, PAA, PAB, PBA, PBB, $PABA$, and $PABB$ in that order. This particular traversal is known as the **Breadth-First Search Algorithm** (BFS). Starting with the root node P in the queue, BFS repeats the following process: add the descendants of the head of the queue to the back of the queue and delete the head of the queue. This is the kind of search an army might employ along the streets of a city, splitting up the troops at a corner to send down each of the streets to broaden their control. Another well-known traversal is called the **Depth-First Search Algorithm** (DFS), which recursively uses DFS on its descendants. In this case, if the tree in Figure 13.6 was the entire tree, then DFS would produce the search order P, PA, PAA, PAB, $PABA$, $PABB$, PB, PBA, and PBB. This is the kind of search a person might use in a city, going as deeply as possible and then retracing steps — in fact, this is how **backtracking** algorithms operate. But Figure 13.6 isn't the entire tree: $PABA$ was pruned instead of branched on because it didn't beat the bound of $PABB$. Notice, however, that DFS doesn't reach $PABB$ until $PABA$ is branched on, and so in this case, BFS saves us extra effort. There are also cases in which DFS saves more effort than BFS, but most evidence suggests that BFS typically performs better than DFS.

Breadth-First Search Algorithm

Depth-First Search Algorithm

backtracking

Workout 13.2.13 *How many extra ILOPs would the DFS implementation of the Branch-and-Bound Algorithm need to solve in Problem 13.1.1?*

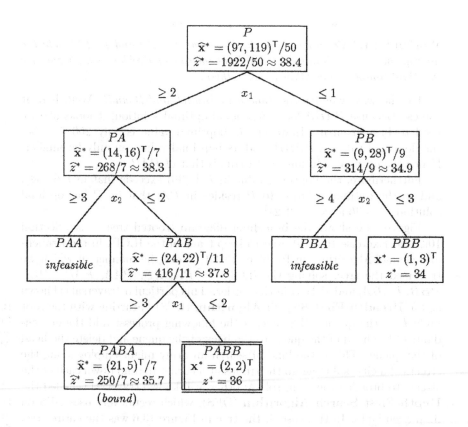

Figure 13.6: Branching tree for Problem 13.1.1

13.3. Последняя Практика

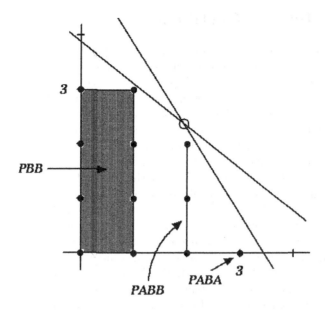

Figure 13.7: Feasible regions of the terminal LOPs for Problem 13.1.1

Recall the definition of a leaf of a tree on page 168. In Figure 13.6 the leaves are the LOPs PAA, PBA, PBB, $PABA$, and $PABB$. We call these the **terminal** LOPs of the Branch-and-Bound Algorithm, and note that they are distinguished by being either infeasible, bound, iloptimal, or unbounded (see Exercise 13.5.28). Figure 13.7 shows the feasible regions of the terminal LOPs of the above. The main point of the Branch-and-Bound Algorithm is that the union of these regions contains all integer feasible points of Problem 13.1.1.

terminal LOP

13.3 Последняя Практика

Consider the following ILOP.

Problem 13.3.1

$$\begin{aligned}
\text{Max.} \quad z &= 62x_1 + 83x_2 \\
\text{s.t.} \quad 11x_1 + 7x_2 &\leq 360 \\
4x_1 + 8x_2 &\leq 275 \\
\& \quad x_1, x_2 &\in \mathbb{N}
\end{aligned}$$

Workout 13.3.2 *Use the Cutting Plane Algorithm to solve Problem 13.3.1.*

Workout 13.3.3 *Use the Branch-and-Bound Algorithm to solve Problem 13.3.1.*

13.4 Integer Certificates

The act of certifying ILOPs is not quite as simple as in the case of FLOPs. However, it is not so complex, either, merely more lengthy. Whereas a FLOP requires only a single multiplier $\mathbf{y}^*$ to certify a given $\mathbf{x}^*$ and z^*, an ILOP requires a sequence or set of multipliers to certify the sequence or set of FLOPs that cutting planes or branching generates to solve it. For this reason we usually use the term proof rather than certificate, as in **cutting plane proof** or **branch-and-bound proof**.

cutting plane/ branch- and-bound proof

In the Branch-and-Bound Algorithm case, the branching tree provides the outline of the proof. It proves that the union of the feasible regions of the terminal FLOPs contains all the integer feasible points of the original ILOP. Thus we only need to certify each terminal FLOP, which is done in the usual way, whether infeasible, optimal or unbounded.

In the Cutting Plane Algorithm case, the dual multiplier y^* of the final LOP will not convince your boss, who will wonder why you added all those extra constraints to his original ILOP. You must therefore justify each cutting plane to her. Thus a cutting plane proof is a sequence of multipliers $\widehat{\mathbf{y}}^i$ that justify each new valid constraint $\mathbf{a}_i \leq b_i$ (in decision form), along with the final y^*.

For example we can see from the work of Section 13.1 and Workout 13.1.12 that the following multipliers generate the following valid inequalities (using the floor function because of the fact that $\mathbf{x}$ is integral).

multipliers $\widehat{\mathbf{y}}^i$	valid inequalities $\mathbf{a}_i \leq b_i$
$(9, 43)^\mathsf{T}/50$	$8x_1 + 9x_2 \leq \lfloor 1847/50 \rfloor = 36$
$(35, 0, 11)^\mathsf{T}/43$	$11x_1 + 8x_2 \leq \lfloor 1726/43 \rfloor = 40$
$(0, 0, 11, 27)^\mathsf{T}/35$	$11x_1 + 9x_2 \leq \lfloor 1476/35 \rfloor = 42$
$(0, 0, 0, 3, 1)^\mathsf{T}/11$	$4x_1 + 3x_2 \leq \lfloor 162/11 \rfloor = 14$
$(0, 0, 5, 0, 0, 5)^\mathsf{T}/12$	$5x_1 + 5x_2 \leq \lfloor 250/12 \rfloor = 20$
$(0, 0, 0, 0, 0, 10, 2)^\mathsf{T}/5$	$10x_1 + 8x_2 \leq 36$

Workout 13.4.1 *Give a cutting plane proof for your solution in Workout 13.3.2.*

Consider the following system of constraints from Problem 13.1.1.

System 13.4.2
$$\begin{array}{rcl} 11x_1 + 7x_2 & \leq & 38 \\ 7x_1 + 9x_2 & \leq & 35 \\ 10x_1 + 8x_2 & \geq & 37 \end{array}$$
$$x_1, \ x_2 \ \in \ \mathbb{N}$$

Notice that System 13.4.2 is fractionally feasible (recall that Problem 13.1.1 has floptimum $\widehat{z}^* = 1922/50 = 38.44$ at $\widehat{\mathbf{x}}^* = (97, 119)^\mathsf{T}/50$) but integer infeasible. Indeed, we just proved that if $11x_1 + 7x_2 \leq 38$, $7x_1 + 9x_2 \leq 35$, and $\mathbf{x} \in \mathbb{N}$, then $10x_1 + 8x_2 \leq 36$. In fact, with $\mathbf{x}' = (2, 2)^\mathsf{T}$

in hand, having $z(\mathbf{x}') = 36$, we have that Problem 13.1.1 is iloptimal at $\mathbf{x}'$ if and only if System 13.4.2 is integer infeasible. This leads us to the following theorem.

Theorem 13.4.3 *Let P be the ILOP*
$$\max\{\mathbf{c}^\mathsf{T}\mathbf{x} \mid \mathbf{A}\mathbf{x} \leq \mathbf{b},\ \mathbf{x} \in \mathbb{N}^n\}\ ,$$
with $\mathbf{c} \in \mathbb{Z}^n$, and S_t be the system
$$\{\mathbf{A}\mathbf{x} \leq \mathbf{b},\ \mathbf{c}^\mathsf{T}\mathbf{x} \geq t,\ \mathbf{x} \in \mathbb{N}^n\}\ .$$
Then P is iloptimal if and only if there is some $k \in \mathbb{Z}$ such that

a. *there is some P-feasible $\mathbf{x}'$ with $\mathbf{c}^\mathsf{T}\mathbf{x}' = k$ and*

b. *the system S_{k+1} is infeasible.*

Workout 13.4.4 *Prove Theorem 13.4.3.*

More interesting is the following sequence of derivations from System 13.4.2 (after putting it into standard form with $\leq$s).

multipliers $\widehat{\mathbf{y}}^i$	valid inequalities $\mathbf{a}_i \leq b_i$
$(8,0,7)^\mathsf{T}/18$	$x_1 \leq \lfloor 45/18 \rfloor = 2$
$(0,0,1,2)^\mathsf{T}/8$	$-x_1 - x_2 \leq \lfloor -33/8 \rfloor = -5$
$(1,2,0,0,25)^\mathsf{T}/17$	$0 \leq -1$

Workout 13.4.5 *Explain how the multipliers above were found.*

Theorem 13.4.6 *If S is a an integer-unsolvable system then there is a cutting plane proof of the inequality $0 \leq -1$.*   Integer Farkas Theorem

Workout 13.4.7 *Prove Theorem 13.4.6.*

13.5 Exercises

Practice

13.5.1 *Give an example of an ILOP that is floptimal but not iloptimal.*

13.5.2 *How many ILOPs were avoided by bounding in Workout 13.3.3?*

13.5.3 *Solve Exercise 1.5.37(a) with*

    a. *the Cutting Plane Algorithm.*

    b. *the Branch-and-Bound Algorithm.*

13.5.4 *Solve Exercise 1.5.37(b) with*

    a. the Cutting Plane Algorithm.

    b. the Branch-and-Bound Algorithm.

13.5.5 *Solve Exercise 1.5.37(c) with*

    a. the Cutting Plane Algorithm.

    b. the Branch-and-Bound Algorithm.

13.5.6 *Solve Exercise 1.5.38 with*

    a. the Cutting Plane Algorithm.

    b. the Branch-and-Bound Algorithm.

13.5.7 *Solve Problem 1.2.5 with*

    a. the Cutting Plane Algorithm.

    b. the Branch-and-Bound Algorithm.

13.5.8 *Solve Problem 1.5.1(a) over the integers.*

    a. Use the Cutting Plane Algorithm.

    b. Use the Branch-and-Bound Algorithm.

13.5.9 *Solve the following ILOP with*

$$\text{Max.} \quad z = 10x_1 + 8x_2$$

$$\text{s.t.} \quad \begin{array}{rcl} 11x_1 + 5x_2 & \leq & 36 \\ 5x_1 + 9x_2 & \leq & 34 \end{array}$$

$$\& \quad x_1, \ x_2 \in \mathbb{N}$$

    a. *the Cutting Plane Algorithm.*

        (i) *write each valid inequality in original and decision form,*

        (ii) *write the sequence of pivots, and*

        (iii) *write the final optimality certificate and use* **MAPLE** *to graph the relevant constraints.*

    b. *the Branch-and-Bound Algorithm.*

        (i) *draw the decision tree,*

        (ii) *include for each node (LOP) its label and floptimal solution, and*

        (iii) *use* **MAPLE** *to graph (and label) the resulting feasible regions of each leaf node.*

13.5.10 Consider the following ILOP.

$$\text{Max. } z = 89x_1 - 66x_2 + 74x_3$$

s.t.
$$\begin{aligned} 6x_1 - 14x_2 + 10x_3 &\geq 13 \\ -4x_1 + 14x_2 - 2x_3 &\geq 31 \\ 2x_1 + x_3 &\leq 9 \\ -10x_1 + 14x_2 + 2x_3 &\leq 53 \end{aligned}$$

& $x_1, x_2, x_3 \in \mathbb{N}$

a. Solve it using the Cutting Plane Algorithm.

b. Solve it using the Branch-and-Bound Algorithm.

13.5.11 Consider the following ILOP.

$$\text{Max. } z = 109x_1 + 116x_2$$

s.t.
$$\begin{aligned} 3x_1 + 4x_2 &\geq 17 \\ 7x_1 + 5x_2 &\leq 43 \\ 6x_1 - 8x_2 &\leq 19 \\ 9x_1 - 6x_2 &\geq 25 \end{aligned}$$

& $x_1, x_2 \in \mathbb{N}$

a. Solve it using the Cutting Plane Algorithm.

b. Solve it using the Branch-and-Bound Algorithm.

13.5.12 Consider the following ILOP.

$$\text{Max. } z = 502x_1 + 628x_2 - 551x_3$$

s.t.
$$\begin{aligned} 3x_1 + 2x_2 - x_3 &\leq 14 \\ -3x_1 + 18x_2 - 4x_3 &\geq 51 \\ 3x_1 - 3x_2 + 4x_3 &\geq 24 \\ -3x_1 + 2x_3 &\leq 6 \end{aligned}$$

& $x_1, x_2, x_3 \in \mathbb{N}$

a. Solve it using the Cutting Plane Algorithm.

b. Solve it using the Branch-and-Bound Algorithm.

13.5.13 For each of the ILOPs below, give a $0 \leq -1$ proof of integer unsolvability.

a. Exercise 13.5.1.

b. Exercise 13.5.10.

c. Exercise 13.5.11.

d. Exercise 13.5.12.

13.5.14 For each of the ILOPs below, provide a certificate of its optimality by proving the integer unsolvability (with a $0 \leq -1$ proof) of its system of constraints after the addition of the constraint $z(\mathbf{x}) \geq z^* + 1$.

a. Exercise 13.5.8.

b. Exercise 13.5.20.

c. Exercise 13.5.19.

d. Exercise 13.5.18.

e. Exercise 13.5.17.

f. Problem 13.3.1.

g. Exercise 13.5.27.

h. Exercise 13.5.9.

Challenges

13.5.15 Describe the branching tree of an infeasible ILOP. Explain. [HINT: See Exercises 13.5.1, 13.5.10, 13.5.11, and 13.5.12.]

13.5.16 Describe the branching tree of an unbounded ILOP. Explain.

13.5.17 Consider the following ILOP.

$$\text{Max. } z = 23x_1 + 60x_2 + 31x_3 + 44x_4$$

$$\text{s.t.} \begin{array}{rcrcrcrcr} 6x_1 & + & 2x_2 & & & + & 3x_4 & \leq & 338 \\ 5x_1 & - & x_2 & + & 9x_3 & + & 4x_4 & \leq & 254 \\ 2x_1 & & & + & 5x_3 & + & 6x_4 & \leq & 238 \\ x_1 & + & 2x_2 & + & 2x_3 & + & 3x_4 & \leq & 170 \\ 3x_1 & + & 5x_2 & + & x_3 & + & 2x_4 & \leq & 210 \end{array}$$

$$\& \quad x_1, \ x_2, \ x_3, \ x_4 \in \mathbb{N}$$

a. Solve it using the Cutting Plane Algorithm.

b. Solve it using the Branch-and-Bound Algorithm.

13.5.18 Solve the following ILOP (as in Exercise 13.5.9) with

$$\text{Max. } z = 6x_1 + 5x_2$$

$$\text{s.t.} \begin{array}{rcrcr} 9x_1 & + & 4x_2 & \leq & 36 \\ 5x_1 & + & 8x_2 & \leq & 40 \end{array}$$

$$\& \quad x_1, \ x_2 \in \mathbb{N}$$

a. the Cutting Plane Algorithm.

b. the Branch-and-Bound Algorithm.

13.5.19 *Consider the following ILOP.*

$$
\begin{array}{rrrrrrrl}
\text{Max.} \ z = & 73x_1 & + & 36x_2 & + & 85x_3 & & \\
\text{s.t.} & 5x_1 & + & 8x_2 & + & 3x_3 & \leq & 140 \\
& x_1 & + & 2x_2 & + & 9x_3 & \leq & 165 \\
& 6x_1 & + & 4x_2 & + & 7x_3 & \leq & 143 \\
& 2x_1 & + & 7x_2 & + & 3x_3 & \leq & 112 \\
\& & x_1, & & x_2, & & x_3 & \in & \mathbb{N}
\end{array}
$$

a. Solve it using the Cutting Plane Algorithm.

b. Solve it using the Branch-and-Bound Algorithm.

13.5.20 *Solve Problem 4.5.8 over the integers.*

a. Use the Cutting Plane Algorithm.

b. Use the Branch-and-Bound Algorithm.

13.5.21 *Solve Problem 12.3.1 over the integers.*

a. Use the Cutting Plane Algorithm.

b. Use the Branch-and-Bound Algorithm.

13.5.22 *Solve Problem 1.2.4 with*

a. the Cutting Plane Algorithm.

b. the Branch-and-Bound Algorithm.

13.5.23 *Solve Exercise 1.5.35(a) with*

a. the Cutting Plane Algorithm.

b. the Branch-and-Bound Algorithm.

13.5.24 *Solve Exercise 1.5.35(b) with*

a. the Cutting Plane Algorithm.

b. the Branch-and-Bound Algorithm.

13.5.25 *Solve Exercise 1.5.36 with*

a. the Cutting Plane Algorithm.

b. the Branch-and-Bound Algorithm.

13.5.26 *Solve Problem 8.5.25 over the integers.*

a. Use the Cutting Plane Algorithm.

b. Use the Branch-and-Bound Algorithm.

13.5.27 Consider the following ILOP.

$$\begin{aligned}
\text{Max.} \quad z = 62x_1 + 59x_2 + 66x_3 & \\
\text{s.t.} \quad 11x_1 + 7x_2 + 2x_3 &\leq 210 \\
4x_1 + 5x_2 + 8x_3 &\leq 185 \\
6x_1 + x_2 + 9x_3 &\leq 173 \\
\& \quad x_1, x_2, x_3 &\in \mathbb{N}
\end{aligned}$$

a. Solve it using the Cutting Plane Algorithm.

b. Solve it using the Branch-and-Bound Algorithm.

13.5.28 Given an ILOP P, prove that a LOP in the branching tree for P is terminal if and only if it is infeasible, bound, integer optimal, or unbounded. [HINT: This is an analogue of the Fundamental Theorem of Linear Optimization 2.9.1.]

13.5.29 Write pseudocode for finding a single cutting plane.

13.5.30 Write pseudocode for the Cutting Plane Algorithm.

13.5.31 Prove that the tableau remains integral throughout the Cutting Plane Algorithm. [HINT: Check the determinant.]

13.5.32 Write pseudocode for traversing a binary tree using BFS.

13.5.33 Write pseudocode for traversing a binary tree using DFS.

13.5.34 Use Exercise 13.5.32 to write pseudocode for the BFS implementation of the Branch-and-Bound Algorithm.

13.5.35 Use Exercise 13.5.33 to write pseudocode for the DFS implementation of the Branch-and-Bound Algorithm.

13.5.36 Devise a tree that could potentially be a branching tree of some ILOP, with the property that the DFS implementation of the Branch-and-Bound Algorithm visits fewer nodes than the BFS implementation; in particular, DFS prunes off ILOPs that BFS would otherwise solve.

13.5.37 Let $O_{n,k}$ be the polytope $\sum_{i=1}^{n} |x_i| \leq k$ (the generalized octahedron).

a. Compute the number of integer points in $O_{3,k}$.

b. Use your answer to part a to calculate the volume of $O_{3,1}$ in $\mathbb{R}^3$.

c. Repeat parts a and b in $\mathbb{R}^n$.

Projects

13.5.38 *Present the finiteness of the Cutting Plane Algorithm.*

13.5.39 *Present the* **Chvátal rank** *of a polytope.*

13.5.40 *Present the* **Branch-and-Cut Algorithm** *for solving ILOPs.*

13.5.41 *Present the the* **Nearest Vector** *and* **Shortest Vector** *Problems.*

13.5.42 *Present the the* **Subset Sum Problem***.*

Chvátal rank

Branch-and-Cut Algorithm

Nearest/Shortest Vector Problem

Subset Sum Problem

Appendix A

Linear Algebra Review

Our purpose here is to review briefly some of the necessary concepts from Linear Algebra that are used in Linear Optimization. This is not meant to be a tutorial but instead a refresher.

We assume familiarity with augmented matrices (which we coin **augmats** for short) and how they represent systems of linear equations. For example, System A.1 is represented by AugMat A.2, below.

augmat

System A.1

$$\begin{aligned}
5x_1 \phantom{{}+{}} - 1x_3 + 2x_4 - 3x_5 &= 19 \\
4x_1 + 6x_2 \phantom{{}-{} 1x_3} - 1x_4 \phantom{{}+{} 9x_5} &= 27 \\
-7x_1 - 3x_2 + 8x_3 \phantom{{}-{} 1x_4} + 9x_5 &= -11
\end{aligned}$$

AugMat A.2

$$\left(\begin{array}{ccccc|c} 5 & 0 & -1 & 2 & -3 & 19 \\ 4 & 6 & 0 & -1 & 0 & 27 \\ -7 & -3 & 8 & 0 & 9 & -11 \end{array} \right)$$

As a method of solving System A.1, one could use Gaussian elimination on AugMat A.2 as follows. First, subtract 4/5 of the first row from the second row and add 7/5 of the first row to the third row, obtaining AugMat A.3, below.

AugMat A.3

$$\left(\begin{array}{ccccc|c} 5 & 0 & -1 & 2 & -3 & 19 \\ 0 & 6 & 4/5 & -13/5 & 12/5 & 59/5 \\ 0 & -3 & 33/5 & 14/5 & 24/5 & 78/5 \end{array} \right)$$

Using the notation R_i (upper case) to denote row i of the resulting augmat and r_j (lower case) to denote row j of the previous augmat, we write the two **row operations** as

row operation

G. H. Hurlbert, *Linear Optimization*, Undergraduate Texts in Mathematics, DOI: 10.1007/978-0-387-79148-7_A, © Springer Science+Business Media LLC 2010

$$\begin{aligned} R_1 &= r_1 \\ R_2 &= r_2 - (4/5)r_1 \\ R_3 &= r_3 + (7/5)r_1 \,. \end{aligned}$$

A handy simplification, visually as well as computationally, involves avoiding fractions. The row operations

$$\begin{aligned} R_1 &= r_1 \\ R_2 &= 5r_2 - 4r_1 \\ R_3 &= 5r_3 + 7r_1 \end{aligned} \tag{A.1}$$

produce AugMat A.4, below.

AugMat A.4

$$\begin{pmatrix} 5 & 0 & -1 & 2 & -3 & | & 19 \\ 0 & 30 & 4 & -13 & 12 & | & 59 \\ 0 & -15 & 33 & 14 & 24 & | & 78 \end{pmatrix}$$

The corresponding system of equations is as follows.

System A.5

$$\begin{aligned} 5x_1 \quad\quad - \quad x_3 + 2x_4 - 3x_5 &= 19 \\ + 30x_2 + 4x_3 - 13x_4 + 12x_5 &= 59 \\ - 15x_2 + 33x_3 + 14x_4 + 24x_5 &= 78 \end{aligned}$$

equivalent system Systems A.1 and A.5 are **equivalent** in the sense that they have the same set of solutions. Indeed, what holds for one set of equations holds for any linear combination of them, so any solution of System A.1 is a solution of System A.5. The reverse is true as well since the row operations (A.1) can be inverted by solving for the r_j:

$$\begin{aligned} r_1 &= R_1 \\ r_2 &= (R_2 + 4R_1)/5 \\ r_3 &= (R_3 - 7R_1)/5 \,. \end{aligned} \tag{A.2}$$

One can check that row operations (A.2) convert AugMat A.4 back to AugMat A.2.

What is nice about the conversion of System A.1 to System A.5 is that the latter is slightly simpler, having eliminated x_1 from two of its equations. One can continue to simplify, as below.

$$\begin{aligned} R_1 &= r_1 \\ R_2 &= r_2 \\ R_3 &= 2r_3 + r_1 \end{aligned} \tag{A.3}$$

AugMat A.6

$$\begin{pmatrix} 5 & 0 & -1 & 2 & -3 & | & 19 \\ 0 & 30 & 4 & -13 & 12 & | & 59 \\ 0 & 0 & 70 & 15 & 60 & | & 215 \end{pmatrix}$$

$$R_1 = 70r_1 + r_3$$
$$R_2 = 70r_2 - 4r_3$$
$$R_3 = r_3$$

AugMat A.7

$$\begin{pmatrix} 350 & 0 & 0 & 155 & -150 & | & 1545 \\ 0 & 2100 & 0 & -970 & 600 & | & 3270 \\ 0 & 0 & 70 & 15 & 60 & | & 215 \end{pmatrix}$$

We can uniformize the sight of AugMat A.7 by multiplying each row appropriately, giving rise to AugMat A.8.

AugMat A.8

$$\begin{pmatrix} 210 & 0 & 0 & 93 & -90 & | & 927 \\ 0 & 210 & 0 & -97 & 60 & | & 327 \\ 0 & 0 & 210 & 45 & 180 & | & 645 \end{pmatrix}$$

This form yields the solution

$$\begin{pmatrix} x_1 \\ x_2 \\ x_3 \end{pmatrix} = \frac{1}{210} \left[\begin{pmatrix} 927 \\ 327 \\ 645 \end{pmatrix} - \begin{pmatrix} 93 \\ -97 \\ 45 \end{pmatrix} x_4 - \begin{pmatrix} -90 \\ 60 \\ 180 \end{pmatrix} x_5 \right] \quad (A.4)$$

for any values of x_4 and x_5 whatsoever.

Other forms of the same solution set can be given as well. For example, one can solve for x_1, x_3 and x_5 in terms of x_2 and x_4. This can be accomplished from AugMat A.8 using the row operations

$$\begin{aligned} R_1 &= (2r_1 + 3r_2)/7 \\ R_2 &= r_2 \\ R_3 &= 2(r_3 - 3r_2)/7 \end{aligned} \quad (A.5)$$

to produce Augmat A.9.

AugMat A.9

$$\begin{pmatrix} 60 & 90 & 0 & -15 & 0 & | & 405 \\ 0 & 210 & 0 & -97 & 60 & | & 327 \\ 0 & -180 & 60 & 96 & 0 & | & -96 \end{pmatrix}$$

This form determines the solution

$$\begin{pmatrix} x_1 \\ x_3 \\ x_5 \end{pmatrix} = \frac{1}{60} \left[\begin{pmatrix} 405 \\ -96 \\ 327 \end{pmatrix} - \begin{pmatrix} 90 \\ -180 \\ 210 \end{pmatrix} x_2 - \begin{pmatrix} -15 \\ 96 \\ -97 \end{pmatrix} x_4 \right] \quad (A.6)$$

for any values of x_2 and x_4.

If it is a mystery how row operations (A.5) were devised so as to make AugMat A.9 uniform, the mystery should be cleared up by rewriting the row operations as

$$\begin{aligned} R_1 &= (60r_1 + 90r_2)/210 \\ R_2 &= r_2 \\ R_3 &= (60r_3 - 180r_2)/210 \end{aligned} \qquad (A.7)$$

and comparing this form with the values of AugMat A.9.

pivot operation The action of clearing all but one of the entries in a column of an augmat is called a **pivot operation**. Writing pivot operations cleverly as in (A.7) is the subject of Section 2.2. Proving that the clever form of them maintains both the uniformity and integrality of augmats is the subject of Section 5.4, and the arguments are based on the knowledge of certain determinants. That knowledge includes the following theorem.

Cramer's Rule **Theorem A.10** *Let* $\mathbf{M}$ *be an* $n \times n$ *invertible matrix and let* $\mathbf{b}$ *be a length* n *vector. Denote by* $\mathbf{M}_j$ *the matrix obtained from* $\mathbf{M}$ *by replacing its* j^{th} *column by* $\mathbf{b}$. *Then the solution to the system* $\mathbf{Mx} = \mathbf{b}$ *is given by*

$$x_j = \frac{|\mathbf{M}_j|}{|\mathbf{M}|}, \qquad (1 \le j \le n).$$

Appendix B

Equivalence of Auxiliary and Shortcut Methods

The goal of this appendix is to compare two methods used in Phase I of the Simplex Algorithm, namely, the Auxiliary Method of Section 2.4 and the Shortcut Method of Section 2.5. The two methods look quite different but turn out to be equivalent in a certain, precise way.

Let's return to Problem 2.5.1 (B.1), below.

Problem B.1

$$\begin{aligned}
\text{Max.} \quad z &= 28x_1 + 21x_2 + 26x_3 \\
\text{s.t.} \quad -7x_1 + 2x_2 + 3x_3 &\leq -210 \\
5x_1 - 8x_2 + x_3 &\leq -305 \\
2x_1 + 4x_2 - 9x_3 &\leq -250 \\
\& \quad x_1, x_2, x_3 &\geq 0
\end{aligned}$$

Its auxiliary is the following LOP.

Problem B.2

$$\begin{aligned}
\text{Max.} \quad v &= -x_0 \\
\text{s.t.} \quad -x_0 - 7x_1 + 2x_2 + 3x_3 &\leq -210 \\
-x_0 + 5x_1 - 8x_2 + x_3 &\leq -305 \\
-x_0 + 2x_1 + 4x_2 - 9x_3 &\leq -250 \\
\& \quad x_0, x_1, x_2, x_3 &\geq 0
\end{aligned}$$

We recall the sequence of tableaux and pivots that solve Problem B.2. Note that we have numbered the tableaux somewhat oddly. We also have left some space where columns may appear or disappear.

Tableaux B.3 (Auxiliary Method)

	x_0	x_1	x_2	x_3	x_4	x_5	x_6	v	
$T0$:	-1	-7	2	3	1	0	0	0	-210
	-1	5	-8	1	0	1	0	0	-305
	-1	2	4	-9	0	0	1	0	-250
	1	0	0	0	0	0	0	1	0

$P1:\ 0 \mapsto 5$

	x_0	x_1	x_2	x_3	x_4	x_5	x_6	v	
$T1$:	0	-12	10	2	1	-1	0	0	95
	1	-5	8	-1	0	-1	0	0	305
	0	-3	12	-10	0	-1	1	0	55
	0	5	-8	1	0	1	0	1	-305

$P3:\ 2 \mapsto 6$

	x_0	x_1	x_2	x_3	x_4	x_5	x_6	v	
$T3$:	0	-114	0	124	12	-2	-10	0	590
	12	-36	0	68	0	-4	-8	0	3220
	0	-3	12	-10	0	-1	1	0	55
	0	36	0	-36	0	4	8	12	-3220

$P5:\ 3 \mapsto 4$

	x_0	x_1	x_2	x_3	x_4	x_5	x_6	v	
$T5$:	0	-114	0	124	12	-2	-10	0	590
	124	274	0	0	-68	-30	-26	0	29930
	0	-126	124	0	10	-12	2	0	1060
	0	-274	0	0	68	30	26	124	-29930

$P6:\ 1 \mapsto 0$

	x_0	x_1	x_2	x_3	x_4	x_5	x_6	v	
$T6$:	114	0	0	274	-36	-32	-46	0	28820
	124	274	0	0	-68	-30	-26	0	29930
	126	0	274	0	-47	-57	-22	0	32755
	274	0	0	0	0	0	0	274	0

Keep in mind that Pivot P1 was performed in order that Tableau T1 might be feasible. By pivoting x_0 into the basis, Tableau T1 will be feasible, according to Workout 2.4.9, if and only if the leaving variable is chosen to be that basic variable which is most negative. From that point on, we resort to the pivoting rules of Phase II, as you can see.

Extended Method Now we wish to take a moment to investigate the effect of these same pivots on the original Problem B.1. To do this, we **extend** the Auxiliary Method as follows. Consider each pivot $x_l \mapsto x_k$ as two pivots $x_l \mapsto x_0$ and $x_0 \mapsto x_k$. These two pivots have the same effect of exchanging x_k and x_l, while leaving x_0 basic. But by splitting each pivot in two we will be able to analyze individually each decision of choosing an entering or leaving variable. In addition, we will include in the tableaux the columns and objective rows of both objective variables z and u. What we produce is the following sequence.

Tableaux B.4 (Extended Method)

	x_0	x_1	x_2	x_3	x_4	x_5	x_6	z	v	
$T0$:	-1	-7	2	3	1	0	0	0	0	-210
	-1	5	-8	1	0	1	0	0	0	-305
	-1	2	4	-9	0	0	1	0	0	-250
	0	-28	-21	-26	0	0	0	1	0	0
	1	0	0	0	0	0	0	0	1	0

$P1: \quad 0 \mapsto 5$

	x_0	x_1	x_2	x_3	x_4	x_5	x_6	z	v	
$T1$:	0	-12	10	2	1	-1	0	0	0	95
	1	-5	8	-1	0	-1	0	0	0	305
	0	-3	12	-10	0	-1	1	0	0	55
	0	-28	-21	-26	0	0	0	1	0	0
	0	5	-8	1	0	1	0	0	1	-305

$P2: \quad 2 \mapsto 0$

	x_0	x_1	x_2	x_3	x_4	x_5	x_6	z	v	
$T2$:	-10	-46	0	26	8	2	0	0	0	-2290
	1	-5	8	-1	0	-1	0	0	0	305
	-12	36	0	-68	0	4	8	0	0	-3220
	21	-28	0	-229	0	-21	0	8	0	6405
	8	0	0	0	0	0	0	0	8	0

$P3: \quad 0 \mapsto 6$

	x_0	x_1	x_2	x_3	x_4	x_5	x_6	z	v	
$T3$:	0	-114	0	124	12	-2	0	0	0	590
	0	-3	12	-10	0	-1	0	0	0	55
	12	-36	0	68	0	-4	12	0	0	3220
	0	-399	0	-522	0	-21	0	12	0	1155
	0	36	0	-68	0	4	0	0	12	-3220

$P4: \quad 3 \mapsto 0$

	x_0	x_1	x_2	x_3	x_4	x_5	x_6	z	v	
$T4$:	-124	-274	0	0	68	30	26	0	0	-29930
	0	-47	68	0	0	-9	-1	0	0	2995
	12	-36	0	68	0	-4	-8	0	0	3220
	522	-3827	0	0	0	-293	-229	68	0	146615
	68	0	0	0	0	0	0	0	68	0

$P5: \quad 0 \mapsto 4$

	x_0	x_1	x_2	x_3	x_4	x_5	x_6	z	v	
$T5$:	124	274	0	0	-68	-30	-26	0	0	29930
	0	-126	124	0	10	-12	2	0	0	1060
	0	-114	0	124	12	-2	-10	0	0	590
	0	-9082	0	0	522	-304	-218	124	0	37600
	0	-274	0	0	68	30	26	0	124	-29930

$P6: \quad 1 \mapsto 0$

	x_0	x_1	x_2	x_3	x_4	x_5	x_6	z	v	
$T6$:	124	274	0	0	-68	-30	-26	0	0	29930
	126	0	274	0	-47	-57	-22	0	0	3275
	114	0	0	274	-36	-32	-46	0	0	28820
	9082	0	0	0	-3827	-2869	-2386	274	0	2275215
	274	0	0	0	0	0	0	0	274	0

Now you can see why we numbered the tableaux in the Auxiliary Method the way we did. Also, you can see that Pivot P3 in the Auxiliary sequence became Pivots P2 and P3 in the Extended sequence. Likewise, Pivot P5 in the Auxiliary sequence became Pivots P4 and P5 in the Extended sequence.

Before proceeding, let's spend a moment translating the rules we used in the Auxiliary Method into rules for the Extended Method. In both cases we start by choosing the row whose **b**-column was most negative, in this case row 2.

In Auxiliary Pivot P3, x_2 was chosen as the entering variable because of the -8 in the objective row of Tableau T1, the first negative number we see when reading left-to-right (Least Subscript rule). This corresponds to the 8 in row 2 of Tableau T1, or the -8 in row 2 of Tableau T0. Neglecting the auxiliary variable x_0, this translates into a rule which chooses the variable with the least subscript whose coefficient in the prior row is negative.

Next, in Auxiliary Pivot P3, x_6 was chosen as the leaving variable because it was the variable which placed the greatest restriction on x_2. That is, its **b**-ratio of $\frac{55}{12}$ was smallest and nonnegative. As we will soon see, $\frac{55}{12}$ was the smallest for the same reason that -3220 was the most negative in Tableau T2. That is to say, x_6 was chosen as the leaving variable in Extended Pivot P3 because -3220 was most negative in Tableau T2. From there, we argue as before.

Hence we have developed rules of pivoting which are always in terms of the even numbered tableaux. This suggests that we can skip the odd numbered tableaux by combining the two pivot operations $x_0 \mapsto x_k$ and $x_l \mapsto x_0$ into the single exchange $x_l \mapsto x_k$. The method for finding a pivot in Phase I is now the reverse of that which finds the pivot in Phase II in the following sense. Instead of deciding upon an entering variable first, followed by a leaving variable, we now choose a leaving variable first, then an entering variable. The rule for choosing the leaving variable is to pick that basic variable whose value is most negative. The entering variable is that with the least subscript whose coefficient in the pivot row is negative. It is these rules, translated from the Auxiliary Method, which give us the Shortcut Method for Phase I, below. Notice that we no longer need a column or row for the auxiliary objective variable u since these are (virtually) unchanged throughout the sequence of even tableaux. Also, we do not need a column for x_0 since it is never basic.

Tableaux B.5 (Shortcut Method)

	x_1	x_2	x_3	x_4	x_5	x_6	z	
$T0$:	-7	2	3	1	0	0	0	-210
	5	-8	1	0	1	0	0	-305
	2	4	-9	0	0	1	0	-250
	-28	-21	-26	0	0	0	1	0

$P2: \quad 2 \mapsto 5$

$T2:$

−46	0	26	8	2	0	0	−2290
−5	8	−1	0	−1	0	0	305
36	0	−68	0	4	8	0	−3220
−28	0	−229	0	−21	0	8	6405

$P4: \quad 3 \mapsto 6$

$T4:$

−274	0	0	68	30	26	0	−29930
−47	68	0	0	−9	−1	0	2995
−36	0	68	0	−4	−8	0	3220
−3827	0	0	0	−293	229	68	146615

$P6: \quad 1 \mapsto 4$

$T6:$

274	0	0	−68	−30	−26	0	29930
0	274	0	−47	−57	−22	0	3275
0	0	274	−36	−32	−46	0	28820
0	0	0	−3827	−2869	2386	274	2275215

Having taken it on faith that achieving a minimum of $\frac{55}{12}$ was equivalent to having -3220 be most negative, we would like now to justify that claim in the general setting.

Let's say that at some stage of the Auxiliary Method we have a feasible tableau, a portion of which, including the incoming variable and the b-column, is shown below.

$$\begin{array}{cc|c} a & \cdot & b \\ c & \cdot & d \\ e & \cdot & f \end{array}$$

For example, it could be the first three rows of Tableau T1. Since this is a feasible tableau, we know that $b, d, f \geq 0$. Let's suppose that the Auxiliary Method tells us to pivot on the entry a, although c and e were also under consideration. Thus $a, c, e > 0$ and $0 \leq \frac{b}{a} < \frac{d}{c}$.

This portion of the corresponding tableau during the Extended Method is identical, and let's say that the bottom row includes the basic auxiliary variable x_0. Since x_0 is basic, $-f$ was most negative in the previous Extended tableau, and so $f > 0$ as well. The Extended Method mandates that we pivot on the entry e, resulting in the following tableau.

$$\begin{array}{cc|c} 0 & \cdot & be - af \\ 0 & \cdot & de - cf \\ e & \cdot & f \end{array}$$

Suppose now that $de - cf < 0$. Then the Extended Method would tell us to pivot in that row unless $be - af$ were more negative. This would be in contrast to the Auxiliary choice of pivoting in the top row. But $de - cf < 0$ implies that $\frac{d}{c} < \frac{f}{e}$, and $\frac{b}{a} < \frac{d}{c}$ implies that $b < \frac{ad}{c}$. Hence $b - d < \frac{d}{c}(a - c) < \frac{f}{e}(a - c)$, which implies that $e(b - d) < f(a - c)$, or

$be - af < de - cf$. Since this whole argument is reversible, we can make the following statement. Let $b, d \geq 0$, $a, c, e, f > 0$, and $\frac{d}{c} < \frac{f}{e}$. Then $be - af < de - cf$ if and only if $\frac{b}{a} < \frac{d}{c}$. This means that the assertion that the Auxiliary and Shortcut Methods make the same choice of leaving variable is a correct one. (The proof is not entirely complete. One must allow for the possibility of equality with the tie-breaking decision made by the Least Subscript rule.) We record this as the following theorem.

Theorem B.6 *The Auxiliary and Shortcut Methods are equivalent in the sense that, ignoring the auxiliary variable x_0, they make the same sequence of decisions for incoming and outgoing variables.* ◇

Workout B.7 *Complete the proof of Theorem B.6, taking into account equality and tie-breaking considerations.*

Appendix C

Complexity

Our aim here is to give the reader a brief and informal introduction to the running time complexity of algorithms. There are excellent places to learn about the subject more formally and in greater detail, so we'll just offer a taste as it relates to LO.

C.1 P versus NP

The question of whether or not P = NP is something every undergraduate mathematician or computer scientist should know about.[1] While the famous million-dollar question is whether or not P = NP, the infamous hundred-forint joke is that P = NP if and only if N = 1 or P = 0. While a bit of an oversimplification, one can think of modern day cryptosystems as being dependent on the assumption that P $\neq$ NP. So what are P and NP?[2]

Without getting into discussions of Turing machines and models of computation, we simply describe NP as the set of computational problems whose solutions can be verified in time that is bounded by a polynomial in the size of its input.[3] This is something we are quite familiar with by now, as our certificates for the infeasibility, unboundedness, or optimality of a particular LOP are as fast as could be, putting LO in NP. Many computational problems have three versions, namely, the **recognition**, **evaluation**, and **optimization** versions. The recognition version asks whether there exists an object satisfying certain conditions (a feasible solution) — it is a Yes-No question. The evaluation version finds the value of the optimal feasible so-

NP problem

recognition/ evaluation/ optimization versions

---

[1] It is considered the *Holy Grail* of theoretical computer science, and the Clay Mathematical Institute offers $1 million for its solution, as one of its seven Millennium Problems (see http://www.claymath.org/millennium).

[2] While most people guess correctly that P stands for *polynomial*, it is not the case that NP stands for *not polynomial* — instead it means *nondeterministic polynomial*, which comes from an alternative definition that allows for machines to make random choices.

[3] The phrase *in the size of the input* will be dropped hereafter, although we will continue to mean it.

G. H. Hurlbert, *Linear Optimization*, Undergraduate Texts in Mathematics, DOI: 10.1007/978-0-387-79148-7_C, © Springer Science+Business Media LLC 2010

lution, given some objective function, and the optimization version finds the optimal feasible solution. With regard to a LOP or ILOP, the objects are real or integer vectors, the constraints are linear, and the objective function is linear.

<small>complexity</small>   When we speak of the **complexity** of a problem, we mean (usually the best) upper bound on the running time of the best algorithm that solves every instance of the problem, with polynomial (in particular linear, quadratic, etc.) and exponential being the main distinctions, although there are certainly others (e.g., subexponential, factorial). It turns out that the three versions above have the same complexity — in the sense that the ratio of the running times of any pair of them is bounded by a polynomial — provided that the number of digits required to represent the optimal cost is bounded by a polynomial. The provision pretty well takes care of everything we care about, but for LOPs again points to the importance of the floptimal value's denominator (optimal basic coefficient); that is, we care that none of the basic determinants has exponentially many digits. We see immediately that LINEAR FEASIBILITY and INTEGER FEASIBILITY are in NP because, for any solution $\mathbf{x}$, $\mathbf{Ax}$ can be calculated in polynomial time (notice here that the size of the input must include the size of the entries of $\mathbf{A}$, not merely its dimensions), and then trivially compared to $\mathbf{b}$ to determine feasibility. It follows by duality that LO is in NP. However, note that it does not follow that IO is in NP, since optimality and infeasibility certificates might not always be polynomial in length (many of the Challenge exercises from

<small>NP-hard</small>   Chapter 13 exemplify this!). In fact, IO is an example of an **NP-hard** problem, one that is at least as hard (see reduction, below) as any in NP.

<small>P problem</small>   Next we describe P as the set of computational problems whose solutions can be calculated in polynomial time (certainly $P \subseteq NP$). For example, the multiplication of two n-digit integers is calculated by the usual algorithm in at most $2n^2$ steps, counting single-digit multiplications and additions. Thus, MULTIPLICATION $\in$ P. Since MULTIPLICATION is verification for FACTORING, we have FACTORING $\in$ NP. Interestingly, it is not known if FACTORING $\in$ P.[4] Here, the input size is the number of binary digits, so if n has k digits, then the naïve algorithm of checking every integer up to $\sqrt{n}$ for divisibility could need to make $2^{k/2}$ tests (or roughly $(\frac{2}{k})2^{k/2}$ if restricting its domain to primes, supposing it had such a list). More interesting, it was discovered recently[5] that RECOGNIZING COMPOSITES $\in$ P — that is, it was discovered how to recognize in polynomial time that a number can be factored without finding a factor.

There are certain problems in NP that are hardest in the class. To
<small>oracle</small>   describe these we introduce **oracles**, which are like little black boxes that give answers in unit time. We say that a problem B **reduces** to problem
<small>reduction</small> A if B can be solved in polynomial time using an oracle for A. Thus the complexity of B is not harder than that of A (it is possible that a fast

---

[4] The seeming computational difficulty of FACTORING is the basis for the security of the RSA cryptosystem.

[5] M. Agrawal, N. Kayal, and N. Saxena, Primes is in P, *Ann. Math.* **160** (2004), 781793.

algorithm for solving B exists that doesn't use the oracle). The problem A is NP-**complete** if every problem in NP reduces to A. In other words, it is an NP-hard problem that is also in NP. It is one of the remarkable beauties of complexity theory that such problems exist. Of course, if one can show that some NP-complete problem is in P, then P = NP.

NP-complete

C.2 Examples

Given a graph G, a **Hamilton cycle** in G is a cycle that contains all of the vertices of G. If the edges of G are weighted, then the **weight** of a subgraph H is the sum of the weights of the edges of H. The MWHC Problem (also called the TRAVELING SALESPERSON Problem) asks if there is a Hamilton cycle of weight less than w. The problem is in NP because a set of edges of G can be verified to be a Hamilton cycle or not and its weight can be calculated, both in linear time. However, there are potentially $(n-1)!/2$ Hamilton cycles in G, so checking each one of them for weight less than w is not a polynomial algorithm. Unfortunately, there is no better idea at present.

Hamilton cycle

weight

A similar sounding problem is MWST, which asks if there is a spanning tree of G having weight less than w. Since there are potentially n^{n-2} spanning trees of G,[6] a number which far exceeds $(n-1)!/2$ (the ratio is close to $e^n/n\sqrt{2\pi n}$ for large n), one might expect that MWST fares no better than MWHC. It is in NP for the same reason, but it turns out that MWST $\in$ P, while MWHC is NP-complete! In fact, both problems are special cases of INTEGER FEASIBILITY (forcing INTEGER FEASIBILITY, and hence IO, to be NP-hard), with the total unimodularity of MWST making it a special case of LINEAR FEASIBILITY, which we will see is in P in Section C.3.

Another classic problem is k-COLORING, in which one asks if it is possible to assign to each vertex of a graph one of k colors so that the colors of adjacent vertices differ. For 2-COLORING, one shouldn't simply test all 2^n possible colorings; instead it is not hard to show that the following (quadratic) algorithm decides if G can be 2-colored. Pick any vertex of G to be the only vertex of the queue Q, color it red (as opposed to blue), and repeat the following task. Take the first vertex $v \in Q$, remove v from Q, and for each of its adjacent vertices u, add u to the end of Q and either color u the opposite of v if it doesn't conflict with the other colored vertices that are adjacent to u, or halt and return that G is not 2-colorable. If every vertex gets colored then halt and return that G is 2-colorable. Thus 2-COLORING $\in$ P. Unfortunately, no analogous algorithm is known for 3-COLORING; in fact, 3-COLORING is NP-complete. It is also another special case of IO.

As noted, IO is NP-hard. In fact, even its restriction to binary variables (BO) is NP-hard. However, 2-COLORING is a special case of BO that is in P. We see in Chapter 10 another case of IO that is in P, namely, NETWORK

---

[6]A result of Cayley — see Exercise 10.5.37.

OPTIMIZATION. Therefore someone who builds a cryptosystem based on an NP-complete problem must be careful not to accidentally use one of its instances that lies in P.[7] As a final example, we mention the SUBSET SUM Problem, which asks if some subset of a set of n integers adds up to k. This is an NP-complete problem, also a special case of IO. But suppose each number on the list is at least k/t, for some fixed t. Then there are at most a polynomial number $(\sum_{i=1}^{t-1} \binom{n}{i} = O(n^t))$ of subsets that need checking, a disaster for security.

There are abundantly more interesting and natural problems to include, and a vast literature on modern developments in the field. Papadimitriou[8] is a good place to start.

C.3 LO Complexity

Is the Simplex Algorithm polynomial? Let's suggest not. Consider the 3-dimensional cube, the convex hull of the eight binary triples. The edges of this polytope form the graph Q^3, whose vertices are the extreme points of the cube, and it is not difficult to construct a Hamilton cycle H in it. Imagine now a plane tangent to the cube at the origin gets pushed gently through the cube. If the cube is a kind of elastic clay, then the topologist in you might be able to stretch it in such a way that the next vertex that the plane touches is the one that follows the origin along H. Really, you may be able to continue this argument, stretching carefully in order to preserve the corresponding vertex orders. If so, you have constructed an instance of a LOP having 3 variables and 3 constraints on which Simplex makes $2^3 - 1$ pivots in finding the optimum. Indeed, one can carefully extend the argument with 3 replaced by n everywhere, showing that Simplex can take exponential time. Theoretically, then, it is a horrible failure.

What makes it such a practical success is that such pathological examples as these seem to be rare, apparently never experienced in applications. As mentioned in Chapter 5, Simplex tends to run very fast on average, often in linear time.

In terms of theoretical complexity, it wasn't until Khachian's Ellipsoid Algorithm in 1979 that we learned that LO $\in$ P. The Ellipsoid Algorithm solves LINEAR FEASIBILITY in polynomial time by surrounding the feasible region by a sequence of ever shrinking ellipsoids. It halts either when an ellipsoid's center is feasible or when polynomially many ellipsoids have been constructed. In the latter case, the final ellipsoid has too small a volume to contain a feasible point, showing that the feasible region is empty. The technical calculations again involve the largest possible basic determinant the given constraint matrix could have. Unfortunately, this great theoretical success is a practical flop — its typical running time compares very

---
[7]See D. Gutfreund, R. Shaltiel and A. Ta-Shma, If NP languages are hard on the worst-case, then it is easy to find their hard instances, *Comput. Complexity* **16** (2007), no. 4, 412–441.

[8]C. Papadimitriou, *Computational Complexity*, Addison-Wesley, 1994.

poorly against Simplex.

A different polynomial algorithm that continues to compete with Simplex was invented in 1984 by Karmarkar. Unlike the exterior approach of Khachian, Karmarkar's Algorithm is an interior approach. It now belongs to a more general class of interior-point methods called *central path following* algorithms that solve more general problems including CONVEX OPTIMIZATION (okay, add COP to the acronym list) and more. Karmarkar solves LO by first converting a given LOP to one with a special form that is feasible and which has minimum 0 if and only if the original problem is feasible. The algorithm generates a sequence of feasible points whose objective values don't always shrink, but do so over many iterations. After polynomially many iterations the resulting value is either small enough that it must be 0, since all basic determinants are bounded by some constant, or returns that the optimum is positive.

More details and precise descriptions of both of these polynomial algorithms can be found in a number of other texts. Neither algorithm pays attention to the combinatorial structure of the feasible region, giving running times that depend on the entries. Roughly, an algorithm is **strongly polynomial** if it is polynomial in the number, not the size of the entries. In that respect it is more of a *combinatorial* algorithm, and neither Khachian's nor Karmarkar's qualifies. It is still unknown whether such an algorithm exists for LO.

strongly polynomial algorithm

The wonderful behavior exhibited by the supposedly bad Simplex Algorithm begs the question about problems in NP: are they almost always polynomial, or polynomial on average, in the same sense that Simplex is? Levin[9] and others[10] explore this direction.

---

[9] L. Levin, Average case complete problems, *SIAM J. Computing* **15** (1986), 285–286.

[10] Special issue on worst-case versus average-case complexity (O. Goldreich and S. Vadhan, eds.), *Comput. Complexity* **16** (2007), no. 4.

Appendix D

Software

In this appendix we discuss various software that is useful for the course. Many instructors like using professional solvers such as CPLEX, LINGO, QSopt, CLP, and others, but we do not address them here. Instead, we restrict our attention to what we're using to supplement this text, namely WebSim and MAPLE, and refer you to three outstanding web sites for learning more about the incredible variety of excellent software available for solving optimization problems of all types.

- INFORMS (The Institute for Operations Research and the Management Sciences):
  www.informs.org

- COIN-OR (Computational Infrastructure for Operations Research):
  www.coin-or.org

- Hans Mittelmann's Decision Tree for Optimization Software:
  plato.asu.edu/guide.html (also includes his fantastic Benchmarks:
  plato.asu.edu/bench.html)

Outside of solving LOPs, one may be interested in visualizing the polyhedra that form their feasible regions. We show some of the ways this can be done with MAPLE in Section D.3. However, a very nice software designed specifically for this purpose is POLYMAKE.[1]

D.1 WebSim

WebSim is merely a web-based tool for performing integer pivots on an array. Because we use it to solve LOPs, as well as systems of linear

---

[1] See www.math.tu-berlin.de/polymake.

G. H. Hurlbert, *Linear Optimization*, Undergraduate Texts in Mathematics,
DOI: 10.1007/978-0-387-79148-7_D, © Springer Science+Business Media LLC 2010

(in)equalities, there are a few bells and whistles included to facilitate its use for this purpose.

First, from my web site (search for `Hurlbert math homepage`), click on the WebSim logo. There you will find everything you need regarding how to use it. The best approach is to click on `Download` and use it on your own machine, so that you don't have to worry about an Internet connection while using it. You will find advice on which browser works best and how to set its preferences, some tips on how to use WebSim efficiently (which you can contribute to if you have some neat ideas), and a list of known bugs (that you should also contribute to if you find troubles).

There are several ways to enter a tableau into WebSim. One can enter each entry by hand into the cells (using tabs, arrows, or your mouse) of an array whose size you define, or paste the entire tableau at once (from the `Empty` option), having copied it from MAPLE, LaTeX, or Excel, for example. In the former case you will have had the opportunity to tell WebSim whether you are solving a LOP or system, and whether the contraints are inequalities or equalities (with a mixture of both, pick the closest option and modify the tableau later). WebSim will then include separating lines in the appropriate places; in the latter case you can insert them yourself by clicking on the lines between the buttons. Pivoting is most easily accomplished by double-clicking on the cell of the pivot entry.

Every cell, row button, column button, and tableau button has a menu that can be accessed by right-clicking on it. These are explained fully on the web site, and include such things as adding, deleting, and highlighting rows and columns, filling them with 0s or 1s, copying a tableau to a new tab or window or to Excel, and converting a tableau to MAPLE or LaTeX. One can also change the widths of the cells to accommodate fitting large tableaux (a frequent occurrence with the networks of Chapter 10, for example) on your screen or seeing large numbers in the cells.

Hopefully, you will find WebSim useful in remembering how the Simplex Algorithm works, merely by recalling the infamous bumper sticker WWWD.[2]

D.2 Algorithms

In this section we review basic algorithmic constructions that can be useful in the course for solving certain exercises or exploring interesting questions. pseudocode We use a version of what is often called **pseudocode**, which concentrates on making the instructions clear without regard to the actual syntax of a particular language, such as Fortran, C++, Java, or MAPLE. The next section will illustrate how MAPLE writes some of the following algorithms using its own symbols and structure.

---

[2] What Would WebSim Do?

D.2. Algorithms

Algorithm D.1

$s \leftarrow 0$
for i **from** 1 **to** 20 **by** 3 **do**
  if $i \mod 2 = 0$
    then $s \leftarrow s + i$
    else $s \leftarrow s - i$
  end if
end do
print s

Algorithm D.1 illustrates a `for loop` and conditional `if` statement, calculating and printing the difference between the even and odd multiples of 3 among the first 20 natural numbers: $9 (= 0 - 3 + 6 - 9 + 12 - 15 + 18)$.

Procedure D.2

procedure modprime(m)
  $count \leftarrow \mathbf{array}(m-1)$
  $p \leftarrow 2$
  while $p < 1000$ **do**
    if $p \neq m$
      then $count(p \mod m) \leftarrow count(p \mod m) + 1$
    end if
    $p \leftarrow \text{nextprime}(p)$
  end do
  output(argmax($count$))
end procedure

Procedure D.2 illustrates a `while loop`, taking as input the positive integer m and returning as output the integer $1 \leq k < m$ that is congruent modulo m to the most primes under 1000. Note that the pseudocode leaves out considerations like how the array is initialized to zero, what type of variables m and p are, and that $count$ and p are local, rather than global variables. These would all need to be specific in your language of choice. The functions nextprime and argmax are assumed here to be predefined functions of the language or procedures you've already written.

Procedure D.3

procedure expo(r, e, m)
  if $e = 0$
    then output(1)
    elseif $e = 1$ **then output**($r \mod m$)
    else output(expo($r, e \mod 2, m$) * expo($r, \lfloor e/2 \rfloor, m$))
  end if
end procedure

Procedure D.3 illustrates recursion, quickly calculating the value of r^e mod m for any nonnegative integer e. Notice that the case $e = 0$ is included to insure that the procedure halts, and that the case $e = 1$ is unnecessary. However, we included it only to show the use of the **elseif** construction, which saves nested **if** statements.

There is obviously much more to know about algorithms,[3] including various data structures, floating point arithmetic, parallel architectures, etc., but really, this is about all you need for this course.

D.3 MAPLE

MAPLE is a very well known computer algebra software, available at many universities for use by faculty and students in varying degrees of licensing. A student version is available for reduced price for personal use; usually most of those in my course want the software on their own laptop instead of having to always go to the school's computer labs. The student version is more than sufficient for our purposes.

The use of MAPLE is, of course, completely up to your instructor. In the text I have included workouts and exercises that allow one to explore the geometry, algebra, combinatorics, and probability related to LO by using a computer algebra system with a visual component. It is certainly reasonable to use something different than MAPLE, such as MATHEMATICA or MATLAB, or example. The choice of MAPLE over MATHEMATICA , MATLAB , and others was made after an extensive and thorough comparison that revealed the most serious flaw with other systems —I don't know how to use them.

While buying a book[4] on how to learn MAPLE can be quite useful, it might be overkill for this particular course. I have found that the combination of MAPLE's excellent help functions and my occasional Internet queries have been sufficient to handle everything I've needed so far. Do what works for you; the following may help you get started. Help on a particular word like `Vector` can be accessed by typing `?Vector`. One can also click on the Help menu to take a tour, browse the table of contents, search key words, read the Quick Reference Card, or see the lists of packages and commands. There are also interactive tools for plotting and building mathematical objects, tutors in various subjects, and other dropdown menus for editing, formatting, and other operations. Surf the joint.

MAPLE can act as a calculator, and it requires punctuation to execute code. While hitting the `Enter` key executes the code, hitting `Shift+Enter` gives a carriage return to the next screen line without executing. Thus one can write

---

[3] Read the classic A. Aho, J. Hopcroft, and J. Ullman, *The Design and Analysis of Computer Algorithms*, Addison-Wesley (1974) if you like.

[4] Such as F. Wright, *Computing with Maple*, Chapman & Hall/CRC (2001) or M. Abell and J. Braselton, *Maple by Example*, Academic Press (2005).

D.3. MAPLE

```
y := 7:
x := 2*y;
x = %-1;
```

to obtain the following output.

$$x := 14$$

$$14 = 13$$

Notice how the colon suppresses output, while the semicolon displays it. Also see that := makes assignments from right to left, while = takes no action. The % sign stands for the most recently calculated value, which in this case is 14. Thus x=%-1 is a statement, in this case a false one. This can be used as a test in an if-then statement, for example.

The MAPLE code

```
S := [1,1]:
for i from 1 to 10 do
  S := [op(S),S[-1]+S[-2]]
od:
print(S);
r := (1+sqrt(5))/2:
[seq(round((r^i)/sqrt(5)),i=1..nops(S))];
```

produces

$$[1, 1, 2, 3, 5, 8, 13, 21, 34, 55, 89, 144]$$

$$[1, 1, 2, 3, 5, 8, 13, 21, 34, 55, 89, 144]$$

as output. The list S is initialized to $[1, 1]$. Then op turns it into the sequence $1, 1$. While $S[k]$ denotes the kth term in the list S, counting from the left (MAPLE always starts its counting with 1 instead of 0), $S[-k]$ denotes the kth term in S, counting from the right. Thus the sum of the two rightmost terms are appended to S. The brackets turn the sequence back into the list $[1, 1, 2]$ and the process is repeated. Finally, the command seq creates a sequence with the given formula for the range given, where round rounds a number to its nearest integer and nops(S) counts the number of terms in S.

It is important to remember that lists and sequences are ordered, whereas sets are not. The following commands and output highlight the difference.

```
L := [5,4,6,3,7]; L[3]; S := {5,4,6,3,7}; S[3];
```

$$L := [5, 4, 6, 3, 7]$$

$$6$$

$$S := 3, 4, 5, 6, 7$$

5

Whenever possible, use MAPLE's sequence-like constructions instead of an iterative algorithm. For example, instead of writing

```
s := 0:
for k from 1 to 10 do
   s := s+k
od:
s;
```

to obtain the result

$$55$$

write the following single line.

```
add(k,k=1..10);
```

Remember also that MAPLE can do algebra, after all. The command

```
simplify(sum(k,k=1..n));
```

produces the following output.

$$\frac{1}{2}n^2 + \frac{1}{2}n$$

This capability means that one should save all evaluations until the last moment, since MAPLE performs exact arithmetic until you ask it to evaluate a result numerically. For example,

```
y := exp(2);
evalf(y);
```

produces the following result.

$$y := e^2$$

$$7.389056099$$

There are a number of variations of **eval** that apply in different circumstances.

Here is the way that Procedure D.2 would be written in MAPLE.

Procedure D.4

```
modprime := proc(m)
  local count, p, x, k;
  count := [seq(0,i=1..m-1)];
  p := 2;
  while p<1000 do
    if p<>m
      then count[p mod m] := count[p mod m]+1
```

D.3. MAPLE

```
        fi;
        p := nextprime(p)
      od;
      x := max(op(count));
      for k from 1 to m-1 while count[k]<>x do od;
      return k
   end:
```

The command modprime(5) returns the value 2. Indeed, count ends at the value $[40, 47, 42, 38]$. I personally enjoy writing if and do backwards to end them. Talk about good times ....

When writing programs, it can be useful to execute them line by line before grouping them together in one procedure. The Split or Join action in the Edit menu can be useful in this regard. Also, debugging can be facilitated with interactive debugger (usually an icon of a little bug), or commands like printlevel and trace. Most of the time, little errors can be found by setting printlevel:=10. When its value is 0 no output is shown (the default value is 1), and higher and higher values show more and more of the internal workings of your algorithm.

You may have the need at times to use functions instead of expressions. For example, the code

```
      z := (3*x-7*y)/5:   eval(z,[x=10,y=3]);
```

can be written in terms of functions as follows

```
      f := (x,y) -> (3*x-7*y)/5;   f(10,3);
```

(If you have been testing out all these commands while reading, you will discover that your results don't match mine. The problem stems from your prior values of x and y, which can be cleared with the command unassign('x','y').)

Many times you will need to load various packages into MAPLE before being able to compute things. For example, for most things in this course you will need LinearAlgebra. Here is an example for building a random tableau.

Algorithm D.5

```
      with(LinearAlgebra):
      vars := 5:  cons := 4:   Max := 100:
      rnd := rand(-Max..Max):
      A := Matrix([seq([seq(rnd(),j=1..vars)],i=1..cons)]):
      B := Matrix([seq([rnd()],i=1..cons)]):
      C := Matrix([[seq(rnd(),j=1..vars)]]):
      R := Matrix([[A],[C]]):  J := IdentityMatrix(cons+1):
      S := Matrix([[B],[0]]):  T := Matrix([[R,J,S]]):
      Tab := convert(T,array);
```

You can see how to glue matrices together to form larger ones. The conversion to an array at the end allows you to copy the result into an **Empty** window of WebSim.

The package **combinat** can be useful at times as well. In the next example we calculate every feasible basis of a LOP.

Algorithm D.6

```
with(combinat,choose):
A := Matrix(rows,cols,
      [ [98, 45, 26, 19, 84],
        [55, 88, 90,  8, 34],
        [ 1,  1,  1,  1,  1] ]);
rows := RowDimension(A):
cols := ColumnDimension(A):
b := Vector[column]([55, 55, 1]);
BETA := choose([seq(i,i=1..cols)],rows):
tetra := {}:
for j from 1 to binomial(cols,rows) do
  beta := BETA[j];
  B := A[1..rows,beta];
  t := B^(-1).b;
  if min(seq(t[i],i=1..rows))>=0
    then tetra := tetra union {beta}
  fi;
od:
tetra;
nops(tetra);
```

Of course, LinearAlgebra still needs to be loaded for this to work. But if you have loaded it already in Algorithm D.5, you don't need to do so again. For those MAPLE commands you don't recognize in Algorithm D.6, look them up in MAPLE Help.

Probably, you will use the package plots frequently. Here are three examples to give you a feel for your array of options. Each is a different picture of the same LOP, although Algorithm D.9 doesn't show the objective function.

Algorithm D.7

```
with(plots):
plot( [(210-5*x)/7, (420-12*x)/7, 25, (30-x)/2, 33-x],
x=0..35, color=[red, red, red, blue, green],
linestyle=[solid,solid,solid,solid,dash],
thickness=[1,1,1,1,3]);
```

Algorithm D.8

```
a := (210-5*x)/7:  b := (420-12*x)/7:  c:= 25:
f := min(a,b,c): d := (30-x)/2:  g := max(d,0):
e := 33-x: plot([f,g,e], x=0..35,
color=[red,blue,green], axes=boxed);
```

Algorithm D.9

```
inequal({5*x+7*y<=210, 12*x+7*y<=420, y<=25,
x+2*y>=30, x>=0, y>=0}, x=-1..36, y=-1..26,
optionsclosed=(color=red),
optionsfeasible=(color=blue),
optionsexcluded=(color=white));
```

In three dimensions you can write something like

```
a:=[4,5,13]:   b:=[9,4,2]:
c:=[3,11,5]:   d:=[0,2,1]:
polygonplot3d([a,b,c,a,d,b,c,d], axes=boxed);
```

to draw some polytopes you know the vertices of (but you need to make sure you traverse every one of its edges in your plot list). The feasible region of a 3-variable LOP can be displayed as follows.

Algorithm D.10

```
plot3d([max(0,(146*x+112*y)/18),
max(0,(-135*x+161*y)/19),
max(0,(149*x-122*y)/17),
max(0,(53*x+40*y)/21),0], x=0..20,
y=0..20, color=[red,yellow,blue,green,black],
axes=boxed);
```

With plot3d one can use the cursor to rotate the results in $\mathbb{R}^3$, as well as zoom in and out, or slide the figure around. You can also change how the figure is drawn, with grid lines or dots, etc. But the absolute most fun comes from animations. The following algorithm builds a movie that shows an objective function of a 3-variable LOP sliding through the optimal solution.

Algorithm D.11

```
T := 2:
a := max(min((56-31*x-37*y)/11,T),0):
b := max(min((52-11*x-34*y)/39,T),0):
c := max(min((60-35*x-10*y)/34,T),0):
d := max(min((t-9*x-7*y)/8,T),0):
animate(plot3d,[ [0, a, b, c, d],
x=0..T, y=0..T, axes=boxed,
color=[black,red,blue,green,yellow]], t=10..20);
```

One can change the speed of the animation, pause it, set it to replay continuously, and so on.

With all the above commands and others you may use, check their Help pages for usage, variations, options, and links to related commands. Your suggestions on more efficient or creative MAPLE programming with regard to this course are welcome.

Index

Regarding the fonts included below, theorems (in their full names) are in *italics*, and page numbers that refer to definitions are in **bold**. (You'll notice that theorems and definitions find their way into the margins for easy location.) *Italicized* page numbers denote appearances in theorems, while sans serif numbers signal inclusion in workouts and exercises, and roman fonts cite regular occurrences of the term. Regarding the order of terms, mathematical symbols come first (numbers, then capital letters, then lower case letters), followed by acronyms, and then standard words.

2-Coloring Problem, **243**
3-Coloring Problem, **243**
A, **11**, 89
B, **90**
B', **91**
J, 160
J$_k$, **20**, 69, 116, 118, 142, 148, 149, 154
$\mathbb{N}$, 209
Π, **90**
X-
    simplex, **71**
    tetrahedron, **69**, 142
    triangle, **69**, 138
$\mathbb{Z}$, 209
$a_{i,j}$, **11**
β, **34**, 90
b, **11**, 89
    -column, **33**, 43, 203, 205, 239
    -ratio, **36**, 44, 47, 61, 112, 205, 238
b_i, **11**
c, **11**, 89
    -ratio, **205**, 213
c$_\beta$, **90**
c_j, **11**
c$_\pi$, **90**
d_β, **91**

deg, **178**
det, 91
ϵ-column, **201**
f-vector, 143
k-
    Coloring Problem, **243**
    face, **30**
    intersecting family, 69, 71, **129**, *129*
    regular graph, **183**
k-regular graph, 192
n-person game, 162
π, **34**, 90
vhull, 69, 70, 72, 137, 138, 142
vspan, 70
w, **11**
x, **11**, 89
x$_\beta$, 90
x_j, **11**
x$_\pi$, 90
y, **11**, 89
y_i, **11**
z, **11**
BFS, 181, **219**
BLOP, **viii**, 23, 24, 26
CLP, 247
COIN-OR, 247
COP, **245**

CPLEX, 247
DFS, 181, **219**
DS, **132**, *132*, 139
FLOP, viii, **209**
GI, **37**, 57
GLOP, viii, **148**
IHOP, **209**
ILOP, **6**
INFORMS, 247
JLOP, **209**
LG, 48
LINGO, 247
LOP, **2**
    degenerate, **49**
    terminal, **221**
LS, **37**, 48
MN, **37**, 57
MWHC, **243**
MWST, **243**
NP
    -complete, **243**
    -hard, **242**, 243
    problem, 27, **241**
P problem, 27, **242**
QSopt, 247
SDR, **189**
SDR blocker, **189**, *190*, 193
SLOP, **viii**
TU, **99**, *100*, 100, 104, 105, 173, 243
WWWD, 248
ahull, **68**, 68
aspan, **67**
glb, 148
lhull, **68**, 68
lspan, 66, **67**
nhull, **68**, 68
nspan, **67**
per, **192**
proj, 70
snr, **36**
vhull, **64**, *65*, 68, 140, 143
vspan, **64**, *65*, 66

activity, **196**, 200, 203
acyclic
    graph, **168**
    network, **168**

affine
    combination, **62**, 67
    hull, **68**, 68, 71
    position, **71**
    region, **68**
    span, **67**
Aging Lappers, 26
Aguilera, 4
air, bottled, 18, 163
Algorithm,
    Best Response, **151**, 154
    Branch-and-Cut, 229
    Breadth-First Search, 181, **219**
    Depth-First Search, 181, **219**
    Elimination, Fourier–Motzkin, 57
    Ellipsoid, 87
    Ford–Fulkerson, 181
    Fourier–Motzkin Elimination, 57
    Karmarkar's, 87, 245
    Search,
        Breadth-First, 181, **219**
        Depth-First, 181, **219**
algorithm,
    backtracking, **219**
    central path following, 245
    combinatorial, 245
    polynomial, **242**
    strongly, **245**
alternating path, **193**
alternative system, **122**, 123, 124, 125
Alternative, Theorem of the, *123*, 127, 128, 190
analysis,
    post-optimality, **195**
    sensitivity, **195**
Annette, 19, 183
Anthony, 25
anti-symmetric matrix, **146**
ants, 20, 201
Approximation Theorem, Shadow, *198*
arc, 161, **163**, 168–170
    pendant, **168**
Arope Trucking, 18, 163
Arrow, Kenneth, 26, 208
arugula, 18

Assignment Problem, 194
association, homeowners, 26
Athens, 18, 163
augmat, **231**
augmented
    network, **166**
    tableau, **165**
augmenting path, **193**, 193
Aussie Foods Co., 20, 202
Auxiliary
    Method, **39**, 43, *44*
    Problem, **39**, *39*

Backman, 4
backtracking algorithm, **219**
balanced digraph, **180**
    nearly, **180**
Barbie, 145–151, 153–156, 160, 161
basic
    coefficient, 33, **35**, 38, 56, 91, 205, 212, 215
    solution, 34, **34**, 35, 38, *49*, 59–61, 66, 90, 99, 124, 173, 205
    variable, **33**, 34, 35, 43, 90, 92, 96, 110, 111, 202, 236, 238
    basis, 33, **33**, 34–37, 38, 43, 44, 55, 57, 60, 66, 77, 90, 91, 96, 98, 111, 136, 166, 166, 167, *168*, 169, 188, 197
    convex, **66**
    degenerate, **49**
    feasible, **34**, 37, 55, 85, 253
    infeasible, **34**, *39*, 45, 45, 55
    initial, **33**, 39, *39*, 45, 90, 91
    linear, **66**
    network, 167
    optimal, **34**, 104, 105, 200, 201, 202, 207, 208
    partial, 110, 166
    partner, dual, 85, **204**, 204
Basis Theorem, Network, *168*, 173, 188
Battleship, 158, 158, 161, 190
Beale, Evelyn Marin Lansdowne, 26
Ben, 19, 189
Benannaugh, Anna, 19, 185
Best Response Algorithm, **151**, 154

Biff, 20, 201
Big M Method, 57
Bill, 19, 183
bimatrix, **153**, 154, 155, 155, 159
Binghamton, Captain, 158
bipartite graph, **5**, 5, 183, *184*, *186*, 188, 190, 190–192
Birkhoff, Garrett, 26
Birkhoff–von Neumann Theorem, *132*, 191, 193
Bixby, Robert, 26, 105
Björn, 157, 159
black widow spiders, 20, 201
Blake, James, 157
Bland's Theorem, *49*, 57
Bland, Robert, 26
blocker, SDR, **189**, *190*, 193
Blough, Joseph, 19, 185
boab seeds, 20, 202, 203, 207
Boosch Oil, 21
Boosters, Dolphin, 26
Boris, 152
bottled air, 18, 163
bound proof, branch-and-, **222**
boundary point, **29**, 31
bounded region, **29**
branch-and-bound proof, **222**
Branch-and-Cut Algorithm, 229
Breadth-First Search Algorithm, 181, **219**
breast, 21
bridge, **168**
broccoli, 18
brother Darryl, 160
Brouwer Fixed Point Theorem, 155, *159*, 159, 160
Brown, Farmer, 18
Bunny, 153
butcher, Kingsbury, 21

Café Barphe, Le, 22
Calcium, 1
calf, 21
Calvin, 22
Captain Binghamton, 158
Carathéodory's Theorem, *65*, 142, 193
Carathéodory, Constantin, 26

CarbonDating.com, 19, 183
Carlos, 20
Carter, 4
cash flow, 24
castles, sand, 21
Catalan number, 143
Cayley's Theorem, 181, 243
Center, Spaced Out, 24
central path following algorithm, 245
Central Station, 22
certificate, 17, 222, 241
    infeasible, 46, 97, 114, 121, *121*, 122, 170
    optimal, **4**, **12**, 98, 117, 196, 224, 226
    unbounded, 47, 48, 114, 122, 171
    unsolvable, 124
chads, 18, 207
chain, **167**
    oriented, **167**
Chamique, 24
Charnes, Abraham, 26
Chayni Oil, 21
cheeseburger, deluxe, xi, 1
chicken sandwich, 1
Cholesky factorization, 105
Christmas, 23
chromatic number, fractional, 194
Chvátal rank, 229
circuit, **167**, 169, 171, 179
    oriented, **167**, 171, 187
Clay Mathematical Institute, 241
Clifford, 19, 189
Clive, 153
Club, Inclusivity, 26
cockroaches, 20, 201
Code, Prufer's, 181
coefficient, basic, 33, **35**, 38, 56, 91, 205, 212, 215
column,
    b-, 33
    ϵ-, **201**
    dominated, **150**, 151
    pivot, **36**, 47
combination,
    affine, **62**, 67

conic, **62**, 67, 136, 141
convex, **62**, 63, 64, 66, 67, 132, *132*, 134, 142, 193
linear, 46, 46, 56, **62**, 67, *121*, 122
combinatorial algorithm, 245
Commerce Secretary, 19, 185
Complementary Slackness Theorem, *79*, 80, 80, 84, 85, 148, 155
    General, 115
complete, NP-, **243**
complexity, **242**
component, **167**
    strong, **167**
Composites Problem, Recognizing, **242**
cone, **134**, 139
    extreme ray of, **136**, 136, 139
    polyhedral, **134**
conic
    combination, **62**, 67, 136, 141
    hull, **68**, 68, *136*
    region, **68**
    span, **67**
conjecture, Hirsch, 143
connected
    graph, **167**
    network, **167**
    network,
        strongly, **167**
constraint, **2**
    nonnegativity, **2**
contour lines, **16**
convex
    basis, **66**
    combination, **62**, 63, 64, 66, 67, 132, *132*, 134, 142, 193
    hull, **64**, 64, *65*, 68, 69, 71, 137
    independent, **66**
    position, **71**, 138, 142
    region, **61**
    span, **64**, *65*
Convex Optimization Problem, 245
Cooper, William, 26
cooperative game, **154**
Correspondence Theorem, Primal-Dual, 205, *205*

cost, **163**
  increased, 202, **202**, 202
  reduced, **201**, 201, 202
Cost Theorem,
  Increased, *202*
  Reduced, *201*
cover, **5**
  minimal, **193**
  minimum, **186**, 190
Cramer's Rule, *234*
crickets, 20, 201
Curly, 23
curves, level, **16**
cutting plane, **203**, **209**, **228**
  decision form, **212**, 215
  original form, **212**, 215
cutting plane proof, 222, **222**, *223*
cycle, **167**
  Hamilton, **243**
  negative, **171**, 187

Dantzig, George, vii, 26
Darryl
  brother, 160
  other brother, 160
David, 19, 183
Davydenko, Nikolay, 157
decision form, **212**, 215
Decision Tree, 247
degenerate
  LOP, **49**
  basis, **49**
  extreme point, **49**
  pivot, **49**
  polyhedron, **49**
  tableau, **49**
degree, **178**, 180, 183
deluxe cheeseburger, xi, 1
demand, **163**, 169, 170, 172, 173, 180, 183, 191
Depth-First Search Algorithm, 181, **219**
Descartes, René, 26
Dewey, 23
diamond weevil, 20, 202, 203, 207
Diana, 24
dictionary, **33**

feasible, **34**
infeasible, **34**
optimal, **34**
unbounded, **47**
digraph, **163**
  balanced, **180**
  nearly, **180**
Dilemma, Prisoner's, **153**
Dilworth's Theorem, 194
Dilworth, Robert, 26
dinner, 24
directed graph, **163**
direction, extreme, **139**, 140
distinct representatives, system of, **189**, 189
Djokovic, Novak, 157
Dolphin Boosters, 26
dominated
  column, **150**, 151
  row, **150**, 151
Don, xii
Donyell, 19, 189
doubly stochastic matrix, **132**, 184, 191–193
dual
  basis partner, 85, **204**, 204
  multipliers, **10**, 15, 222
  problem, **11**, *75*, *113*, 149
  variable, **11**, **62**, 197, 206
Duality Theorem,
  General, *113*, 124, 149
  Strong, 3, 5, *75*
  Weak, *11*, 78, 119, 125
Dykstra, 4

edge, **5**
  pendant, **168**
Eduard, Helly, 26
efficiency, Pareto, 208
Egerváry, Jenő, 26
Elimination Algorithm, Fourier–Motzkin, 57
Ellipsoid Algorithm, 87
Elster, 4
emu, 20, 202
entering variable, **36**, 37, 44, 47, 48, 238, *240*

entry, pivot, 50, 53, 99, 248
Environment,
    Matrix, 91, 95, 96, 97, 97, 98, 100, 105, 164, 166
    Network, 164, 166, 190
    Tableau, 36, 91, 92, 95, 96, 97
equilibrium, 153, **153**, 154, 155, 155, 159, 162
equivalent system, **232**
Erdős–Ko–Rado Theorem, 143
Euler tour, **178**, *180*, 180
Euler's Theorem, 180
Eumerica, 18, 163
Eurodollar, 18, 163
evaluation version, **241**
even, **180**
Evens, Odds and, 24
expectation, **145**
expected value, **145**
Extended Method, **236**
exterior point, **29**
extreme
    direction, **139**, 140
    point, **29**, 59, 140
        degenerate, **49**
    ray
        of cone, **136**, 136, 139
        of polycone, 142
        of polyhedron, **139**

face, **30**, 55
    k-, **30**
facet, **30**, 134, 137
Factoring Problem, **242**
factorization, Cholesky, 105
fair game, **146**, 157, 160, 161
family,
    k-intersecting, 69, 71, **129**, *129*
    full-intersecting, **129**, *129*
Family, Gambino, 26
fantasy tennis, 157
Farkas Theorem, Integer, 223
Farkas's Theorem, 125, 137
Farkas, József, 26
Feasibility Problem,
    Integer, **242**, 243
    Linear, **242**, 243, 244

feasible
    basis, **34**, 37, 55, 85, 253
    dictionary, **34**
    network, 183, 191
    objective value, **10**
    point, **10**, *11*, 13, 17, 245
    problem, **34**, 38
    region, **29**, 30, 31, 50, 54, 59, 61, 132, 142, 215, 222, 244, 247
    set, **29**
    solution, 14, 47, *49*, 59, 80, 81, 109, 210, 241
    tableau, **34**, 43, 236
    tree, 170
Federer, Roger, 157
Fernandez, 4
fingers, 24
Finite Basis Theorem, 143
fish sandwich, 1
Fixed Point Theorem, Brouwer, 155, *159*, 159, 160
floptimal
    solution, viii, 215, 224
    tableau, 216
    value, 242
floptimal value, **209**
Florida, State of, 18
Flow Problem, Network, 181
flow, cash, 24
Ford, Lester, 26
Ford–Fulkerson Algorithm, 181
forearm, 21
form,
    decision, **212**, 215
    general, **107**, *112*
    Hermite normal, 128
    original, **212**, 215
    standard, **8**, 9, 9, 12, *39*, *49*, 83, 118, 164, 212
    network, **165**, *168*
Fourier, Jean, 26
Fourier–Motzkin Elimination Algorithm, 57
fractional chromatic number, 194
Frank, 159

Frankfurt, 18, 163
Freddy, 210
free variable, **8**, 111, 118, 122
Frish, Ragnar, 26
Fulkerson Algorithm, Ford–, 181
Fulkerson, Ray, 26
full-intersecting family, **129**, *129*
function,
    improvement, **155**, 159
    objective, **2**
Fundamental Theorem, 49
    General, 112

Gale, David, 26
Gambino Family, 26
game value, 24, **146**
game,
    n-person, 162
    cooperative, **154**
    fair, **146**, 157, 160, 161
    general-sum, **153**, 159
    noncooperative, **154**
    symmetric, **146**, 157
    zero-sum, 153, **153**, 154
Gardeners Guild, 26
General Duality Theorem, 124, 149
general form, **107**, *112*
general-sum game, **153**, 159
Gomory, Ralph, 26
Gooden, 4
Graham's scan, 72
Grail, Holy, 241
graph, **5**
graph,
    k-regular, **183**, 192
    acyclic, **168**
    bipartite, **5**, 5, 183, *184*, 186,
        188, 190, 190–192
    connected, **167**
    directed, **163**
    regular, **183**, *184*
Graphic Method, **31**
Greatest Implementation, Lowest-, **48**
Greatest Increase Implementation, **37**, 57
Gridburg, 19, 185
Guild, Gardeners, 26

Gumbo's Restaurant, 24

Hal's Refinery, 21
half-space, **29**, 61, 65, 130, 134
Hall's Theorem, 190
Hall, Philip, 26
hamburger, 1
Hamilton cycle, **243**
    Minimum Weight Problem, **243**
Handshaking Lemma, 178
hard, NP-, **242**, 243
head, **164**
Helly's Theorem, 129
Hendrix, 21
Hermite normal form, 128
Hernandez, 4
Hilton–Milnor Theorem, 143
Hirsch conjecture, 143
Hitchcock, Frank, 26
Hoffman, Alan, 26
Hoffman–Kruskal Theorem, 100
Holy Grail, 241
homeowners association, 26
homogenization, **127**
hot start, 200, **203**, 207, 208, 212
Hotdogger's Union, 19, 185
Huey, 23
hull,
    affine, **68**, 68, 71
    conic, **68**, 68, *136*
    convex, **64**, 64, *65*, 68, 69, 71, 137
    linear, **68**, 68
Hungarian Method, 194
Hyperplane Separation Theorem, 143
hyperplane, supporting, **30**
iloptimal
    problem, 223
    solution, viii
    value, **209**
Implementation,
    Greatest Increase, **37**, 57
    Least Subscript, **37**, 48
    Lowest-Greatest, **48**
    Most Negative, **37**, 57
improvement function, **155**, 159

Inclusivity Club, 26
inconsistent system, **123**, *123*
Increase Implementation, Greatest, **37**, 57
increased
    cost, 202, **202**, 202
    profit, **203**, 203
Increased Cost Theorem, *202*
indegree, **178**, 180
independent, convex, **66**
Indifference, Principle of, **148**, 155, 159
inequality, valid, **209**
    decision form, **212**, 215
    original form, **212**, 215
infeasible
    basis, **34**, *39*, 45, 45, 55
    certificate, 46, 97, 114, 121, *121*, 122, 170
    dictionary, **34**
    point, **10**
    problem, **10**, **34**, *49*, *112*, 119
    solution, *121*
    tableau, **34**
infinity node, **165**
initial basis, **33**, 39, *39*, 45, 90, 91
Innis, 4
Institute, Clay Mathematical, 241
Integer Farkas Theorem, *223*
Integer Feasibility Problem, **242**, 243
integral polyhedron, **100**, *100*
Integrality Theorem, *173*
interior point, **29**
    methods, 245
intersecting family,
    k-, 69, 71, **129**, *129*
    full-, **129**, *129*
interval matrix, **105**
Iron, 1

J. J., 129
Jake, 19, 189
Jason, 210
Jody, 152
John, 19, 157, 159, 183
Johnson, 4
Johnson, Anders, 26

Joni, 197
judge, 24

König's Theorem, *184*
König, Dénes, 26
König–Egerváry Theorem, *186*, 190, 190, 193
Kantorovich, Leonid, 26
Kara, 19, 189
Karmarkar's Algorithm, 87, 245
Karmarkar, Narendra, viii, 26
Karush, William, 26
Kate, 22
Kathy, 19, 183
Ken, 145–151, 153–156, 160, 161
Khachian, Leonid, viii, 26
Kingsbury butcher, 21
kiwi fruit, 20, 202, 203, 207
Klee, Victor, 26
Koopmans, Tjalling, vii, 26
Kruskal, Joseph, 26
Kuhn, Harold, 26

label, pivot, 60
labeling, Sperner, **140**
Lagrange multipliers, 87
Lappers, Aging, 26
Large Numbers, Weak Law of, *145*
Large World Park, 24
Larry, 23
Law of Large Numbers, Weak, *145*
leaf, **168**, 168, 169, 173, 178, 180, 221
Least Subscript Implementation, **37**, 48
Leatherface, 210
leaves, wattle, 20, 202, 203, 207
leaving variable, **36**, 37, 43, 44, *44*, 47, 236, 238, *240*
Lemke, Carlton, 26
Lemma,
    Handshaking, *178*
    Sperner's, *140*, 141, 143
length, **167**
Leontief, Wassily, 26
level curves, **16**
Lieutenant McHale, 158

Index 265

linear
    basis, **66**
    combination, 46, 46, 56, **62**, 67, *121*, 122
    hull, **68**, 68
    region, **68**
    span, **67**
Linear Feasibility Problem, **242**, 243, 244
lines, contour, **16**
Lisa, 24
Log Barrier Method, 105
Lopez, Jennifer, 209
Louie, 23
Lowest-Greatest Implementation, **48**

MacDonald, Öreg, 20
marginal value, **197**, 200
matching, **5**, 5
    maximal, **190**
    maximum, **186**, 190
    perfect, **183**, *184*, 186, 192, 193
Matching polytope, 72
    Perfect, 72
Mathematical Institute, Clay, 241
Matrix Environment, 91, 95, 96, 97, 97, 98, 100, 105, 164, 166
Matrix Tree Theorem, 181
matrix,
    anti-symmetric, **146**
    doubly stochastic, **132**, 184, 191–193
    interval, **105**
    payoff, **145**, 146, 149, 160
    permutation, **132**, *132*, 139, 184, 191–194
    sparse, **96**, 105
    stochastic, **132**
        doubly, **132**, 184, **191**, 191–193
    totally unimodular, **99**, *100*, 100, 104, 105, 173, 243
maximal matching, **190**
maximum matching, **186**, 190
Mayor, 19, 185
McHale, Lieutenant, 158
Melissa, 23

Merrill, Carol, 1
Method,
    Auxiliary, **39**, 43, *44*
    Big M, 57
    Extended, **236**
    Graphic, **31**
    Hungarian, 194
    Log Barrier, 105
    Perturbation, 57
    Primal-Dual, 128
    Shortcut, **43**, *44*
method, interior-point, 245
Mikey Moose Square, 24
Millenium Problems, 241
minimal cover, **193**
Minimax Theorem, *149*
minimum cover, **186**, 190
Minimum Weight Hamilton Cycle Problem, **243**
Minimum Weight Spanning Tree Problem, **243**
Minkowski sum, **69**, 137
Minkowski, Hermann, 26
Minkowski–Weyl Theorem, *136*
Minty, George, 26
Mittelmann, Hans, 247
mixed strategy, **147**
Moe, 23
Monica, 19, 183
Monty, 160
Moose, Mikey, 24
Morganstern, Oscar, 26
Moscow, 18, 163
Most Negative Implementation, **37**, 57
moths, 20, 201
Motzkin, Theodore, 26
Muffy, 20, 201
Multiplication Problem, **242**
multipliers,
    dual, **10**, 15, 222
    Lagrange, 87
Murray, Andy, 157

Nadal, Rafael, 157
Nalbandian, David, 157
Nash's Theorem, *162*

Nash, John, 26
Natasha, 152
Nearest Vector Problem, 229
nearly balanced digraph, **180**
neck, 21
negative cycle, **171**, 187
Negative Implementation, Most, **37**, 57
Nemirovsky, Arkady, 26
network, 99, **163**
    acyclic, **168**
    augmented, **166**
    basis, 167
    connected, **167**
        strongly, **167**
    feasible, 183, 191
    form, standard, **165**, *168*
Network Basis Theorem, *168*, 173, 188
Network Environment, 164, 166, 190
Network Flow Problem, 181
Neumann, John von, 27
Nobel Prize, vii, 155, 208
node, **163**
    infinity, **165**
nonbasic variable, **33**
noncooperative game, **154**
nonnegativity constraint, **2**
normal form, Hermite, 128
number,
    Catalan, 143
    fractional chromatic, 194
Nykesha, 19, 189

objective
    function, **2**
    row, **33**
    value, **10**
Odds and Evens, 24
Oil,
    Boosch, 21
    Chayni, 21
Old Yorktown, 22
operation,
    pivot, **35**, 36, 37, 38, 40, 43, 45, 53, 56, 90, 95, 111, 111,
166, 202, 205, 212, **234**, 236, 244, 247
    row, **231**
optimal
    basis, **34**, 104, 105, 200, 201, 202, 207, 208
    certificate, 4, **12**, 98, 117, 196, 224, 226
    dictionary, **34**
    pair, *79*, 79, 84, 85, 87
    problem, **34**, *49*, *112*, 119
    solution, 4, *49*, 75, 81, 85, 112, 120, 132, 195, *198*, 202
    strategy, 148, 149, 149, 151, 152, 155, 160
    tableau, **34**
optimization version, **241**
oracle, **242**
Orchard-Hayes, William, 27
oriented
    chain, **167**
    circuit, **167**, 171, 187
original form, **212**, 215
other brother Darryl, 160
outdegree, **178**, 180

pair, optimal, *79*, 79, 84, 85, 87
Panch, 152
Paper-Scissors, Rock-, 157
parameter, **33**
Pareto efficiency, 208
Paris, 18, 163
Park, Large World, 24
partial basis, 110, 166
partner, dual basis, 85, **204**, 204
Pat, 24
path, **167**
    alternating, **193**
    augmenting, **193**, 193
payoff matrix, **145**, 146, 149, 160
pendant
    arc, **168**
    edge, **168**
perfect matching, **183**, *184*, 186, 192, 193
Perfect Matching polytope, 72
permanent, **192**

permutation matrix, **132**, *132*, 139, 184, 191–194
Perturbation Method, 57
Pete, 24
Phase
    0, 111, **111**, 117, 166
    I, **34**, 39, 44, 45, **45**, 57, 111, 120, *121*, 169, 235
    II, **34**, 36, 37, 38, 40, 44, **44**, 47, 55, 111, 122, 136, 170, 199, 205, 236
pivot
    column, **36**, 47
    degenerate, **49**
    entry, 50, 53, 99, 248
    label, 60
    operation, **35**, 36, 37, 38, 40, 43, 45, 53, 56, 90, 95, 111, 111, 166, 202, 205, 212, **234**, 236, 244, 247
    row, **36**, 43, 45, 238
    rule, 43, 48, 91, 166, 215, 238
plane,
    cutting, **203**, **209**, **228**
        decision form, **212**, 215
        original form, **212**, 215
        proof, 222, **222**, *223*
Planet, Sidney, 24
point,
    boundary, **29**, 31
    exterior, **29**
    extreme, **29**, 59, 140
        degenerate, **49**
    feasible, **10**, *11*, 13, 17, 245
    infeasible, **10**
    interior, **29**
    saddle, **152**, 153, 155, 158–160
points, separable, **127**
Police Chief, 19, 185
polycone, **134**, 139
    extreme ray of, 142
polyhedral cone, **134**
Polyhedral Verification Problem, 72
polyhedron, **29**, 65, 68, 69, 96, *100*, *132*, 137, 140
    degenerate, **49**

extreme ray of, **139**
integral, **100**, *100*
polynomial algorithm, **242**
    strongly, **245**
polytope, **29**, 65, 68, 70, 100, 130, 132, 137, 141, 143, 228, 229
    Matching, 72
    Perfect, 72
position,
    affine, **71**
    convex, **71**, 138, 142
post-optimality analysis, **195**
Poussin, Charles de la Vellé, 27
President, 26
price, shadow, **197**, 198, 203, 206
primal
    problem, **11**, *75*, 204
    variable, **11**
Primal-Dual Correspondence Theorem, 205, *205*
Primal-Dual Method, 128
Principle of Indifference, **148**, 155, 159
Prisoner's Dilemma, **153**
Prize, Nobel, vii, 155, 208
problem variable, **8**, 29, 33, 74
Problem,
    2-Coloring, 243
    3-Coloring, 243
    k-Coloring, **243**
    Assignment, 194
    Auxiliary, **39**, *39*
    Composites, Recognizing, **242**
    Convex Optimization, 245
    Factoring, **242**
    Feasibility,
        Integer, **242**, 243
        Linear, **242**, 243, 244
    Hamilton Cycle, Minimum Weight, **243**
    Integer Feasibility, **242**, 243
    Linear Feasibility, **242**, 243, 244
    Minimum Weight
        Hamilton Cycle, **243**
        Spanning Tree, **243**
    Multiplication, **242**

Nearest Vector, 229
Network Flow, 181
Polyhedral Verification, 72
Recognizing Composites, **242**
Salesperson, Traveling, **243**
Shortest Vector, 229
Spanning Tree, Minimum Weight, **243**
Subset Sum, 229
Transshipment, 163
Traveling Salesperson, **243**
Vector,
    Nearest, 229
    Shortest, 229
problem,
    NP, 27, **241**
    P, 27, **242**
    constraint, **2**
    dual, **11**, *75*, *113*, 149
    feasible, **34**, 38
    iloptimal, 223
    infeasible, **10**, **34**, *49*, *112*, 119
    optimal, **34**, *49*, *112*, 119
    primal, **11**, *75*, 204
    unbounded, **10**, **47**, *49*, *112*, 119, *136*
Problems, Millennium, 241
product, **196**
profit, 196, 197, 200
    increased, **203**, 203
    reduced, **203**, 203
proof,
    branch-and-bound, **222**
    cutting plane, 222, **222**, *223*
Prufer's Code, 181
pseudocode, **248**
pull, theological tractor, xi
pure strategy, **147**

rank, Chvátal, 229
ratio,
    b-, **36**, 44, 47, 61, 112, 205, 238
    c-, **205**, 213
    smallest nonnegative, **36**
ray, **134**
    extreme
        of cone, **136**, 136, 139

        of polycone, 142
        of polyhedron, **139**
Rebecca, 19, 189
recognition version, **241**
Recognizing Composites Problem, **242**
reduced
    cost, **201**, 201, 202
    profit, **203**, 203
    tableau, **165**
Reduced Cost Theorem, *201*
reduction, 242
redundant system, **86**
    strongly, **86**
region,
    affine, **68**
    bounded, **29**
    conic, **68**
    convex, **61**
    feasible, **29**, 30, 31, 50, 54, 59, 61, 132, 142, 215, 222, 244, 247
    linear, **68**
    unbounded, **29**
Regis, 19, 183
regular graph, **183**, *184*
relaxation, **209**
Repete, 24
representatives, system of distinct, **189**, 189
resource, **196**
Response Algorithm, Best, **151**, 154
Restaurant, Gumbo's, 24
restricted variable, **8**
rib, 21
Rock-Paper-Scissors, 157
Rockafellar, Tyrell, 27
Roddick, Andy, 157
rooted spanning tree, **178**
rough green snakes, 20, 201
row operation, **231**
row,
    dominated, **150**, 151
    objective, **33**
    pivot, **36**, 43, 45, 238
Rule, Cramer's, *234*
rule, pivot, 43, 48, 91, 166, 215, 238

rump, 21

saddle point, **152**, 153, 155, 158–160
Salesperson Problem, Traveling, **243**
Sallie, 25
Sammy, 25
Samuelson, Paul, 27
sand castles, 21
sandwich,
    chicken, 1
    fish, 1
scan, Graham's, 72
Scissors, Rock-Paper-, 157
scorpions, 20, 201
Search Algorithm,
    Breadth-First, 181, **219**
    Depth-First, 181, **219**
Secretary, Commerce, 19, 185
seeds, boab, 20, 202, 203, 207
Senate, 19, 189
sensivity analysis, **195**
separable points, **127**
Separation Theorem, Hyperplane, *143*
set, feasible, **29**
Shadow Approximation Theorem, *198*
shadow price, **197**, 198, 203, 206
Shapley value, 162
Shapley, Lloyd, 27
Sheryl, 24
Shor, Naum, 27
Shortcut Method, **43**, *44*
Shortest Vector Problem, 229
Sidney Planet, 24
simplex, X-, **71**
sink, **180**
slack variable, **8**, 32, 40, 60, 74, 76,
    99, 112, 113, 121, 164, 199,
    205, 212
Smale, Stephen, 27
smallest nonnegative ratio, **36**
snakes, rough green, 20, 201
solution, **34**
    basic, 34, **34**, 35, 38, *49*, 59–61,
        66, 90, 99, 124, 173, 205
    feasible, 14, 47, *49*, 59, 80, 81,
        109, 210, 241
    floptimal, viii, 215, 224

iloptimal, viii
infeasible, *121*
optimal, 4, *49*, 75, 81, 85, 112,
    120, 132, 195, *198*, 202
songs, 25
source, **180**
space, half-, **29**
Spaced Out Center, 24
span,
    affine, **67**
    conic, **67**
    convex, **64**, *65*
    linear, **67**
spanning
    subgraph, **168**
    subnetwork, **168**
    tree, **168**, *168*, 168, 169, 170,
        172, 174, 178, 243
        rooted, **178**
Spanning Tree Problem, Minimum
    Weight, **243**
sparse
    matrix, **96**, 105
    tableau, **96**, 165
Sperner labeling, **140**
Sperner's Lemma, *140*, 141, 143
spiders, black widow, 20, 201
Square, Mikey Moose, 24
Stable Marriage Theorem, 194
standard form, **8**, 9, 9, 12, *39*, *49*,
    83, 118, 164, 212
    network, **165**, *168*
start, hot, 200, **203**, 207, 208, 212
States, United, 163
Station, Central, 22
Stepford Wives, 26
Stigler, George, 27
stochastic
    matrix, **132**
        doubly, **132**, 184, 191–193
    vector, **132**
strategy,
    mixed, **147**
    optimal, 148, 149, 149, 151, 152,
        155, 160
    pure, **147**

strong component, **167**
Strong Duality Theorem, 3, 5, *75*
    General, *113*
strongly
    connected, **167**
    polynomial algorithm, **245**
    redundant system, **86**
subgraph, **168**
    spanning, **168**
subnetwork, **168**
    spanning, **168**
Subscript Implementation, Least, **37**, 48
Subset Sum Problem, 229
subway, 22
Sue, 19, 189
sum,
    game,
        general-, **153**, 159
        zero-, 153, **153**, 154
    Minkowski, **69**, 137
supply, **163**
supporting hyperplane, **30**
symmetric game, **146**, 157
system of distinct representatives, **189**, 189
system,
    alternative, **122**, 123, 124, 125
    equivalent, **232**
    inconsistent, **123**, *123*
    redundant, **86**
        strongly, **86**

tableau, **33**
    augmented, **165**
    degenerate, **49**
    feasible, **34**, 43, 236
    floptimal, 216
    infeasible, **34**
    optimal, **34**
    reduced, **165**
    sparse, **96**, 165
    unbounded, **47**
    unstable, **96**
Tableau Environment, 36, 91, 92, 95, 96, 97
tail, **164**

Tardos, Eva, 27
tennis, fantasy, 157
Teresa, 19, 183
terminal LOP, **221**
tetrahedron, X-, **69**, 142
Thelma, 129
theological tractor pull, xi
Theorem,
    Alternative, of the, *123*, 127, 128, 190
    Approximation, Shadow, *198*
    Birkhoff–von Neumann, *132*, 191, 193
    Bland's, *49*, 57
    Brouwer Fixed Point, 155, *159*, 159, 160
    Carathéodory's, *65*, 142, 193
    Cayley's, 181, 243
    Complementary Slackness, *79*, 80, 80, 84, 85, 148, 155
        General, *115*
    Correspondence, Primal-Dual, 205, *205*
    Cost,
        Increased, *202*
        Reduced, *201*
    Cramer's Rule, *234*
    Dilworth's, 194
    Duality,
        General, *113*, 124, 149
        Strong, 3, 5, *75*
        Weak, *11*, 78, 119, 125
    Erdős–Ko–Rado, *143*
    Euler's, *180*
    Farkas's, *125*, 137
    Farkas, Integer, *223*
    Finite Basis, *143*
    Fixed Point, Brouwer, 155, *159*, 159, 160
    Fundamental, *49*
        General, *112*
    Hall's, *190*
    Helly's, *129*
    Hilton–Milnor, *143*
    Hoffman–Kruskal, *100*
    Hyperplane Separation, *143*

Increased Cost, *202*
Integer Farkas, *223*
Integrality, *173*
König's, *184*
König–Egerváry, *186*, 190, 190, 193
Marriage, Stable, 194
Matrix Tree, 181
Minimax, *149*
Minkowski–Weyl, *136*
Nash's, *162*
Network Basis, *168*, 173, 188
Primal-Dual Correspondence, 205, *205*
Reduced Cost, *201*
Separation, Hyperplane, *143*
Shadow Approximation, *198*
Stable Marriage, 194
Weak Law of Large Numbers, *145*
thigh, 21
tomatoes, 23
totally unimodular matrix, **99**, *100*, 100, 104, 105, 173, 243
tour, Euler, **178**, *180*, 180
tournament, **161**
tractor pull, theological, xi
Transshipment Problem, 163
Traveling Salesperson Problem, **243**
Tree
    Problem, Minimum Weight Spanning, **243**
    Theorem, Matrix, 181
tree, **168**
    feasible, 170
    spanning, **168**, *168*, 168, 169, 170, **172**, 174, 178, 243
        rooted, **178**
Tree, Decision, 247
triangle, X-, **69**, 138
Trucking, Arope, 18, 163
Tucker, Albert, 27

unbounded
    certificate, 47, 48, 114, 122, 171
    dictionary, **47**

problem, **10**, **47**, *49*, *112*, 119, *136*
region, **29**
tableau, **47**
unimodular matrix, totally, **99**, *100*, 100, 104, 105, 173, 243
Union, Hotdogger's, 19, 185
United States, 163
unsolvable certificate, 124
unstable tableau, **96**

valid inequality, **209**
    decision form, **212**, 215
    original form, **212**, 215
value,
    expected, **145**
    floptimal, **209**, 242
    game, 24, **146**
    iloptimal, **209**
    marginal, **197**, 200
    objective, **10**
    Shapley, 162
variable,
    basic, **33**, 34, 35, 43, 90, 92, 96, 110, 111, 202, 236, 238
    dual, **11**, **62**, 197, 206
    entering, **36**, 37, 44, 47, 48, 238, *240*
    free, **8**, 111, 118, 122
    leaving, **36**, 37, 43, 44, *44*, 47, 236, 238, *240*
    nonbasic, **33**
    primal, **11**
    problem, **8**, 29, 33, 74
    restricted, **8**
    slack, **8**, 32, 40, 60, 74, 76, 99, 112, 113, 121, 164, 199, 205, 212
Varyim Portint Co., 81
vector,
    f-, 143
    stochastic, **132**
Venice, 18, 163
Verification Problem, Polyhedral, 72
version,
    evaluation, **241**
    optimization, **241**

recognition, **241**
vertex, **5**, 168, *173*, 178, 183, 243, 244
Victoria, 19, 183
Vienna, 18, 163
Vitamin
    A, 1
    C, 1

Warren, 19, 183
wattle leaves, 20, 202, 203, 207
Weak Duality Theorem, *11*, 78, 119, 125
Weak Law of Large Numbers, *145*
weevil, diamond, 20, 202, 203, 207
weight, **243**
Wets, Roger, 27
Wives, Stepford, 26
Wolfe, Philip, 27
Wood, Marshal, 27
Wright, Susan, 27

Yolanda, 24

zero-sum game, **153**